A TEXT BOOK OF

QUANTITATIVE TECHNIQUES IN PROJECT MANAGEMENT

FOR
SEMESTER – VI

THIRD YEAR (T.Y.) B. TECH COURSE IN MECHANICAL / PRODUCTION / AUTOMOBILE ENGINEERING

Strictly According to New Revised Credit System Syllabus of Babasaheb Ambedkar Technological University (BATU), Lonere, (Dist. Raigad) Maharashtra, (w.e.f. June 2019-20)

Dr. DNYANESHWAR R. WAGHOLE

Ph.D (Mech. Engg.)
Associate Professor,
School of Mechanical Engineering
Maharashtra Institute of Technology Pune.
MIT World Peace University (MITWPU)
Kothrud, PUNE.

N1110

QUANTITATIVE TECH. IN PRO. MGT. (MECH. BATU) ISBN : 978-93-89686-84-5

First Edition : **December 2019**
© : **Author**

Published By :
NIRALI PRAKASHAN
Abhyudaya Pragati, 1312 Shivaji Nagar
Off J.M. Road, PUNE 411005
Tel : (020) 25512336/37/39
Email : niralipune@pragationline.com

➢ **DISTRIBUTION CENTRES**

PUNE

Nirali Prakashan (Local)   : 119 Budhwar Peth, Jogeshwari Mandir Lane, Pune 411002, Maharashtra
Tel : (020) 2445 2044, Mobile : 9657703145, Email : niralilocal@pragationline.com

Nirali Prakashan (Outstation)   : S. No. 28/27 Dhayari, Near Asian College, Dhayari, Pune 411041, Maharashtra
Tel : (020) 2469 0204, Fax : (020) 2469 0316, Mobile : 9657703143
Email : bookorder@pragationline.com

MUMBAI

Nirali Prakashan   : 385 S.V.P. Road, Rasadhara Co-op. Hsg. Society Ltd., Girgaum, Mumbai 400004, Maharashtra
Tel : (022) 2385 6339 / 2386 9976, Fax : (022) 2386 9976, Mobile : 9320129587
Email : niralimumbai@pragationline.com

➢ **DISTRIBUTION BRANCHES**

JALGAON

Nirali Prakashan   : 34 V. V. Golani Market, Navi Peth, Jalgaon 425001, Maharashtra
Tel : (0257) 222 0395, Mob : 94234 91860, Email : niralijalgaon@pragationline.com

KOLHAPUR

Nirali Prakashan   : New Mahadvar Road, Kedar Plaza 1st Floor, Opp. IDBI Bank
Kolhapur 416012, Maharashtra. Mobile : 9850046155
Email : niralikolhapur@pragationline.com

NAGPUR

Nirali Prakashan   : Above Maratha Mandir, Shop No 3, Second Floor,
Rani Jhanshi Square, Sitabuldi, Nagpur 440012, Maharashtra
Tel : (0712) 254 7129, Email : niralinagpur@pragationline.com

DELHI

Nirali Prakashan   : 4593/15 Basement, Agarwal Lane, Ansari Road, Daryaganj
Near Times of India Building, New DelhiV 110002 Mobile : 8505972553
Email : niralidelhi@pragationline.com

BENGALURU

Nirali Prakashan   : Maitri Ground Floor, Jaya Apartments, No. 99, 6th Cross, 6th Main,
Malleswaram, Bengaluru 560003, Karnataka
Mobile : 9449043034, Email : niralibangalore@pragationline.com

niralipune@pragationline.com | www.pragationline.com
Also find us on www.facebook.com/niralibooks

PREFACE

It gives me great pleasure to present the book **"Quantitative Techniques in Project Management"** for the students of **Semester VI Third Year (T.Y.) B. Tech. Course Mechanical / Production / Automobile Engineering of Dr. Babasaheb Ambedkar Technological University (BATU), Lonere, Dist. Raigad (Maharashtra).** This book is strictly as per the new revised syllabus 2019-20 Pattern, effective from the Academic Year July 2019-20.

In New Revised Syllabus, there will Class Assessment (CA) 20 Marks, Mid Sem. Exam. (MSE) 20 Marks and End Sem. Exam. (ESE) 60 Marks. End Sem. Exam. will be based on all Six units and each unit will carry 20 Marks.

The Theory Course will have 3 Credits.

The basic objective of this book is to bridge the gap between the vast contents of the reference books, written by the renowned International Authors and the concise requirements of Undergraduate Students. This book has been written in a comprehensive manner using Simple and Lucid language, keeping in mind students' requirements. The main emphasis has been given on exploring the basic concepts rather than merely the Information. Solved Examples and Questions for Practice have been provided throughout the book and at the end of the Unit. Also, we have given **Model Question Paper** for practice at the end of book.

My special thanks to my family members, students and all those who directly or indirectly supported us in this project.

I also take this opportunity to express my sincere thanks to Shri. Dineshbhai Furia, Shri. Jignesh Furia, Mrs. Nirali Verma, Shri. M. P. Munde and entire team of Nirali Prakashan, namely Mrs. Deepali Lachake (Co-ordinator), and her colleagues who really have taken keen interest and untiring efforts in publishing this text.

The advice and suggestions of my esteemed readers to improve the text are most welcome and will be highly appreciated.

Pune   **Author**

SYLLABUS

Unit 1: Introduction

Introduction to Operations Research, Stages of Development of Operations Research, Applications of Operations Research, Limitations of Operations Research Linear programming problem, Formulation, graphical method, Simplex method, artificial variable techniques.

Unit 2: Assignment and Transportation Models

Transportation Problem, North west corner method, Least cost method, VAM, Optimality check methods, Stepping stone, MODI method, Assignment Problem, Unbalanced assignment problems, Travelling salesman problem.

Unit 3: Waiting Line Models and Replacement Analysis

Queuing Theory: Classification of queuing models, Model I (Birth and Death model) M/M/I (∞, FCFS), Model II - M/M/I (N/FCFS).

Replacement Theory, Economic Life of an Asset, Replacement of item that deteriorate with time, Replacement of items that failed suddenly.

Unit 4: Inventory Models

Inventory Control, Introduction to Inventory Management, Basic Deterministic Models, Purchase Models and Manufacturing Models without Shortages and with Shortages, Reorder level and optimum buffer stock, EOQ problems with price breaks.

Unit 5: Project Management Techniques

Difference between project and other manufacturing systems. Defining scope of a project, Necessity of different planning techniques for project managements, Use of Networks for planning of a project, CPM and PERT.

Unit 6: Time and Cost Analysis

Time and Cost Estimates: Crashing the project duration and its relationship with cost of project, probabilistic treatment of project completion, Resource allocation and Resource leveling.

CONTENTS

UNIT III : WAITING LINE MODELS AND REPLACEMENT ANALYSIS

UNIT IV : INVENTORY MODELS

UNIT V : PROJECT MANGAMENET TECHNIQUES

UNIT VI : TIME AND COST ANALYSIS

CHAPTER 1
INTRODUCTION

1.1 INTRODUCTION TO OPERATIONS RESEARCH [May 18, 8M]

Evolution of Quantitative Methods and Operations Research Techniques

* The origin of Quantitative Methods and Operations Research took place in 1885. In 1885, *F.W. Taylor*, first used scientific technique for production methods. *Henry L. Gantt* applied scientific management tools for product scheduling after F.W. Taylor, A.K. Erlang, a mathematician, generated waiting line theory for telephone traffic situations in 1917. F.W. Harries, developed inventory control model in 1915, *H.C. Levinson*, an Astronomer, applied scientific technique for merchandising problem in 1930's. In short, the evolution of Quantitative Methods and Operations Research took place during World War II.

* In India, the roots of quantitative methods and operations research extends from 1949 at the Regional Research Laboratory at Hyderabad, Professor P.C. Mahalonobis applied operations research methods in national planning and survey in 1953 for second five year plan. The introduction of quantitative methods and operations research took place to academic studies in 1963 by Delhi University to M.Sc. course. During this period, Institute of Management at Kolkata and Ahmedabad, started operations research subject in MBA course knowing the importance of operations research in Accounts and Administrative, Government of India has introduced this subject to CA, ICWA and IAS exam apart from institute and university for student of engineering, management statistic, mathematics. The techniques of quantitative methods and operation research, now-a-days, applied every where in industry such as Telco, BHEL, Defence, Fertilizers, SAIL, ONGC etc. Quantitative methods and operations research are applied for planning, scheduling, inspection, budgeting, maintenance etc.

1.2 DEFINITIONS

[Oct. 15, 4M, Nov./Dec. 06, 12, 6m]

Quantitative method and operations research are used for resource planning such as material, men, machines etc. Hence so many definitions have been developed which are as follows :

* The application of methods of science (mostly mathematical science) to complex problem arising in directing and management of large resource in industry, business and government is known as **Operations Research**.

* A scientific method in which mathematical models are evolved for the real life situation and solved by application of various algorithms is known as *Quantitative Method or **Operation Research***.

* The use of scientific technique to problem associated with operation of system so as to give optimal solution to the problem is known as **Operations Research**.

* **Apart from Above Definitions;** Various Definitions Suggested by Many Inventors are as Follows :

  ➤ **Operations Research**, in most general sense, can be characterized as the use of scientific methods, tools and technique to problem involving the operations of system so as to give these in control of the operation with optimum solution to the problems.

  – AcKoH, ArnoH, Churchman

  ➤ **Operations Research** is a scientific technique to problem solving for executive management.

  – Wagner H.M.

  ➤ **Operation Research** is concerned with scientifically deciding how to best design and operate man, machine systems generally involving the allocation of scarce resources.

  – Operation Research Society of America

➢ **Operations Research** is applied decision tools. It applies any scientific mathematical mean to solve the situation that confront the executive when he work hard to reach a through going rationality in dealing with his decision problem.

– Starr and Miller

➢ **Operations Research** is a scientific technique of providing executive departments with a quantitative basis for decision relating to the operations under their control.

– Kimball and Morse

1.3 CHARACTERISTICS OF OPERATIONS RESEARCH OR FEATURES OF OPERATIONS RESEARCH

(Characteristics, Advantages of Operations Research and Quantitative Method)

The Features or Characteristics of Operation Research are as Follows :

- The use of scientific method to obtain solution of problem under investigation.
- The use of interdisciplinary groups from various scientific and engineering disciplines.
- Development of quality of decisions by using its scientific technique.
- Use of computer for solution of complex mathematical problems.
- System orientation of operation research concern with problem as whole.
- Consideration of human factors for deriving quantitative solutions to the problems.

1.4 STAGES OF DEVELOPMENT OF OPERATIONS RESEARCH   [Dec. 17, May 07]

- **The Stages of Operations Research are as Follows :**
  1. Problem formulation as per situations.
  2. Model development for system.
  3. Finding solution for model.
  4. Solution analysis (Validation).
  5. Deriving controls over the solution of problem.
  6. Implementation of solution of problem.

1. **Problem Formulation as per Situations :**

- It deals with economics of the operations. The specified problem is splitted into four element which are as follows :
  (i) The surrounding
  (ii) System operator
  (iii) The objectives
  (iv) Alternatives and constraints.

(i) The Surrounding :

The surrounding is the outline within which organised activity system is guided to reach the predefined aims or goals. It involves many factors such as economics, social and physical which may change the problem under study. It consists of resources such as men, materials, machines etc.

(ii) System Operator :

System operator is a person who control the operations of system. It is an important element in problem formulation stage.

(iii) The Objectives :

It is applied for analysis of the problem. It takes into account good and worst effect on system.

(iv) Alternatives and Constraints :

These are applied to reach the prescribed objectives. It consists of technical feasibility, previous experience, legal obligations etc. Once the objectives, alternatives and constraints are available, identification of system operator and the surroundings are ready then the problem is considered for further development.

2. **Model Development for System :**

- It is followed after formulation of problem. The model involves set of equations which defines the system or problem. These equations provides the effectiveness function and constraints. Generally, the objective of system is known as effectiveness function and limitation on the satisfaction of the objectives are known as Constraints.

- The general form of model developed is :

$$R = F(x_i, y_i)$$

where,   R = Effectiveness function or resulting function

         x_i = Constraints or controllable variable

         y_i = Constraints or uncontrollable variable

         F = Relationship between R and x_i, y_i

- A model expresses the relationship between variables and resulting function. Once model is developed the selection of inputs are performed. This requires special efforts to keep minimum error.

3. Finding Solution for Model :

- This is achieved by mathematical methods or performing simulations (experiments). The mathematical methods used for finding solution model consists of the following steps :

(i) Analytical steps

(ii) Numerical steps

(i) Analytical Steps : For analytical steps the various concepts of mathematics such as vector algebra or calculus of matrix, differential calculus, finite difference are used to get the optimal solution. It is somewhat difficult than numerical steps.

(ii) Numerical Steps : For numerical steps trial and error method is applied to get optimal solution. The procedure starts with initial (trial) solution and repeated till the optimality of solution is achieved. In finding solution of model another approach is simulation method. It is the process which consists of series of predefined trial and error experiments to decide the behaviour of model. For this, Monte Carlo method is mostly applied to get solution.

4. Solution Analysis (Validation) :

- The validity of solution is checked by comparing the results achieved. It provides effect of variable in constraints on overall system performance. For this, sometime, certain time may elapsed for data to be collected for solution of the problem.

5. Deriving Controls over the Solution of Problem :

- A solution achieved for model may change as per change parameters or variables, hence controls must be derived to express the limit within which the model and its solution can be considered as reliable. For this certain tools needs to be developed.

6. Implementation of Solution of the Problem :

- The solution achieved above should be implemented to operate procedure which can be known and applied by those who controls the operations. The actual response of system is verified by implementing the solution achieved previously. As per performance of system modification or changes should be made on the part of operation research group. The success of operations research study is based on co-ordination received from management at implementation stage.

- The above stages or methodology may be used as per problem situation and may be interchanged as per need.

1.5 OBJECTIVES OR GOALS OF OPERATIONS RESEARCH

The Main Objectives or Goals of Operations Research are as Follows :

- To provide a scientific base for management of an organization for solution of problem involving iteration element of system.

- To find a solution which is in optimum interest of the organization as a whole.

- To employ a system procedure by a team of personnel considered from various disciplines.

1.6 APPLICATIONS OF OPERATIONS RESEARCH   [May 17, Oct. 15, 4M, May 05, 6M]

- Quantitative method and OR are applied for optimization problem of variety of situations such as :

1. Industrial management

2. Development

3. Defence operation

4. Agricultural field

5. Business and society.

1. Industrial Management :

In industry, it is applied for finding optimized solution to problem such as purchase of raw material for finished components. Quantitative method and OR are used for planning, inventory control, repair, sequencing, scheduling, project management etc.

2. Development :

It is used for developing and developed economics. It is used for optimization of problem such as poverty, hunger in various countries.

3. Defense Operations :

It is used to defense field such airforce, army and navy. It is applied for optimization of defense operation to fulfill desired goals.

4. Agricultural Field :

It is used for optimization of agricultural sector for the problem such as shortage of food, land, environmental condition, water resources.

5. Business and Society :

Operation research and quantitative methods are used for optimization of problem associated with small and large scale organization, society and economy, hospitals, transport, LIC etc.

1.7 CLASSIFICATIONS OF QUANTITATIVE TECHNIQUES   [May 16, NOV./DEC. 06, 6M]

- The various operations research models have been developed and used to problems in industries and business. The most commonly used models are as follows :

1. Mathematical models
2. Statistical models
3. Inventory control models
4. Assignment and Transportation models
5. Project management
6. Sequencing models
7. Routing techniques
8. Queuing theory or models
9. Simulation models
10. Competitive technique
11. Decision making technique
12. Replacement models
13. Reliability theory or models
14. Programming models
15. Combined technique.

1. Mathematical Models :

- Mathematical models can be applied as most important tool for operations personnel. The most common mathematical concepts involved are as follows :
  - ➢ Vector algebra
  - ➢ Matrix algebra
  - ➢ Differential equation
  - ➢ Linear differential equation
  - ➢ Partial differential equation
  - ➢ Integration equation
  - ➢ Theory of operator.

2. Statistical Models :

- It involved probability and statistics theory, probabilities are related with event to predict uncertainties and give inputs with accuracy for decision theory. Probability technique is based on historical evidence of experience of personnel for decision-making for special events. These models also includes discrete and continuous probability, theory of renewal, stochastic operation etc.

3. Inventory Control Models :

- Inventory control models are applied for deciding economic order quantities of an item or unit for minimizing cost of the item. The mathematical technique applied for solution of inventory control problems are such as quadratic program, dynamic program, EOQ equation. It involves planning of raw material, finished items, consumable stocks etc.

4. Assignment and Transportation Models :

- Assignment and Transportation models are applied for assigning of personnel/items to the various destination for maximizing profit, sale or minimizing cost. The transportation model is applied when some of items needs more than one resource and vice versa.

5. Project Management :

- It is applied for complex project which need number of resources (such as material, men, machine etc.). The most common project management technique used are Critical Path Method (CPM) and PERT (Programming Evaluation and Review Technique).

6. Sequencing Models :

- It involves sequences of various items, jobs (works) on various machines for maximizing profit, sale or minimizing cost. These consists of :
  - ➢ Sequence of 'n' job on two machines
  - ➢ Sequence of 'n' job on three machines
  - ➢ Sequence of 'n' job on 'm' machines.
- The main objective is to analyse elapsed time and idle time for item/job and workmen respectively.

7. Routing Techniques :

- It is special case of sequencing models. Routing technique involves two problems which are as follows :
  - ➢ Shortest route (smallest path problem).
  - ➢ Travelling salesman problem.
- The concept of assignment and transportation model is used to routing technique.

8. Queuing Theory or Models :

- It is related to service at various destination as per flow of a resource. It involves parallel and series service sources or both for minimizing cost. It also involves waiting line theory.

9. Simulation Models :

- It involves development of analytical method and solving these models which are concerned with complex relationship. It is applied when actual experimentation is not feasible and analytical technique of construction model and solving them are not helpful. It is most important method of operation research. It is used to the problem such as production lines, management of warehouses, queuing theory, inventory problem, decision-making problems.

10. Competitive Techniques :

- It involves decision-making, game theory technique which are used for situation where chance of wining increases but the predicted profit decrease. It is applied for the following problems :
  - ➢ Share market or stock market
  - ➢ Bidding problems
  - ➢ Trade union problem
  - ➢ Bargaining problem
  - ➢ Planning of war tactics.

11. Decision-Making Techniques :

- These models involved tactical decision theory of strategic decision theory. Tactical decision theory is used in situation such as engaging number of salesman, operators number of shifts. Strategic decision theory is used to situation such as launching new product in market, automation, expenses, manufacturing etc. Decision-making models involving the following cases :
  - ➢ Decision-making under situation of risk.
  - ➢ Decision-making under situation of certainty.
  - ➢ Decision-making under situation of uncertainty.

12. Replacement Models :

- It involves replacing machine or equipment or item after periodic use for certain application due to depreciation or decrease in performance.

13. Reliability Theory or Models :

- An ability of a device or equipment to work without any failure for a specific period of time under specified condition is known as *Reliability*. It involves statistical approach for testing of device or equipment. Reliability of complex device depends on reliability of its support or subcomponents. It is useful in design of principle of complex device such as radar, computer, electrical devices etc.

14. Programming Models :

- It involves non-linear programming, dynamic programming, integer programming, goal programming, heuristics programming, quadratic programming, sensitivity analysis, parametric programming, stochastic programming.

15. Combined Technique :

- In this technique, more than one operation research models are combined as per problem to get optimal solution in specific order. It involves, assignment models, inventory control models, queuing models etc. It is used to problem such as production problems, marketing problems etc.

1.8 LIMITATIONS OF OPERATIONS RESEARCH

The Disadvantages or Limitations of Quantitative Method or Operations Research as Follows :

- Quantitative method and OR do not consider qualitative factors or emotional factors which are quite real.
- Mathematical methods are used to only specific cases of problem.
- The decision person is not completely aware of shortcoming of Quantitative method and OR that he is applying to problems.
- It is applicable to repetitive types of problems or situations.
- Management may itself offer a lot of prohibition or restrictions due to conventional thinking.

QUESTIONS FOR PRACTICE

1. Define quantitative methods and operations research. Describe briefly its function. **[May 17, May 08, 4M]**

2. Explain in brief how operations research has been evolved or developed.   **[Nov. 15, May 07, 5M]**

3. Explain in brief, steps in methodology of operations research.   **[Dec. 17, May 07, 5M]**

4. What methods are used to solve operations research problem? Explain.   **[Dec. 17, May 06, 6M]**

5. What are the characteristics of quantitative methods and operations research.

2.1 INTRODUCTION

[May 12, Dec. 18, Nov./Dec. 06, May 13, 4M]

An optimization technique in which the objective function and constraints functions are linear in relationship or linear in nature is known as **Linear Programming**. The objective function may be related to sale, cost, profit, manufacturing ability etc. which are determined in best possible way or in optimal form. The constraints functions consists of resources (machine, men, material etc.). Linear programming was developed by George Dantzig in 1947 for United state air force situations. He generated *'Simplex Method'* to solve linear programming problems.

2.1.1 Requirements or Conditions of Linear Programming Problem

The conditions or requirement of linear programming problems to be satisfied are as follows :

- An objective function must be linear function of decision variables.
- The constraints function must be expressed in linear form.
- The variable must be non-negative.
- The decision variables must form a set of alternative course of action.

2.1.2 Assumptions in Linear Programming Models

The assumptions for linear programming models or problems are as follows :

- There must be linear relationship between objective function and constraints functions.
- The decision variables should be added as per demands.
- The decision variables are permitted to have any non-negative values that satisfies the constraints.
- There must be certainty in constraints functions.
- There must be availability of alternatives for decision maker.

2.1.3 Limitations or Shortcomings of Linear Programming Problem   [May 16, Nov. 12, 4M]

The limitations or shortcomings of linear programming problem are as follows :

- The relationship between objective function and constraints may not be linear in all cases.

- It may provide answer of variable in fraction due to which solution may not be optimal when round-off.
- It is used only for single objective problems.
- Sometime the decision variables are probabilistic in nature which is not desirable.

2.1.4 Applications of Linear Programming Problem

[May 17, Nov. /Dec. 13, Nov. /Dec. 12, 6M]

- Linear programming technique is used for the following fields :
  1. Industry
  2. Business
  3. Agricultural field
  4. Miscellaneous applications.

1. **Industry :**

   In industry, concept of linear programming is applied in production scheduling, assignment problems, transportation problem, personnel allocations etc.

2. **Business :**

   It is applied for advertising areas, economics, marketing fields, military areas etc.

3. **Agricultural Field :**

   Linear programming is used for planning of resources of agricultural field such as water, land, fertilizers, loads, crops etc.

4. **Miscellaneous Applications :**

   It is applied for medical field, light scheduling at airports, environmental situation, service centres etc.

2.2 FORMULATION OF LINEAR PROGRAMMING PROBLEM   [May 17]

- The steps used for mathematical formulation of linear programming problem are as follows :

  Step 1 : Determining the decision variables.

  Step 2 : Representation of objective function.

  Step 3 : Writing equation for constraints functions or constraints.

  Step 4 : Formation of linear programming problem.

Step 1 : Determining the Decision Variables :

As per the situation or problem, determine the number of decision variables desired. This can be obtained by considering question "What is being needed?"

Step 2 : Representation of Objective Function :

The objective function may be minimization or maximization. Write the objective function in linear form in terms of decision variables.

Step 3 : Writing Equation for Constraints Functions or Constraints :

The constraints are generally the resources such as machine, men, materials, methods etc. Write the relation between the constraints in linear form as per case of problem.

Step 4 : Formation of Linear Programming Problems :

All above steps (1), (2) and (3) are expressed combinely as :

> Objective Function (minimize or maximize) $Z = f(x_i)$
>
>   Subject to constraints : $F(x_i)$
>
>   where, $i = 1, 2, 3, \dots\dots n$

The above steps are explained with example 2.1.

SOLVED EXAMPLES

Example 2.1 : An industry is producing two jobs x and y. The manpower requirement per unit of the job x and y are 4 man hours and 3 man hours. The processing time needed per unit is expressed as 1 and 2 machine hours respectively. The raw material demand are as 0.3 and 0.2 kg per unit. The weekly availability of man hours, machine hours and raw materials are 100 man hour, 80 machine hour and 30 kg respectively. It is estimated that the profit per unit of x and y are as ₹5 and ₹3 respectively. Determine how much of jobs x and y are to be produced per week to maximize profit. Develop the linear programming model only.

Solution : Formulation of linear programming problem is as follows :

Step 1 : Determining the Decision Variables :

To get number of units of jobs x and y, ask the question "How much of jobs x and y are to be produced per week ?"

Let x_1 and x_2 be the number of units of x and y required.

Step 2 : Representation of Objective Function :

Since, the profit per unit of jobs x and y are given as ₹ 5 and ₹ 3 respectively and here profit is to be maximized. Let z be the objective function.

The objective function is expressed as :

$$\boxed{\text{Max. } z = 5x_1 + 3x_2}$$

Step 3 : Writing Equation for Constraints or Constraints Functions :

Since, constraints are man hours. machine hours and raw materials.

The equations are expressed as follows :

(a) For Man Hours :

As total availability of man hours is 100 per week. The requirements are 4 man hours per unit of job x and 3 man hours per unit of job y.

Since, x_1 and x_2 are number of units of jobs x and y per week.

Thus, $4x_1 + 3x_2 \leq 100$

(b) For Machine Hours :

Similarly, as per data of given problem, we get

$x_1 + 2x_2 \leq 80$

(c) For Raw Material :

As per given data of problem, we have

$0.3x_2 + 0.2x_2 \leq 30$

OR

$3x_1 + 2x_2 \leq 300$

Table 2.1

Resources	Job → x	y	Availability
Man hour	4	3	100
Machine hour	1	2	80
Raw material	3	2	300
Profit	₹ 5	₹ 3	

Step 4 : Formation of Linear Programming Problem :

All above steps (1), (2) and (3) are expressed combinely as :

Maximize $z = 5x_1 + 3x_2$

subject to : $4x_1 + 3x_2 \leq 100$

     $x_1 + 2x_2 \leq 80$

     $3x_1 + 2x_2 \leq 300$

     $x_1 \geq 0, x_2 \geq 0$ (Since these must be non-negative in nature)

2.3 GRAPHICAL SOLUTION OF LINEAR PROGRAMMING PROBLEM

- Generally, graphical method can be applied for 2 or 3 variables whereas simplex method can be applied for any number of variables.

- **The Steps Used are as Follows :**

  Step 1: Construction of mathematical problem or standard form of LPP.

  Step 2: Draw x_1 and x_2 axes (or x and y axes).

  Step 3: Sketch the constraints on graph.

  Step 4: Locate the feasible solution or region.

  Step 5: Sketch the objective function on graph.

Step 1 : Construction of Mathematical Problem or Standard Form of LPP :

The mathematical problem or standard form of LPP is obtained as per step given in Section 2.2.

Step 2 : Draw x_1 and x_2 Axes (or x and y Axes) :

Since, $x_1 \geq 0$, $x_2 \geq 0$, this shows that value of constraints x_1 and x_2 lies in first quadrant. Ignore the number of infeasible alternation that lies in second, third and fourth quadrant.

Step 3 : Sketch the Constraints on Graph :

Assign any arbitrary value to one variable and obtain the value of other variable from the equations of constraints. Generally assume, arbitrary value as zero for one variable and obtain the value of other variable and vice versa. Locate these points on graph and draw straight line through this points on graph.

Step 4 : Locate the Feasible Solution or Region :

Identify the feasible solution or region that satisfy all the constraints simultaneously. Generally, for $\geq$ type constraints, the area on or above the constraints line i.e. away from origin and for $\leq$ type constraints, the area on or below the constraints line i.e. towards origin will be preferred or considered. Any point on or within the shaded (hatched) region represents a feasible solution to the given problem.

Step 5 : Sketch the Objective Function on Graph :

Assume objective function to be zero. Draw a line passing through origin. As the value of objective function is increased from zero, the straight line starts moving to right parallel to itself. Sketch line parallel to this line till the line is farthest away from origin (for maximization problem). For minimization problem, the line will be nearest to origin. The point of feasible region or area through which this line passes will be the optimal point. Sometime this line may match with one of edges of the feasible area or region. In that situation, every point on that edge will give the some largest/smallest value of z and will be the optimal point. Alternatively, this point at which objective function z is largest/smallest, is the optimal point and its co-ordinates expresses the optimal solution.

The above steps are explained with example 2.2.

Example 2.2 : *A factory manufactures two items A and B on which the profit obtained per unit are ₹ 3 and ₹ 4 respectively. Each item is processed on two machines x and y. Item A requires one minute of processing time on machine x and two minutes on machine y while B requires one minute on machine x and one minute on machine y. Machine x is available for not more than 7 hrs. 30 minutes. While machine y is available for 10 hrs. during any working days. Find the number of units of items A and B to be manufactured to have maximum profit.*

Solution :

Step 1 : Construction of Mathematical Problem or Standard Form of LPP :

The following steps given in Section 2.2, we have

Table 2.2

Machine	Items		Availability (Minutes)
	A	**B**	
x	1	1	450
y	2	1	600
Profit	₹ 3	₹ 4	

Maximize $z = 3x_1 + 4x_2$

subject to : $x_1 + x_2 \leq 450$

$2x_1 + x_2 \leq 600$

$x_1 \geq 0$ $x_2 \geq 0$

Step 2 : Draw x_1 and x_2 Axes :

Since, $x_1 \geq 0$, $x_2 \geq 0$

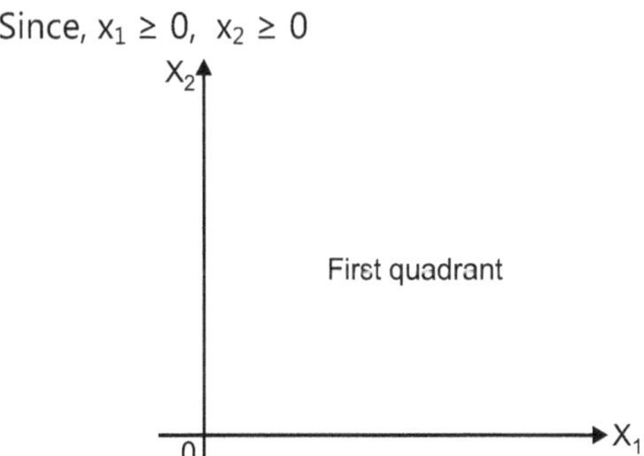

Fig. 2.1

Step 3 : Sketch the Constraints on Graph :

(a) Since, $x_1 + x_2 \leq 450$

    Let $x_1 + x_2 = 450$

    Put $x_1 = 0$, we have $x_2 = 450$, A $(0, 450)$

             $x_2 = 0$, we have $x_1 = 450$, B $(450, 0)$

        $\therefore$ The co-ordinates A and B are shown in Fig. 2.2

(b) Since, $2x_1 + x_2 \leq 600$

    Let $2x_1 + x_2 = 600$

    Put $x_1 = 0$, $x_2 = 600$, C $(0, 600)$

             $x_2 = 0$, $x_1 = 300$, D $(300, 0)$

    $\therefore$ The co-ordinates C and D are shown in Fig. 2.2.

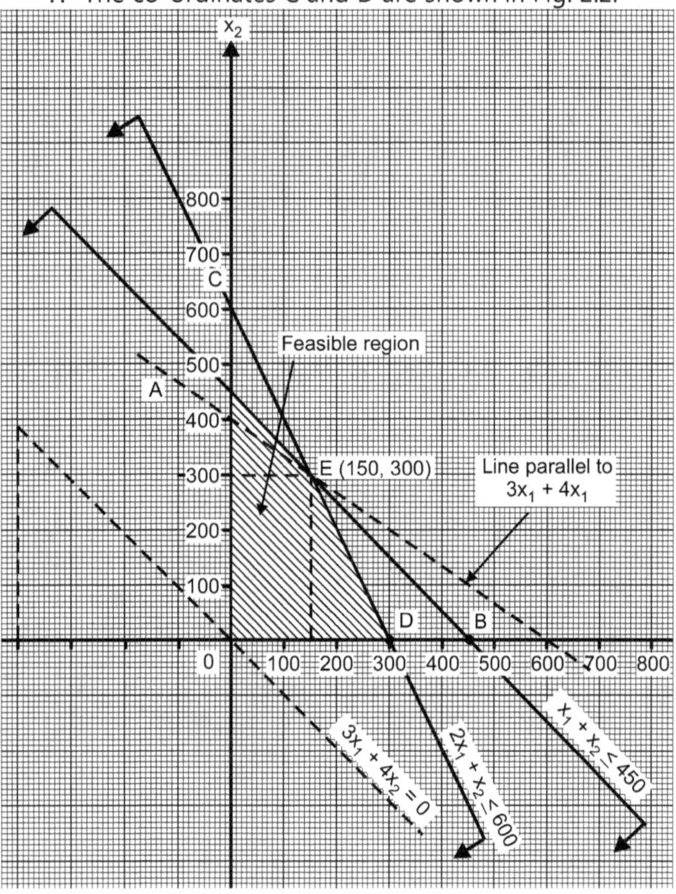

Fig. 2.2

To determine co-ordinate of E (point of intersection) transfer points on x_1 and x_2 axes, i.e. E $(150, 300)$.

To Find Optimal Solution :

Method I :

    Let $z = 0$

    $\therefore$ From objective function, we have

        $3x_1 + 4x_2 = 0$

           $3x_1 = -4x_2$

$$\boxed{\dfrac{x_1}{x_2} = \dfrac{-4}{3}}$$ this is slope of objective function

Draw line with slope $-\dfrac{4}{3}$ from origin. It is as shown in Fig. 2.2.

Draw parallel line to this line till the line is farthest away from origin (touching the feasibles region) and passing through one point of feasible region or area. This is the point where maximum profit is achieved.

Method II : (Alternative Method)

Here, we have four vertices : O, A, E, C of the convex area (shaded region).

Here O$(0, 0)$, C$(300, 0)$, E$(150, 300)$ and A$(0, 450)$, obtain the value of the objective function $z = 3x_1 + 4x_2$ at these vertices.

at O(0, 0)	at C(300, 0)
$z = 3x_1 + 4x_2 z$	$= 3x_1 + 4x_2$
$z = 3(0) + 4(0)$	$= 3(300) + 4(0)$
$\boxed{z = 0}$	$\boxed{z = 900}$

at E(150, 300)	at A(0, 450)
$z = 3x_1 + 4x_2 \ z$	$= 3x_1 + 4x_2$
$z = 3(150) + 4(300)$	$= 3(0) + 4(450)$
$= 450 + 1200$	$= 3(0) + 4(450)$
$\boxed{z = 1650}$	$\boxed{z = 1800}$

Therefore, the maximum value of objective function z is ₹ 1800 and it occurs at vertex A$(0, 450)$.

    $\therefore$ Optimal solution is

$$\boxed{z_{max} = ₹\ 1800 \text{ at } x_1 = 0, \ x_2 = 450}$$

Example 2.3 :

Minimize $z = 2x_1 + x_2$

Subject to :

$$3x_1 + x_2 \geq 3$$
$$4x_1 + 3x_2 \geq 6$$
$$x_1 + 2x_2 \geq 3$$
$$x_1 \geq 0, \qquad x_2 \geq 0$$

Solution : Graphical solution of above problem is given as follows :

(a) Since, $3x_1 + x_2 \geq 3$

    Let $3x_1 + x_2 = 3$

    put $x_1 = 0$, $x_2 = 3$ A$(0, 3)$ ⎤ These are located

             $x_2 = 0$, $x_1 = 1$ B$(1, 0)$ ⎦ on graph (Fig. 2.3)

(b) Since, $4x_1 + 3x_2 \geq 6$

    Let $4x_1 + 3x_2 = 6$

    put $x_1 = 0$, $x_2 = 2$ C$(0, 2)$ ⎤ These are located

             $x_2 = 0$, $x_1 = \dfrac{3}{2}$ D$\left(\dfrac{3}{2}, 0\right)$ ⎦ on graph (Fig. 2.3)

(c) Since, $x_1 + 2x_2 \geq 3$

    Let $x_1 = 0$, $x_2 = \dfrac{3}{2}$ E$\left(0, \dfrac{3}{2}\right)$ ⎤ These are located

             $x_2 = 0$, $x_1 = 3$ F$(3, 0)$ ⎦ on graph (Fig. 2.3)

(d) Since, $z = 2x_1 + x_2$

    Let $z = 0$

    $\therefore$ $2x_1 + x_2 = 0$

$$2x_1 = -x_2$$

$$\boxed{\frac{x_1}{x_2} = -\frac{1}{2}} \quad ....\text{This is slope of objective function.}$$

To find co-ordinates of point G (point of intersection), transfer point on x_1 and x_2 axes. Alternatively point of intersection is obtained, by solving equation of two lines giving point of intersection.

For optimal solution

at G(0.6, 1.2)

$z = 2x_1 + x_2$

$z = 2(0.6) + 1.6$

$\boxed{z = 2.4}$

at A(0, 3)

$z = 2x_1 + x_2$

$= 2(0) + 3$

$\boxed{z = 3}$

at F(3, 0)

$z = 2x_1 + x_2$

$= 2(3) + 0$

$\boxed{z = 6}$

Thus, z (objective function) is minimum at G(0.6, 1.2)

∴   The optimal solution is

$$z_{min} = 2.4 \quad \text{at} \quad x_1 = 0.6, \quad x_2 = 1.2$$

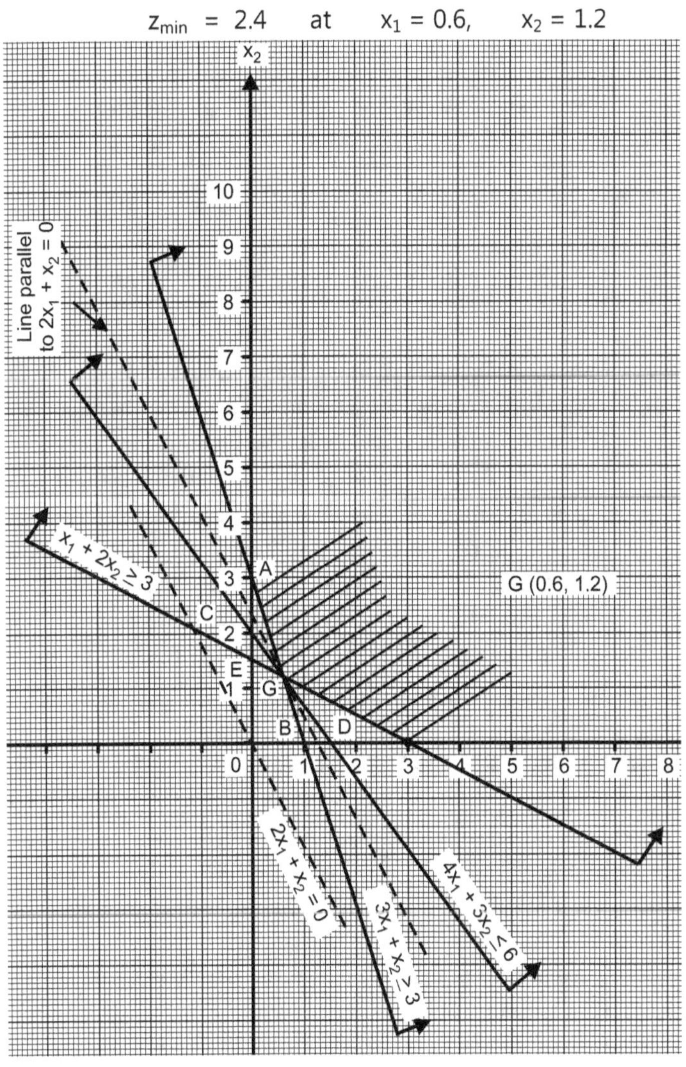

Fig. 2.3

UNIVERSITY QUESTIONS

1. What are the limitation of LPP. (refer Article 2.1.32. Explain graphical method to solve linear programming problem.   **[May 17, Nov 14, 4M]**

2. Write short note on "Optimal Solution on Graph".

4. Maximize

$$z = 5x_1 + 3x_2$$

Subject to :

$$3x_1 + 5x_2 \le 15$$

$$5x_1 + 2x_2 \le 10$$

$$x_1 \ge 0, \quad x_2 \ge 0 \qquad \text{(refer solved example 2.2)}$$

$$\left[\textbf{Ans.:} \; z_{max} = \frac{235}{19} \text{ at } x_1 = \frac{20}{19}, \; x_1 = \frac{45}{19} \right]$$

5. Define linear programming. Give application of linear programming.   **[May 18, Nov./Dec. 06, 6M]**

6. A firm uses lathe, milling and grinding machines to manufactures parts. The given table represents the machining time required for parts. The manufacturing time available on different machines and processing on each machine part. Find graphically number of Part I and Part II to be manufactured per week in order to maximize profit.

Table 2.3

Type of Machine	Machining Time Required for Machined Parts (Minutes)		Machining time Required Per Week (Minutes)
	I	II	
Lathe machine	12	06	3000
Milling machine	04	10	2000
Grinding machine	02	03	900
Profit per unit	₹ 40	₹ 100	

[**Ans. :** Mathematical formulation :

Maximize $z = 40x_1 + 100x_2$

Subject to : $12x_1 + 6x_2 \le 3000$

$4x_1 + 10x_2 \le 2000$] **[May 17, Nov./Dec. 06, 6M]**

7. Two animal feeds A and B are available in market one kg of A contains 0.1 kg of x_1, 0.1 kg of x_3 of 0.2 kg of x_4 and 1 kg of B contains 0.1 kg of x_2, 0.2 kg of x_3 and 0.1 kg of x_4 (x_1, x_2, x_4 are different ingredients). A feed mixing operation can be described in terms of two activities. The daily per head requirement is of atleast 0.4 kg of x_1, 0.6 kg of x_2, 2.0 kg of x_3 and 1.8 kg of x_4. Feed A can be bought for ₹ 0.07 per kg and feed B for ₹ 0.05 per kg. The availabilities, requirements and costs are summarised in the following Table 2.4.

Table 2.4

Ingredient	A (kg)	B (kg)	Requirement (kg)
x_1	0.1	0	0.4
x_2	0	0.1	0.6
x_3	0.1	0.2	2.0
x_4	0.2	0.1	1.8

Cost → ₹ 0.07 / kg, ₹ 0.05 / kg

Determine the quantities of feeds A and B in mixture so that the total cost is minimum (refer solved example 2.2).

8. Solve LPP by Suitable Method.　　　**[Dec 11, 10M]**

　　　Maximize :　$Z = 10x_1 + 20x_2$

　　　Subject to : $5x_1 + 2x_2 \leq 30$

　　　　　　　　$3x_1 + 6x_2 < 36$

　　　　　　　　$2x_1 + 5x_2 \leq 20$

where　　　　　　$x_1 > 0, x_2 > 0$

[Ans. $Z = 90.5$, $X_1 = 5.2$, $X_2 = 1.9$] [For graphical method error of 6% is allowed]

9. A firm manufactures two products $P_1 P_2$ on which the profits earned are ₹ 5 and ₹ 8 respectively. Each product is prepared on two machines M_1 and M_2. The machine time required for these products on the two machines and their availability is as shown below.

	Product P_1	Product P_2	Availability of Machines (Mins.) Per Day
Machine M_1	2	1	400
Machine M_2	4	1	600

Find the number of units of products P_1 and P_2 to be manufactured per day to get maximum profits?

Ans. : Refer Section 2.2 [Example 2.1].

[Ans. $P_1 = 0$, $P_2 = 400$, $Z = 3200$

[May 15, Dec. 11, 10M]

10. A firm manufacturers three products S_1, S_2 and S_3 on which the profits earned are ₹ 2, ₹ 5 and ₹ 4 respectively. Each product need two types of raw materials R_1 and R_2 which the firm can purchase upto a maximum of 500 and 400 units respectively. Design production plan so as to maximize the profit.

Raw Material	Consumption of Raw Materials Per Unit Product		
	S_1	S_2	S_3
R_1	0.5	1	1
R_2	2	0.5	0.5

Ans. : Refer Section 2.2 [Example 2.2].

[Dec. 17, May 10, 16M]

11. A factory has decided to diversify its activities. The data collected for the sales and production departments are summarized below :

[Nov. / Dec. 12, 12M]

Potential demand exists for two products A and B. Market can absorb, any quantity of A, whereas the share of B for this organization is expected to be not more than 400 units per month. Contribution per unit of product A and B is expected to be ₹ 6 and ₹ 8 respectively. These products, require three different processes and the time required per unit of product is given in the table below, :

Process	Product A	Product B	Available Hours
1	2	3	900
2	1	2	600
3	2	2	1200

Find the product mix to optimize the contribution.

Ans. : Refer Section 2.2　　　**[Dec. 16, May 12, 10M]**

12. A person requires 10, 12 and 12 units of chemicals A, B and C respectively per Jar. A dry product contains 1, 2 and 4 units of A, B and C per cartoon. If the liquid product sells for ₹ 3 per Jar and dry product sell for ₹ 2 per cartoon, how many of each should be purchased to minimize the cost and meet the requirements.

Ans. Refer Section 2.2 [Example 2.2].　**[May 13, 12M]**

2.4 SIMPLEX METHODS

[May 17, Nov./Dec. 06, 10, 07, May 12, 6M]

- The simplex method was developed by Dantzig G.B. an American mathematician in 1947. Simplex method is applied for solving linear programming problem having any number of variables (two or more than two variables).

- In simplex method, an initial basic feasible solution can be improved in consecutive iteration until a final optimal solution is achieved. It can provide a systematic reduction from an infinite number of solution to a definite number of promising solutions.

2.4.1 Requirements or Conditions of Simplex Method

- Simplex method consists of the following two fundamental requirements or conditions :

  1. Condition of feasibility
  2. Condition of optimality.

1. Condition of Feasibility :

- When the starting solution is basic feasible then only basic feasible solution will be obtained during iteration.

2. Condition of Optimality :

- Simplex method ensures that only better solution (with respect to current solution) will be encountered. Basically, it can be applied for searching better corner point with number of repetitive iterations.

2.4.2 Simplex Algorithms or Steps for Solution of L.P.P. [May 18, Nov. 12, 6M]

The Steps (Algorithms) of Simplex Method are as Follows :

 Step 1 : Write the problem in standard mathematical form.

 Step 2 : Find an initial basic feasible solution for the problem.

 Step 3 : Preparation of simplex table and check for optimality.

 Step 4 : Application of simplex criterion (I and II).

 Step 5 : Preparation of next iteration table or format.

 Step 6 : Continue successive iteration until optimal solution is achieved.

Step 1 : Write the Problem in Standard Mathematical Form :

- When the decision variables are non-negative, right hand side of constraints are non-negative and constraints are expressed as equation then the given problem is said to be represented in standard mathematical form.

- Let $s_1, s_2, s_3, ... , s_n$ be slack variables (as per constraints equation). Add this slack variable to the left hand side of first, second, third ... n^{th} constraints respectively to put them into equation. The values of $s_1, s_2, s_3, ... , s_n$ depends on values of $x_1, x_2, x_3, ... , x_n$ i.e. slack variable vary with $x_1, x_2, x_3, ... , x_n$ variables.

- Slack variable represents **Unutilised Resources** or **Availability** i.e. it represents idle time, men, material etc. Since, slack variable represents an idle resource, they contribute zero in objective function. Therefore, they are added with zero coefficient in the specified objective function.

- Thus, the standard form of specified problem is represented as follows :

Table 2.5

Max. or Min. $z = a_1x_1 + a_2x_2 + a_3x_3 + ... + a_nx_n + 0s_1 + 0s_2 + 0s_3 + ... + 0s_n$

Subject to : $a_{11}x_1 + a_{12}x_2 + a_{1n} x_n + s_1 + 0s_2 + 0s_3 + ... + 0s_n = b_1$

$a_{21}x_1 + a_{21}x_2 + ... + a_{2n}x_n + 0s_1 + s_2 + ... + 0s_n = b_2$

$a_{31}x_1 + a_{32}x_2 + ... + a_{3n}x_n + 0s_1 + 0s_2 + s_3 + ... + 0s_n = b_3$

and so on

where, $x_1 \geq 0$ $x_2 \geq 0$ $x_3 \geq 0$ $x_n \geq 0$, $s_1 \geq 0$, $s_2 \geq 0$, $s_3 \geq 0$, $s_n \geq 0$

Step 2 : Find an Initial Basic Feasible Solution for the Problem :

- Put $x_1 = 0$, $x_2 = 0$, $x_3 = 0$, ... , $x_n = 0$ in constraints equation of step (1) and find values of $s_1, s_2, s_3, ... , s_n$. Values of $s_1, s_2, s_3, ... , s_n$ are referred as initial basic feasible solution.

Step 3 : Preparation of Simplex Table and Check for Optimality :

The simplex table is prepared as follows :

Table 2.6 : Initial Simplex Table

	c_j	a_1	a_2	a_3	....	a_n	0		0	0	....	0	
C_b	Basis	X_1	X_2	X_3	....	X_n	S_1		S_2	S_3	....	S_n	B
0	s_1	a_{11}	a_{12}	a_{13}	....	a_{1n}	1		0	0	....	0	b_1
0	s_2	a_{21}	a_{22}	a_{23}	....	a_{2n}	0		1	0	....	0	b_2
0	s_3	a_{31}	a_{32}	a_{33}	....	a_{3n}	0		0	1	....	0	b_3
.	.	.	.	.		.	.		.	.		.	.
.	.	.	.	.		.	.		.	.		.	.
.	.	.	.	.		.	.		.	.		.	.
0	s_n	a_{n1}	a_{n2}	a_{n3}	....	a_{nn}	0		0	0	....	1	b_n
	z_j	0	0	0	....	0	0	....	0	0	....	0	b
	Δ_j	a_1	a_2	a_3	....	a_n	0		0	0	....	0	

In above simplex table (Table 2.6), the interpretation is as follows :

- The coefficient of c_j (contribution/unit) of the variable in the objective function equation (in Table 2.5) is given in first row. These coefficient remain unchanged in further simplex table. The variable in the problem for which c_j coefficient have already been expressed is represented in second row.

- The coefficient of current basic variable in the objective function is represented in first column (column c_b). The basic variable (slack variable) of the current solution are represented in second column (basis).

- The coefficient matrix under non-basic variables x_1, x_2, x_3, ..., x_n expresses their coefficient in constraints equation (as in Table 2.5). These coefficient expresses the quantity of resource required to make a unit of product.

- The coefficient of slack variable in the constraints are represented in coefficient of slack variable. It is always unit matrix or identity matrix [I_r where r, is order of matrix].

- The column B (last column of Table 2.6) is known as quantity column or solution value column. This column represents the quantities of available resource or right hand side value of constraints or value of basic variables s_1, s_2, s_3, ... , s_n in the initial basic feasible solution achieved earlier. Variables not written under basic column are hence basic variable and their values are zero.

- **Check for Optimality :** It is used for checking improved solution. In (Table 2.6) row coefficient (z_j) under any column is estimated by adding the products of element under that column with respective c_b value or from objective function. i.e. $z_j = \sum cb \cdot a_{ij}$ where a_{ij} are element of matrix in the ith row and j^{th} column. In Table 2.6, Δ_j is referred as net evaluation row or index row. It is expressed as $\boxed{\Delta_j = c_j - z_j}$. The value of Δ_j expresses whether solution is optimal or not.

[**Note :** The current solution is said to be optimal if Δ_j is negative or zero. If Δ_j is positive then current solution is not optimal and therefore there is a need of improvement.]

Step 4 : Application of Simplex Criterion (I and II) :

The simplex criterion are as follows :

1. Simplex criterion (I) : Decision of variable to be entered (entering variable).

2. Simplex criterion (II) : Decision of variable to be leaved (leaving variable).

1. Simplex Criterion (I) : Decision of Variable to be Entered (Entering Variable) :

To decide the entering variable, choose the column having maximum value of Δ_j. The variable leading that column is the variable to be entered (entering variable). This variable is known as variable to be entered (entering variable or incoming variable). The corresponding column is known as key-column and marked as k_c. When there are more than one variable with maximum value of Δ_j then choose the variable arbitrarily. If there is no more positive value in Δ_j row, the profit achieved is maximum or optimal solution is achieved.

2. Simplex Criterion (II) : Decision of Variable to be Leaved (Leaving Variable) :

The entering variable will replace the current basic variable s_1 or s_2 or s_3 or s_n. To decide which of slack variable to be leaved (removed or made zero or made non-basic) (leaving variable), divide elements under B-column (quantity column) by respective element of key column (k_c) i.e. determine minimum ratio.

$$\text{Minimum ratio} = \frac{\text{Element under B-column}}{\text{Respective element of key column } (k_c)}$$

Choose the minimum non-negative ratio (i.e. ignore negative ratio) and mark the column containing these ratio as ϕ – column. The row so chosen (marked) is known as key-row and represented by k_R. The element laying at the intersection of key-row and key-column is known as **Key Element** and is enclosed by( ).

Step 5 : Preparation of Next Iteration Table or Format :

After decision of entering and leaving variable, replace slack variable (s) by decision variable (x) and change corresponding c_b coefficient. If key element is other than '1' then make it one by dividing that element as well as all elements in the row of key element. New element in rows are obtained as :

$$\begin{bmatrix} \text{New element} \\ \text{in respective row} \end{bmatrix} = \begin{bmatrix} \text{Old element of} \\ \text{respective row} \end{bmatrix} - \begin{bmatrix} \text{Element} \\ \text{below or above} \\ \text{key element of} \\ \text{respective row} \end{bmatrix} \times \begin{bmatrix} \text{Corresponding} \\ \text{element} \\ \text{[element of} \\ \text{respective row} \\ \text{i.e. obtained by} \\ \text{dividing key element]} \end{bmatrix}$$

Step 6 : Continue Successive Iteration until Optimal Solution is Achieved :

If Δ_j is positive then next successive iteration are needed. Continue successive iteration until Δ_j is negative or zero. All the above steps can be explained with example 2.4.

Example 2.4 : *Solve the following linear programming problem by simplex method.*

[May 16, June 09, 8M, Nov. 14, 16M]

Maximize $\quad z = 3x_1 + 4x_2$

Subject to : $x_1 + x_2 \le 450$

$\quad\quad 2x_1 + x_2 \le 600$

$\quad\quad x_1 \ge 0, \quad x_2 \ge 0$

Solution : Using steps as per Section 2.4.2.

Step 1 : Write the Problem in Standard Mathematical Form :

When the decision variables are non-negative, right hand side constraints are expressed as equation then the given problem is said to be represented in standard mathematical form.

Let s_1 and s_2 be the two slack variables (as per constraints equations). Add this slack variable to the left handside of first and second constraints respectively to put them in equation. The values of s_1 and s_2 depends on values of x_1 and x_2 i.e. slack variables vary with x_1 and x_2 variables.

Therefore, the standard mathematical form of specified problem is represented as follows :

> Min. $z = 3x_1 + 4x_2 + 0s_1 + 0s_2$
> Subject to   : $x_1 + x_2 + s_1 + 0s_2 = 450$
> $2x_1 + x_2 + 0s_1 + s_2 = 600$
> where, $x_1 \geq 0,\ x_2 \geq 0,\ s_1 \geq 0,\ s_2 \geq 0$

Step 2 : Find an Initial Basic Feasible Solution for the Problem :

Put $x_1 = 0$, $x_2 = 0$ in constraints equation of step (1) and find values of s_1 and s_2. Values of s_1 and s_2 are referred as initial basic feasible solution.

Therefore,

$x_1 + x_2 + s_1 + 0s = 450$	$2x_1 + x_2 + 0s_1 + s_2 = 600$
$0 + 0 + s_1 + 0 = 450$	$2(0) + 0 + 0 + s_2 = 600$
$\boxed{s_1 = 450}$	$\boxed{s_2 = 600}$

Step 3 : Preparation of Simplex Table and Check for Optimality :

The simplex table is prepared as follows :

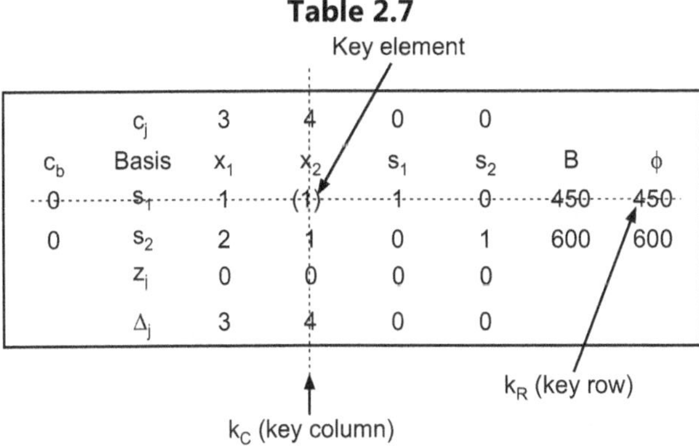

Table 2.7

c_b	Basis	c_j				B	ϕ
		3	4	0	0		
		x_1	x_2	s_1	s_2		
0	s_1	1	(1)	1	0	450	450
0	s_2	2	1	0	1	600	600
	z_j	0	0	0	0		
	Δ_j	3	4	0	0		

Key element — k_C (key column), k_R (key row)

Since, Δ_j is positive, the current solution is not optimal and therefore there is need for improvement. In above simplex table, the interpretation is as follows :

- The coefficient of c_j of variable in the objective function equation is given in first row. These coefficient remains unchanged in further simplex table. the variable in the problem for which c_j coefficient have already been expressed is represented in second row.

- The coefficient of current basic variable in the objective function is represented in first column [Columb c_b]. The basic variables (slack variable) of the current solution are represented in second column (basis).

- The coefficient matrix under non-variables x_1 and x_2 expresses their coefficient in constraints equation. These coefficient expresses the quantity of resource required to make a unit of product.

- The coefficient of slack variable in the constraints are represented in coefficient of slack variable. It is always **Unit Matrix** or **Identity Matrix** $[I_r]$ (where r is order of matrix).

- The column B (last column of Table 2.7) is known as quantity column or solution value column. This column represents the quantities of available resources or right hand side value of constraints or value of basic variables s_1 and s_2 in the initial basic feasible solution achieved earlier. Variables not written under basic column are non-basic variables and their values are zero.

- **Check for Optimality :** It is used for checking improved solution. In Table 2.7 row coefficient (z_j) under any column is estimated by adding the product of element under that column with respective c_b value or from objective function.

i.e.

$$z_j = \sum c_b \cdot a_{ij}$$ where c_{ij} are element of matrix in the ith row and j^{th} column. In Table 2.7, Δj is referred as net evaluation row or index row. It is expressed as $\boxed{\Delta_j = c_j - z_j}$. The value of Δ_j expresses whether the solution is optimal or not.

Step 4 : Application of Simplex Criterion (I and II)

The simplex criteria are as follows :

1. Simplex Criterion (I) : Decision of Variable to be Entered (Entering Variable) :

To decide the entering variable, choose the column having maximum value of Δ_j. The variable leading that column is the variable to be entered (entering variable). This variable is known as *Variable to be entered* (entering variable or incoming variable). The corresponding column is known as key column and marked as k_c. When there are more than one variable with maximum value of Δ_j then choose the variable arbitrarily. If there is no more positive value in Δ_j row, the profit achieved is maximum or optimal solution is achieved. In Table 2.7, x_2 i.e. 4 is maximum value, hence this column is selected as key column.

2. Simplex Criterion (II) : Decision of Variable to be Leaved (Leaving Variable) :

The entering variable will replace the current basic variable s_1 or s_2. Since, variable x_2 is to be entered, it will replace the current basic variable s_1 or s_2. To decide which of slack variable to be leaved (removed or made zero or made non-basic) (leaving variable), divide elements under B-column (quantity column) by respective element of key-column (k_c) i.e. determine minimum ratio.

$$\text{Minimum ratio} = \frac{\text{Element under B-column}}{\text{Respective element of key column } (k_c)}$$

Choose the minimum non-negative ratio (i.e. ignore negative ratio) and mark the column considering these ratio as ϕ column. The row so chosen (marked) is known as key-row and represented by k_R. The element lying at the intersection of key-row and key-column is known as **Key Element** and it is enclosed by ( ). Here s_1 is variable to be leaved (leaving variable).

Step 5 : Preparation of Next Iteration Table or Format :

After decision of entering variable and leaving variable, replace slack variable (s) by decision variable (x) and change corresponding c_b coefficient. Since, x_2 is entering variable, replace s_1 by x_2 i.e. $\boxed{s_1 \rightarrow x_2}$, corresponding c_b coefficient is changed from 0 to 4. i.e. if key element is other than '1' then make it '1' by dividing that element as well as all element in the row of key element. Here key element is (1) accordingly element 1, (1), 1, 0, 450 of s_1 row are retained as such as the element of 1, 1, 1, 0, 450 of x_2 row of Table 2.8.

The new element in row, s_2 row of Table 2.8 is obtained as :

$$\begin{aligned}\text{New element of } s_2 \text{ row} &= \text{Old element of } s_2 \text{ row of Table 2.7} \\ &- \left[\begin{array}{l}\text{Element above or below key element of respective } s_2 \text{ row of Table 2.7}\end{array} \times \begin{array}{l}\text{Corresponding element of } x_2 \text{ row in Table 2.8}\end{array}\right]\end{aligned}$$

Table 2.8

c_b	Basis	c_j					
		3	4	0	0		
		x_1	x_2	s_1	s_2	B	ϕ
4	x_1	1	1	1	0	450	
0	s_2	1	0	−1	1	150	
	z_j	4	4	4	0	1800	
	Δ_j	−1	0	−4	0		

Second feasible solution

Second feasible solution

$$\begin{aligned}\text{New element of } s_2 \text{ row} &= \text{Old element of } s_2 \text{ row of Table 2.7} \\ &- \left[\begin{array}{l}\text{Element above or below key element of respective } s_2 \text{ row of Table 2.7}\end{array} \times \begin{array}{l}\text{Corresponding element of } x_2 \text{ row in Table 2.8}\end{array}\right]\end{aligned}$$

Therefore,

(i) New element of s_2 row $= 2 - (1 \times 1) = 2 - 1 = 1$
(ii) New element of s_2 row $= 1 - (1 \times 1) = 0$
(iii) New element of s_2 row $= 0 - (1 \times 1) = 1$
(iv) New element of s_2 row $= 1 - (1 \times 0) = 1$
(v) New element of s_2 row $= 600 - (1 \times 450) = 150$

All these elements are shown in s_2 row of Table 2.8

In Table 2.8, element of zj are obtained from objective function or by adding the product of element under that column with respective c_b value i.e. $z_j = \Sigma c_b\, a_{ij}$ where a_{ij} are elements of matrix in the ith row and j^{th} column.

i.e. $z_j = 3x_1 + 4x_2 + 0s_2 + 0s_2$ In this, consider value of x_2 from x_2 row and s_2 from s_2 row

(i) $z_j = 3(0) + 4(1) + 0 + 0 = 4$
(ii) $z_j = 3(0) + 4(1) + 0 + 0 = 4$
(iii) $z_j = 3(0) + 4(1) + 0 + 0 = 4$
(iv) $z_j = 3(0) + 4(0) + 0 + 0 = 0$
(v) $z_j = 3(0) + 4(450) = 1800$

All these are shown in Table 2.8

In Table 2.8, Δ_j is not evaluative row or index row. It is expressed as $\boxed{\Delta_j = c_j - z_j}$

Step 6 : Continue Successive Iteration Until Optimal Solution is Achieved :

If Δ_j is zero or negative, the above solution is optimal solution. Need not to proceed further. If Δ_j is positive then next successive iteration are required.

Therefore optimum solution is :

$$\boxed{z_{max} = ₹\ 1800 \text{ at } x_1 = 0,\ x_2 = 450}$$

Example 2.5 : *Solve by simplex method*

[Dec. 17, May 12, 10M]

Maximize $z = 3x_1 + 2x_2$

Subject to : $x_1 + x_2 \leq 4$

$x_1 - x_2 \leq 2$

$x_1 \geq 0,\ x_2 \geq 0$

Solution : Using steps of simplex method, we have.

Step 1 : Write the Problem in Standard Mathematical Form :

Let s_1 and s_2 be two slack variables. Add this variable to left hand side of first and second constraints equation.

Therefore,

$$\text{Max. } z = 3x_1 + 2x_2 + 0s_1 + 0s_2$$

Subject to :
$$x_1 + x_2 + s_1 + 0s_2 = 4$$
$$x_1 - x_2 + 0s_1 + s_2 = 2$$
$$x_1 \geq 0,\ x_2 \geq 0,\ s_1 \geq 0,\ s_2 \geq 0$$

Step 2 : Find an Initial Basic Feasible Solution for the Problem :

Let $x_1 = 0$, $x_2 = 0$ in constraints equation

Thus,

$$x_1 + x_2 + s_1 + 0s_2 = 4 \qquad x_1 - x_2 + 0s_1 + s_2 = 2$$
$$0 + 0 + s_1 + 0 = 4 \qquad 0 - 0 + 0s_1 + s_2 = 2$$

$$\boxed{s_1 = 4} \qquad\qquad \boxed{s_2 = 2}$$

Step 3 : Preparation of Simplex Table and Check for Optimality :

Table 2.9

c_b	Basis	c_j 3 x_1	2 x_2	0 s_1	0 s_2	B	ϕ
0	s_1	1	1	1	0	4	
0	s_2	(1)	1	0	1	2	
	z_j	0	0	0	0		
	Δ_j	3	2	0	0		

(k_C under x_1; k_R pointing at s_2 row)

Step 4 : Application of Simplex Criterion (I and II) :

By applying simplex criterion (I and II)

$$\text{Here } \boxed{s_2 \to x_1}$$

Leaving variable   Entering variable

Step 5 : Preparation of Next Iteration Table or Format :

New element of s_1 row in Table 2.10 is calculated as :

Table 2.10

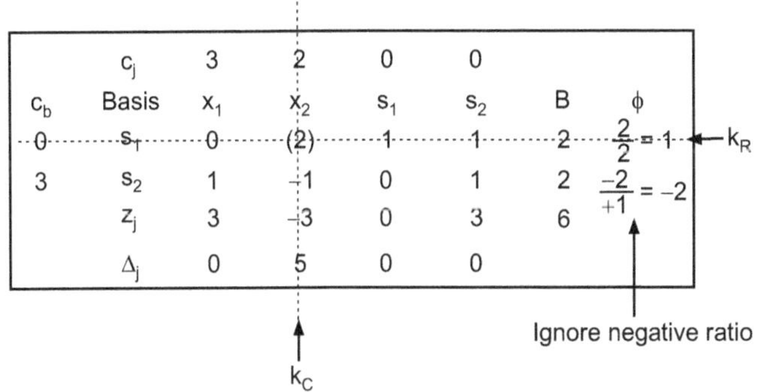

c_b	Basis	c_j 3 x_1	2 x_2	0 s_1	0 s_2	B	ϕ
0	s_1	0	(2)	1	1	2	$\frac{2}{2} = 1 \leftarrow k_R$
3	s_2	1	−1	0	1	2	$\frac{-2}{+1} = -2$
	z_j	3	−3	0	3	6	
	Δ_j	0	5	0	0		

(k_C under x_2; Ignore negative ratio)

New element of s_1 row $= \dfrac{\text{Old element of } s_1 \text{ row}}{} -\begin{bmatrix}\text{Element above or below key element of respective row}\end{bmatrix} \times \begin{bmatrix}\text{Corresponding new element of } x_1 \text{ row}\end{bmatrix}$

(i) New element of s_1 row $= 1 - (1 \times 1) = 0$

(ii) New element of s_1 row $= 1 - (1 \times -1) = 2$

(iii) New element of s_1 row $= 1 - (1 \times 0) = 1$

(iv) New element of s_1 row $= 0 - (1 \times 1) = -1$

(v) New element of s_1 row $= 4 - (1 \times 2) = 2$

These are expressed in s_1 row in Table 2.10

Step 6 : Continue Successive Iteration until Optimal Solution is Achieved :

Since in Table 2.10, Δ_j is positive, solution is not optimal,

Now $\boxed{s_1 \to x_2}$

Table 2.11

c_b	Basis	c_j 3 x_1	2 x_2	0 s_1	0 s_2	B	ϕ
2	x_1	0	1	$\frac{1}{2}$	$-\frac{1}{2}$	1	
3	x_2	1	0	$\frac{1}{2}$	$\frac{3}{2}$	3	
	z_j	3	2	$\frac{5}{2}$	$\frac{7}{2}$	11	
	Δ_j	0	0	$-\frac{5}{2}$	$-\frac{7}{2}$		

$$\therefore\ z_{max} = 11 \text{ at } x_1 = 3,\ x_2 = 1$$

Since in table 2.11, Δ_j is zero or negative, the given solution is optimal.

Example 2.6 : *Solve by simplex method*   **[May 15]**

$$\text{Maximize } z = 3x_1 + 2x_2 + 5x_3$$

Subject to :
$$x_1 + 2x_2 + x_3 \leq 430$$
$$3x_1 + 2x_3 \leq 460$$
$$x_1 + 4x_2 \leq 420$$
$$x_1 \geq 0,\ x_2 \geq 0,\ x_3 \geq 0$$

Solution : Using steps of simplex method.

Step 1 : Write the Problem in Standard Mathematical Form

Let s_1, s_2 and s_3 be slack variables as per constraints in given problem.

Thus,

$$\text{Maximize } z = 3x_1 + 2x_2 + 5x_3 + 0s_1 + 0s_2 + 0s_3$$

Subject to :
$$x_1 + 2x_2 + x_3 + s_1 + 0s_2 + 0s_3 = 430$$
$$3x_1 + 0x_2 + 2x_3 + 0s_1 + s_2 + 0s_3 = 460$$
$$x_1 + 4x_2 + 0x_3 + 0s_1 + 0s_2 + s_3 = 420$$
$$x_1 \geq 0,\ x_2 \geq 0,\ x_3 \geq 0,\ s_1 \geq 0,\ s_2 \geq 0,\ s_3 \geq 0$$

Step 2 : Find an Initial Basic Feasible Solution for the Problem :

Let $\left.\begin{array}{l} x_1 = 0 \\ x_2 = 0 \\ x_3 = 0 \end{array}\right\}$ In constraints equation,

e get

$$x_1 + 2x_2 + x_3 + s_1 + 0s_2 + 0s_3 = 430$$
$$0 + 0 + s_1 + 0 + 0 = 430$$
$$\boxed{s_1 = 430}$$

$$3x_1 + 0x_2 + 2x_3 + 0s_1 + s_2 + 0s_3 = 460$$
$$0 + 0 + 0 + 0 + 0s_2 + 0 = 460$$
$$\boxed{s_2 = 460}$$

$$x_1 + 4x_2 + 0x_3 + 0s_1 + 0s_2 + s_3 = 420$$
$$0 + 0 + 0 + 0 + 0 + s_3 = 420$$
$$\boxed{s_3 = 420}$$

Step 3 : Preparation of Simplex Table and Check for Optimality :

Here simplex table (Table 2.12) is prepared from step (1).

Table 2.12

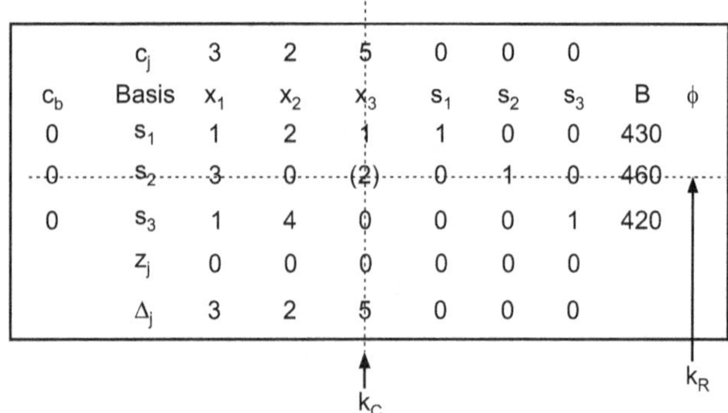

c_b	c_j Basis	3 x_1	2 x_2	5 x_3	0 s_1	0 s_2	0 s_3	B	ϕ
0	s_1	1	2	1	1	0	0	430	
0	s_2	3	0	(2)	0	1	0	460	
0	s_3	1	4	0	0	0	1	420	
	z_j	0	0	0	0	0	0		
	Δ_j	3	2	5	0	0	0		

(k_C points to column x_3; k_R points to row s_2)

Step 4 : Application of Simplex Criterion (I and II) :

By applying simplex criterion (I and II)

$$\boxed{s_2 \rightarrow x_3}$$

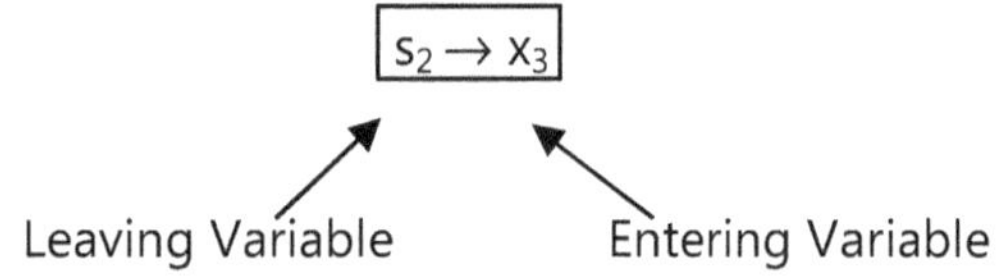

Step 5 : Preparation of Next Iteration Table or Format :

In Table 2.13, new element of s_1 and s_3 row are obtained as follows :

$$\begin{aligned} \text{New element of respective row} &= \text{Old element of respective row} \\ &- \begin{bmatrix} \text{Element above or below key element of respective row} \times \text{Corresponding new element of } x_3 \text{ row} \end{bmatrix} \end{aligned}$$

Table 2.13

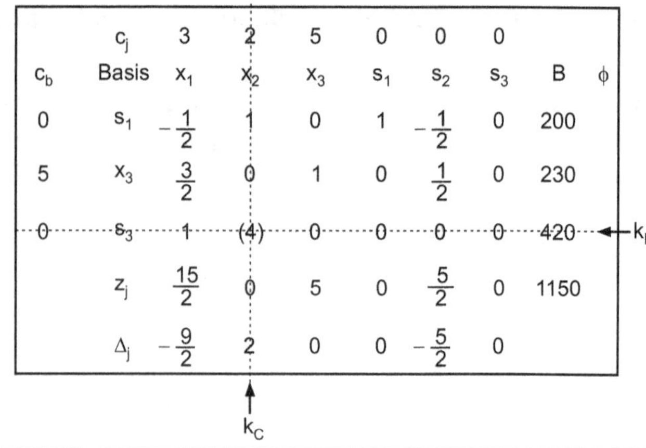

c_b	c_j Basis	3 x_1	2 x_2	5 x_3	0 s_1	0 s_2	0 s_3	B	ϕ
0	s_1	$-\dfrac{1}{2}$	1	0	1	$-\dfrac{1}{2}$	0	200	
5	x_3	$\dfrac{3}{2}$	0	1	0	$\dfrac{1}{2}$	0	230	
0	s_3	1	(4)	0	0	0	0	420	$\leftarrow k_R$
	z_j	$\dfrac{15}{2}$	0	5	0	$\dfrac{5}{2}$	0	1150	
	Δ_j	$-\dfrac{9}{2}$	2	0	0	$-\dfrac{5}{2}$	0		

(k_C points to column x_2)

New Element of s_1 Row	New Element of s_3 Row
(i) $1 - \left(1 \times \dfrac{3}{2}\right) = 1 - \dfrac{3}{2} = -\dfrac{1}{2}$	(i) $1 - \left(0 \times \dfrac{3}{2}\right) = 1$
(ii) $2 - (1 \times 0) = 1$	(ii) $4 - (0 \times 0) = 4$
(iii) $1 - (1 \times 1) = 0$	(iii) $0 - (0 \times 1) = 0$
(iv) $1 - (1 \times 0) = 1$	(iv) $0 - (0 \times 1) = 0$
(v) $0 - \left(1 \times \dfrac{1}{2}\right) = -\dfrac{1}{2}$	(v) $0 - \left(0 \times \dfrac{1}{2}\right) = 0$
(vi) $0 - (1 \times 0) = 0$	(vi) $0 - (0 \times 0) = 0$
(vii) $430 - (1 \times 230) = 200$	(vii) $420 - (0 \times 230) = 420$

In above (Table 2.13), z_j is obtained by objective function with respective value of x_3 row.

Step 6 : Continue Successive Iteration Until Optimal Solution is Achieved :

Since, in Table 2.13 Δ_j, all values are not zero or negative hence it is not optimal solution, successive iteration are shown in further tables.

Table 2.14

c_b	c_j Basis	3 x_1	2 x_2	5 x_3	0 s_1	0 s_2	0 s_3	B	ϕ
0	s_1	$-\dfrac{3}{2}$	0	0	1	$-\dfrac{1}{2}$	0	95	
5	x_3	$\dfrac{3}{2}$	0	1	0	$\dfrac{1}{2}$	0	230	
2	x_2	$\dfrac{1}{4}$	1	0	0	0	0	105	
	z_j	8	2	5	0	$\dfrac{5}{2}$	0	1360	
	Δ_j	-5	0	0	0	$-\dfrac{5}{2}$	0		

Optimal solution

In Table 2.14

New Element of s_1 Row	New Element of x_3 Row
(i) $-\dfrac{1}{2} - \left(1 \times \dfrac{1}{4}\right) = -\dfrac{1}{2} - \dfrac{1}{4}$ $= \dfrac{-4-2}{8} = -\dfrac{6}{8} = -\dfrac{3}{2}$	(i) $\dfrac{3}{2} - \left(0 \times \dfrac{1}{4}\right) = \dfrac{3}{2}$
(ii) $1 - (1 \times 1) = 0$	(ii) $0 - (0 \times 1) = 0$
(iii) $0 - (1 \times 0) = 0$	(iii) $1 - (0 \times 0) = 1$
(iv) $1 - (1 \times 0) = 1$	(iv) $0 - (0 \times 0) = 0$
(v) $\dfrac{-1}{2} - (1 \times 0) = \dfrac{-1}{2}$	(v) $\dfrac{1}{2} - (0 \times 0) = \dfrac{1}{2}$
(vi) $0 - (1 \times 0) = 0$	(vi) $0 - (0 \times 0) = 0$
(vii) $200 - (1 \times 105) = 95$	(vii) $230 - (0 \times 105) = 230$

In Table 2.14, z_j is obtained from objective function with corresponding value of s_1, x_3 and x_2 rows.

Since Δ_j is negative or zero, the solution is optimal.

$\therefore$ $\boxed{z_{max} = 1360 \text{ at } x_1 = 0, x_2 = 105, x_3 = 230}$

Example 2.7   *[May 18, Dec. 13, 10M]*

$Minimize\ z = 60x_1 + 48x_2$

$Subject\ to:\ 4x_1 + 2x_2 \geq 8$

$2x_1 + 4x_2 \geq 6$

$x_1 \geq 0,\ x_2 \geq 0$

Solution : Using steps of simplex method.

Step 1 : Write the Problem in Standard Mathematical Form :

Let s_1 and s_2 be slack variables as per constraints equations.

Thus,

Minimize $z = 60x_1 + 48x_2 + 0s_1 + 0s_2$

Subject to : $4x_1 + 2x_2 + s_1 + 0s_2 = 8$

$2x_1 + 4x_2 + 0s_1 + s_2 = 6$

$x_1 \geq 0, x_2 \geq 0, s_1 \geq 0, s_2 \geq 0$

Step 2 : Find an Initial Basic Feasible Solution for the Problem :

Let $x_1 = 0$, $x_2 = 0$ in constraints equation.

$\therefore$ $\boxed{s_1 = 8}$ and $\boxed{s_2 = 6}$

Step 3 : Preparation of Simplex Table and Check for Optimality :

Table 2.15

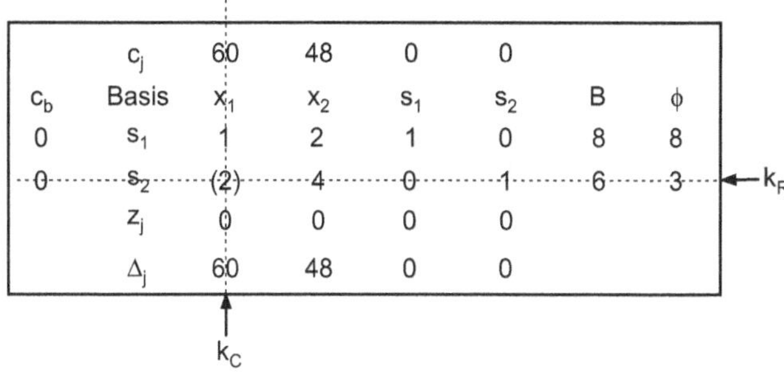

c_b	Basis	x_1	x_2	s_1	s_2	B	ϕ
c_j		60	48	0	0		
0	s_1	1	2	1	0	8	8
0	s_2	(2)	4	0	1	6	3	$\leftarrow k_R$
	z_j	0	0	0	0		
	Δ_j	60	48	0	0		

(k_C under x_1 column)

Step 4 : Application of Simplex Criterion (I and II) :

By applying simplex criterion

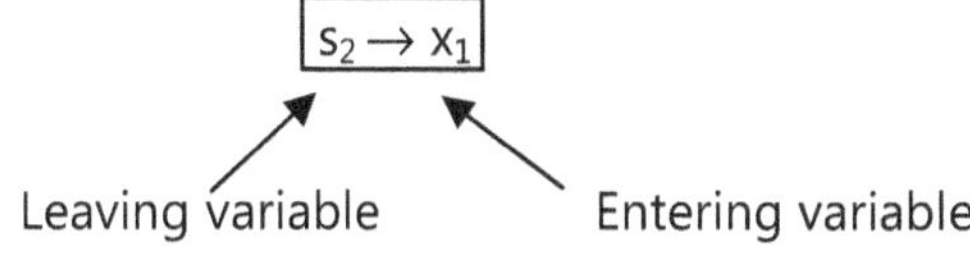

Leaving variable   Entering variable

Step 5 : Preparation of Next Iteration Table or Format :

Table 2.16

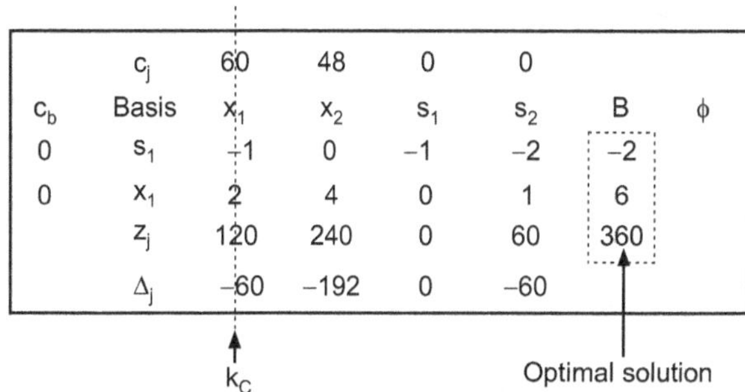

c_b	Basis	x_1	x_2	s_1	s_2	B	ϕ
c_j		60	48	0	0		
0	s_1	-1	0	-1	-2	-2	
0	x_1	2	4	0	1	6	
	z_j	120	240	0	60	360	
	Δ_j	-60	-192	0	-60		

(k_C under x_1 column; Optimal solution at B = 360)

Step 6 : Continue Successive Iteration Until Optimal Solution is Achieved :

Since in Table 2.16, Δj is zero or negative, the above solution is optimal, need not to proceed further.

$\therefore$ $\boxed{z_{min} = ₹ \ 360 \text{ at } x_1 = 6,\ x_2 = 0}$

UNIVERSITY QUESTIONS

1. Define the following terms in LP :
   (a) Slack variable
   (b) Redundant constraints
   (c) Minimum ratio or replacement ratio
   (d) Degenerative solution.
          [May 18, June 09, 4M, Nov. /Dec. 12]

2. Explain simplex method for solving LPP.
          [Dec. 16, May 08, 9M]

3. Solve by simplex method to maximize $10x_1 + 20x_2$
   Subject to :
   $5x_1 + 3x_2 \leq 30$
   $3x_1 + 6x_2 \leq 36$
   $2x_1 + 5x_2 \leq 20$
   $x_1 \geq 0,\quad x_2 \geq 0$
          [May 15, Nov./Dec. 06, 6M]

4. Minimize :

$$z = x_1 - 3x_2 + 2x_3$$

Subject to :

$$3x_1 - x_2 + 2x_3 \leq 7$$
$$- 2x_1 + 4x_2 \leq 12$$
$$- 4x_1 + 3x_2 + 8x_3 \leq 10$$
$$x_1, x_2, x_3 \geq 0$$

(Refer solved example 2.6)

5. Maximize $z = 3x_1 + 5x_2 + 4x_3$

Subject to :

$$2x_1 + 3x_2 \leq 8$$
$$2x_2 + 5x_3 \leq 10$$
$$3x_1 + 2x_2 + 4x_3 \leq 15$$
$$x_1 \geq 0, \ x_2 \geq 0, x_3 \geq 0$$

[May 16, Nov. / Dec. 14, 10M]

$$\left[\textbf{Ans.:} \ x_1 = \frac{89}{41}, \ x_2 = \frac{50}{41}, \ x_3 = \frac{62}{41}, \ z_{max} = \frac{765}{41} \right]$$

6. Maximize

$$z = 2x_1 + 5x_2$$

Subject to :

$$x_1 + 4x_2 \leq 24$$
$$3x_1 + x_2 \leq 21$$
$$x_1 + x_2 \leq 9$$
$$x_1 \geq 0 \ \ x_2 \geq 0$$

(Refer solved example 2.4)

[**Ans. :** $x_1 = 4$, $x_2 = 5$, $s_2 = 4$, $s_1 = 0$, $s_2 = 0$, $z_{max} = 33$]

7. Maximize

$$z = 5x_1 + 7x_2 \quad \textbf{[May 17, Nov. /Dec. 14, 10M]}$$

Subject to :

$$x_1 + x_2 \leq 4$$
$$3x_1 - 8x_2 \leq 24$$
$$10x_1 + 7x_2 \leq 35$$
$$x_1 \geq 0 \ \ \ x_2 \geq 10$$

[**Ans. :** $x_1 = 0$, $x_2 = 4$, $z_{max} = 28$]

8. Maximize

$$z = 2x_1 + 5x_2$$

Subject to :

$$x_1 + 3x_2 \leq 3$$
$$3x_1 + 2x_2 \leq 6$$
$$x_1 \geq 0 \ \ \ x_2 \geq 10$$

[**Ans. :** $x_1 = 2$, $x_2 = 0$, $z_{max} = 4$]

9. $$z_{max} = 3x_1 + 2x_2$$

Subject to :

$$2x_1 + x_2 \leq 5$$
$$x_1 + x_2 \leq 5$$
$$x_1 \geq 0 \ \ \ x_2 \geq 10$$

[**Ans. :** $x_1 = 0$, $x_2 = 12$, $z_{max} = 60$]

10. Maximize

$$z = 2x_1 + 4x_2 + x_3 + x_4$$

Subject to :

$$x_1 + 3x_2 + x_4 \leq 4$$
$$2x_1 + x_2 \leq 3$$
$$x_2 + 4x_3 + x_4 \leq 3$$
$$x_1 \geq 0, \ x_2 \geq 10, \ x_3 \geq 0, \ x_4 \geq 0$$
$$x_1 \geq 0 \ \ x_2 \geq 0$$

11. Maximize

$$z = x_1 - 3x_2 + 2x_3$$

Subject to :

$$3x_1 - x_2 + 2x_3 \geq 7$$
$$- 2x_1 + 4x_2 \geq 12$$
$$- 4x_2 + 3x_2 + 8x_3 \geq 10$$
$$x_1 \geq 0 \ \ \ x_2 \geq 0$$

[**Ans. :** $x_1 = 4$, $x_2 = 3$, $x_3 = 0$, $z_{min} = 11$]

12. Solve LPP by Suitable Method

Maximize :

$$Z = X_1 - 3X_2 + X_3$$

Subject to :

$$3X_1 - X_2 + 2X_3 \leq 7$$
$$2X_1 + 4X_2 \geq - 2$$
$$- 4X_1 + 3X_2 + 8X_3 \leq 10$$

13. Solve LPP by Suitable Method

Maximize :

$$Z = 2X_1 + 5X_2$$

Subject to :

$$X_1 + 4X_2 \leq 24$$
$$3X_1 + X_2 \leq 21$$
$$X_1 + X_2 \leq 9$$
$$X_1, X_2 \geq 0$$

[May 15, Nov. /Dec. 12, 10M]

14. Solve LPP by suitable Method

Maximize :

$$Z = 3X_1 + 5X_2 + 4X_3$$

Subject to :

$$2X_1 + 3X_2 \leq 8$$
$$2X_2 + 5X_3 \leq 10$$
$$3X_1 + 2X_2 + 4X_3 \leq 15$$
$$X_1, X_2, X_3 \geq 0$$

[Dec. 17, May /June 13, 12M, Nov. 13, 12M]

15. A company manufacturers three products namely X, Y and Z. Each of the products requires processing on three machines, Turning, Milling and Grinding. Product X requires 10 hours of turning, 5 hours of milling and 1 hour of grinding. Product Y requires 5 hours of turning, 10 hours of milling and 1 hour of grinding and Product Z requires 2 hours of turning, 4 hours of milling and 2 hours of grinding. In the coming planning period, 2700 hours of turning, 2200 hours of milling and 500 hours of grinding are available. The profit contribution of X, Y and Z are ₹ 20, ₹ 15 and ₹ 20 per unit respectively. Find the optimal product mix to maximize the profit.

Ans.: $X = 0$

$Y = 150$

$Z = 174.4$

$Z_{max} = ₹\ 5738$

2.5 DUALITY AND SENSITIVITY ANALYSIS

[May 17, Dec. 11 6m, May 12, 6M]

Dual of linear programming problem is very important interesting feature of linear programming. The actual linear programming is known as primal programming. When this programming can be rewritten by transposing (changing) rows or columns of the algebraic statement of problem them it is known as Dual programming and Duality. Solution of dual programming is obtained by the same way as primal programming.

2.5.1 Mathematical Formulation of Dual Problem

The following points should be considered for converting primal problem into dual problem :

- When the primal consists of 'n' variable and 'm' constraints then dual will have 'm' variables and 'n' constraints.

- The maximization problem in primal changes to minimization problem in dual programming.

- The maximization problem have ($\le$) constraints whereas the minimization problem have ($\ge$) constraints.

- Similarly constraints of ($\le$) type in primary programming changes to constraints of ($\ge$) type of dual programming.

- The coefficient matrix of constraints of dual is the reverse of the primal programming.

- The right hand side constraints of primal programming will become constraints coefficient of dual and vice versa i.e. ['Bi's is will become 'cj's].

- In both programming problem, the variables are non-negative. In general, mathematical formulation of problems are as follows :

Primal Programming Problem	Dual Programming Problem
Max. $z = \sum\limits_{j=1}^{n} x_j\, c_j$	Min. $z' = \sum\limits_{i=1}^{m} v_i\, b_i$
Subject to : $\sum\limits_{j=1}^{n} a_j\, x_j \le b_j$	Subject to : $\sum\limits_{i=1}^{n} a_{ij}\, v_i \le c_j$
where, $i = 1, 2, 3, \ldots, m$	where, $j = 1, 2, 3, \ldots, n$
$x_j \ge 0,\ j = 1, 2, 3, \ldots, m$	$v_i \ge 0,\ j = 1, 2, 3, \ldots, m$
$\boxed{\begin{array}{l} x_j \to v_i \\ b_i \to c_j \end{array}}$	

Example 2.8 : *For the following primal programming problem, write its dual or construct the dual of the following primal problem.*

$$Max.\ z = c_1 x_1 + c_2 x_2 + c_3 x_3 + c_4 x_4$$

$$Subject\ to: a_{11} x_1 + a_{12} x_2 + a_{13} x_3 + a_{14} x_4 \le b_1$$

$$a_{21} x_1 + a_{22} x_2 + a_{23} x_3 + a_{24} x_4 \le b_2$$

$$a_{31} x_1 + a_{32} x_2 + a_{33} x_3 + a_{34} x_4 \le b_3$$

$$a_{41} x_1 + a_{42} x_2 + a_{43} x_3 + a_{44} x_4 \le b_4$$

$$x_1 \ge 0,\ x_2 \ge 0,\ x_3 \ge 0,\ x_4 \ge 0$$

Solution : Primal problem

Write the system as $\boxed{AX = B}$

where,   A = Matrix of coefficient of constraints

X = Matrix of constraints

B = R.H.S. of constraints

$\therefore$ AX = B

$$\begin{bmatrix} a_{11} & a_{12} & a_{13} & a_{14} \\ a_{21} & a_{22} & a_{23} & a_{24} \\ a_{31} & a_{32} & a_{33} & a_{34} \\ a_{41} & a_{42} & a_{43} & a_{44} \end{bmatrix} \begin{bmatrix} x_1 \\ x_2 \\ x_3 \\ x_4 \end{bmatrix} = \begin{bmatrix} b_1 \\ b_2 \\ b_3 \\ b_4 \end{bmatrix}$$

Dual problem :

Write the system in $\boxed{A'V = C}$

where,   A' = Transpose of A

V = Matrix of V_i

C = Matrix of c_j

$$\begin{bmatrix} a_{11} & a_{12} & a_{13} & a_{14} \\ a_{21} & a_{22} & a_{23} & a_{24} \\ a_{31} & a_{32} & a_{33} & a_{34} \\ a_{41} & a_{42} & a_{43} & a_{44} \end{bmatrix} \begin{bmatrix} v_1 \\ v_2 \\ v_3 \\ v_4 \end{bmatrix} = \begin{bmatrix} c_1 \\ c_2 \\ c_3 \\ c_4 \end{bmatrix}$$

Therefore, min. $z' = \sum b_i v_j = b_1 v_1 + b_2 v_2 + b_3 v_3 + b_4 v_4$

Subject to :

$$a_{11} v_1 + a_{12} v_2 + a_{13} v_3 + a_{14} v_4 \leq c_1$$
$$a_{21} v_1 + a_{22} v_2 + a_{23} v_3 + a_{24} v_4 \leq c_2$$
$$a_{31} v_1 + a_{32} v_2 + a_{33} v_3 + a_{34} v_4 \leq c_3$$
$$a_{41} v_1 + a_{42} v_2 + a_{43} v_3 + a_{44} v_4 \leq c_4$$
$$v_1 \geq 0,\ v_2 \geq 0,\ v_3 \geq 0,\ v_4 \geq 0$$

Example 2.9 : *Construct the dual of the problem*

[Dec 18, May 14, 10M]

$$\text{Min. } z = 3x_1 - 2x_2 + 4x_3$$

Subject to :
$$3x_1 + 5x_2 + 4x_3 \geq 7$$
$$6x_1 + x_2 + 3x_3 \geq 4$$
$$-7x_1 - 2x_2 - x_3 \leq 10$$
$$x_1 - 2x_2 + 5x_3 \geq 3$$
$$4x_1 + 7x_2 - 2x_3 \geq 2$$
$$x_1 \geq 0,\ x_2 \geq 0,\ x_3 \geq 0$$

Solution :

Since, the specified problem is of minimization, all constraints should be of ($\geq$) type. Here 3rd constraints is multiply by -1 to have ($\geq$) type constraints.

i.e. $+7x_1 + 2x_2 + x_3 \geq -10$

Therefore, the problem becomes

$$\text{Min. } z = 3x_1 - 2x_2 + 4x_3$$

Subject to :
$$3x_1 + 5x_2 + 4x_3 \geq 7$$
$$6x_1 + x_2 + 3x_3 \geq 4$$
$$7x_1 + 2x_2 + x_3 \geq -10$$
$$x_1 + 2x_2 + 5x_3 \geq 3$$
$$4x_1 + 7x_2 - 2x_3 \geq 2$$
$$x_1 \geq 0,\ x_2 \geq 0,\ x_3 \geq 0$$

Primal Problem	Dual Problem
$AX = B$	$A'V = C$
$\begin{bmatrix} 3 & 5 & 4 \\ 6 & 1 & 3 \\ 7 & 2 & 1 \\ 1 & -2 & 5 \\ 4 & 7 & -2 \end{bmatrix} \begin{bmatrix} x_1 \\ x_2 \\ x_3 \end{bmatrix} = \begin{bmatrix} 7 \\ 4 \\ -10 \\ 3 \\ 2 \end{bmatrix}$	$\begin{bmatrix} 3 & 6 & 7 & 1 & 4 \\ 5 & 1 & 2 & -2 & 7 \\ 4 & 3 & 1 & 5 & -2 \end{bmatrix} \begin{bmatrix} v_1 \\ v_2 \\ v_3 \\ v_4 \\ v_5 \end{bmatrix} = \begin{bmatrix} 3 \\ -2 \\ 4 \end{bmatrix}$

$\therefore$ Dual problem is as follows :

$$\text{Max. } z' = 7v_1 + 4v_2 - 10v_3 + 3v_4 + 2v_5$$

Subject to :
$$3v_1 + 6v_2 + 7v_3 + v_4 + 4v_5 \leq 3$$
$$5v_1 + v_2 + 2v_3 - 2v_4 + 7v_5 \leq -2$$
$$4v_1 + 3v_2 + v_3 + 5v_4 - 2v_5 \leq 4$$
$$v_1 \geq 0,\ v_2 \geq 0,\ v_3 \geq 0,\ v_4 \geq 0,\ v_5 \geq 0$$

Example 2.10 : *Write the dual of the problem.*

$$\text{Max. } z = 3x_1 + 10x_2 + 2x_3$$

Subject to :
$$2x_1 + 3x_2 + 2x_3 \leq 7$$
$$3x_1 - 2x_2 + 4x_3 \leq 3$$
$$x_1 \geq 0,\ x_2 \geq 0,\ x_3 \geq 0$$

Solution : The given problem is maximization, all the constraints should be q ($\leq$) type. Here in equation of constraints convert all other to ($\leq$) type by multiplying -1 to the respective equation.

Here the equation $3x_1 - 2x_2 + 4x_3 = 3$ is to be expressed as a pair of inequalities.

i.e. $3x_1 - 2x_2 + 4x_3 \leq 3$ and $3x_1 - 2x_2 + 4x_3 \geq 3$

or $3x_1 - 2x_2 + 4x_3 \leq 3$ and $-3x_1 + 2x_2 - 4x_3 \leq -3$

multiply by -1

Thus, the problem becomes

$$\text{Max. } z = 3x_1 + 10x_2 + 2x_3$$

Subject to :
$$2x_1 + 3x_2 + 2x_3 \leq 7$$
$$-3x_1 - 2x_2 + 4x_3 \leq -3$$
$$x_1 \geq 0,\ x_2 \geq 0,\ x_3 \geq 0$$

Primal Problem	Dual Problem
$AX = B$	$A'V = C$
$\begin{bmatrix} 2 & 3 & 2 \\ -3 & 2 & -4 \end{bmatrix} \begin{bmatrix} x_1 \\ x_2 \\ x_3 \end{bmatrix} = \begin{bmatrix} 7 \\ -3 \\ 0 \end{bmatrix}$	$\begin{bmatrix} 2 & -3 \\ 3 & 2 \\ 2 & -4 \end{bmatrix} \begin{bmatrix} v_1 \\ v_2 \end{bmatrix} = \begin{bmatrix} 3 \\ 10 \\ 2 \end{bmatrix}$

$\therefore$ The dual problem is written as :

$$\text{Max. } z' = 7v_1 - 3v_2$$

Subject to :
$$2v_1 - 3v_2 \geq 3$$
$$3v_1 + 2v_2 \geq 10$$
$$2v_1 - 4v_2 \geq 2$$
$$v_1 \geq 0,\ v_2 \geq 0$$

UNIVERSITY QUESTIONS

1. Define the following term of LP
   - (i) Basic solution
   - (ii) Basic variable
   - (iii) Non-basic variable
   - (iv) Feasible solution
   - (v) Artificial variable
   - (vi) Slack variable.

   [May 17, Nov./Dec. 06, 6M]

2. Write the dual of the following problem and write the value of decision variables from primal simplex table.

   Maximize :
   $$10x_1 + 20x_2$$
   Subject to :
   $$5x_1 + 3x_2 \leq 30$$
   $$3x_1 + 6x_2 \leq 36$$
   $$2x_1 + 5x_2 \leq 20$$
   $$x_1 \geq 0, \ x_2 \geq 0$$

   [Dec. 17, May 08, 6M]

3. In above problem (Q_2) which constraints can be deleted without affecting final solution ? What is the name of such type of constraints ?

   [Dec. 16, May 08, 4M]

4. Using the dual, solve the following linear programming problem.

   Max. $Z_{max} = 2x_1 + 2x_2 + 4x_3$

   Subject to : $2x_1 + 3x_2 + 5x_3 \geq 2$
   $$3x_1 + x_2 + 7x_3 \leq 3$$
   $$x_1 + 4x_2 + 6x_3 \leq 5$$
   $$x_1 \geq 0, \ x_2 \geq 0, \ x_3 \geq 0$$

   [May 17, Dec. 10, 16M]

5. From dual of the following problem and write the values of dual decision variables from the final simplex table.

   Max. $z = 3x_1 + 5x_2$

   Subject to : $x_1 \leq 14$
   $$2x_2 \leq 12$$
   $$3x_1 + 2x_2 \leq 18$$
   $$x_1 \geq 0, \ x_2 \geq 0$$

   (Refer solved examples 2.8 and 2.6)

   [Dec. 16, May/June 11, 6M]

6. Write short note on sensitivity analysis of LPP.

2.6 SIMPLEX METHOD

2.6.1 Artificial Variable Techniques
(Big m Method) (Penalty Method)

[Dec. 10, 6M, Nov. /Dec. 12, 8M]

If at least one constants is of [(=) or (≥)] type then artificial variable techniques are applied for determining solution of LPP. For this Big 'm' method and two phase method are generally applied for finding optimal solution. Artificial variables are fictitious and have no physical meanings. Artificial variable performs role of slack variable in the most iteration, only to replaced at a next iteration or further iteration. Thus, they are applied to get initial basic feasible solution so that simplex algorithm can be applied to get optimal solution. The various steps involved in Big 'm' method are as follows :

Step 1 : Write the LPP in standard form by adding slack variables. Generally, slack variable are added to left hand side constraints of (≤) type and subtracted from the constraints of (≥) type.

Step 2 : Add artificial variable (non-negative) to left hand side of all the constraints of initially (≥) or (=) type. Assign a very large per unit penalty in the objective function. This penalty is expressed by – m (for maximization problem) and + m (for minimization problem) where m > 0.

Step 3 : Apply simplex method to solve the modified LPP. The following situation may occur during iterations :

- (i) When two artificial variables remains in the basis and the condition of optimality is satisfied, then solution is optimal feasible solution to the given problem.
- (ii) When atleast one artificial variable appears in the basis at zero level (i.e. with zero value in right hand side B column) and condition of optimality is satisfied then the solution is optimal feasible (though degenerate) solution to the specified problem.
- (iii) When atleast one artificial variable remains in the basis at non-zero level (i.e. with positive value in right hand side B column) and the condition of optimality is satisfied there original problem has no feasible solution existence of feasible solution, then artificial variable should be assigned zero value.

Notes :

1. Artificial variables are added to the constraints of (≥) type and (=) type. Equality constraints needs neither slack variable or surplus variables.
2. Other variable (except artificial) once driven out in as iteration, may re-enter in a subsequent iteration. But, an artificial variable, once driven out, can never re-enter due to the larger penalty coefficient in associated with it in the objective function.
3. Slack variables are added to left hand side constraints of (≤) type and subtracted from constraints of (≥) type.
4. By use of computer, generally, the largest value that can be represented in computer is used.

Example 2.11 : *Solve by using big 'm' method the following LPP.*   **[Dec. 17, May 10, 8M]**

Maximize $z = x_1 + 2x_2 + 3x_3 - x_4$

Subject to : $x_1 + 2x_2 + 3x_3 = 15$
$$2x_1 + x_2 + 5x_3 = 20$$
$$x_1 + 2x_2 + x_3 + x_4 = 10$$
$$x_1 \geq 0, \ x_2 \geq 0, \ x_3 \geq 0, x_4 \geq 0$$

Solution : Using steps big 'm' method.

Step 1 : Write the LPP in standard form by adding slack variable. Here these are not required (since constraints of ($\leq$) and ($\geq$) are not used).

Step 2 : Add artificial variable to left hand side of all constraints of initially ($\geq$) type or (=) type. Assign a large per unit penalty in the objective function. This penalty is expressed by $- m$ (for maximisation problem, where $m > 0$).

Let a_1, a_2, a_3 be the artificial variables. Thus, the given problem in standard form is expressed as :

Maximize $z = x_1 + 2x_2 + 3x_3 - x_4 - ma_1 - ma_2 - ma_3$

Subject to :

$$x_1 + 2x_2 + 3x_3 + 0x_4 + a_1 + 0a_2 + 0a_3 = 15$$
$$2x_1 + x_2 + 3x_3 + 0x_4 + 0a_1 + a_2 + 0a_3 = 20$$
$$x_1 + 2x_2 + x_3 + x_4 + 0a_1 + 0a_2 + a_3 = 10$$
$$x_1 \geq 0, \ x_2 \geq 0, \ x_3 \geq 0, \ a_1 \geq 0, \ a_2 \geq 0, \ a_3 \geq 0$$

Step 3 : Apply simplex method to solve modified L.P.P.

Let $x_1 = 0$, $x_2 = 0$, $x_3 = 0$, $x_4 = 0$ in constraints equation. Therefore, $a_1 = 15$, $a_2 = 20$, $a_3 = 10$, $z = - 45\,m$ (Initial basic flexible solution).

Use simplex method (as previous steps or examples).

Table 2.17

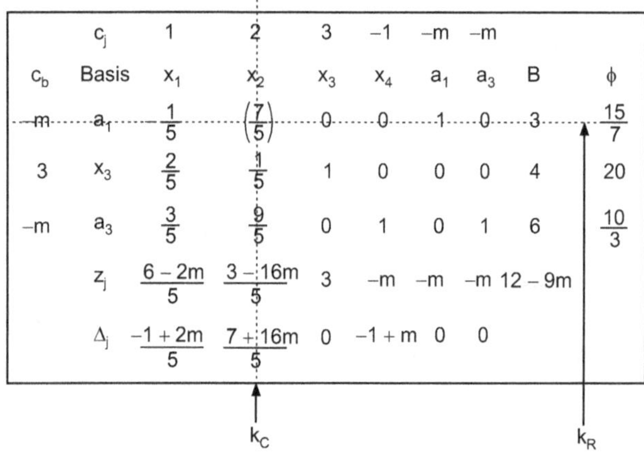

c_b	Basis	c_j 1 x_1	2 x_2	3 x_3	-1 x_4	$-m$ a_1	$-m$ a_2	$-m$ a_3	B	ϕ
0	s_1	1	2	3	0	1	0	0	15	5
5	x_3	2	1	(5)	0	0	1	0	20	4	$\leftarrow k_R$
2	x_2	1	2	1	1	0	0	1	10	10
	z_j	$-4m$	$-5m$	$-9m$	$-m$	$-m$	$-m$	$-m$	-45	
	Δ_j	$1 + 4m$	$2 + 5m$	$3 + 9m$	$-1 + m$	0	0	0		

k_C

Since, Δ_j is positive under some variable column, the solution is not optimal. Therefore, performing iteration to get an optimal solution in the following tables.

Table 2.18

c_b	Basis	c_j 1 x_1	2 x_2	3 x_3	-1 x_4	$-m$ a_1	$-m$ a_3	B	ϕ
$-m$	a_1	$\dfrac{1}{5}$	$\left(\dfrac{7}{5}\right)$	0	0	1	0	3	$\dfrac{15}{7}$
3	x_3	$\dfrac{2}{5}$	$\dfrac{1}{5}$	1	0	0	0	4	20
$-m$	a_3	$\dfrac{3}{5}$	$\dfrac{9}{5}$	0	1	0	1	6	$\dfrac{10}{3}$
	z_j	$\dfrac{6 - 2m}{5}$	$\dfrac{3 - 16m}{5}$	3	$-m$	$-m$	$-m$	$12 - 9m$	
	Δ_j	$\dfrac{-1 + 2m}{5}$	$\dfrac{7 + 16m}{5}$	0	$-1 + m$	0	0		

k_C k_R

Here a_2 basis column is neglected or deleted.

Table 2.19

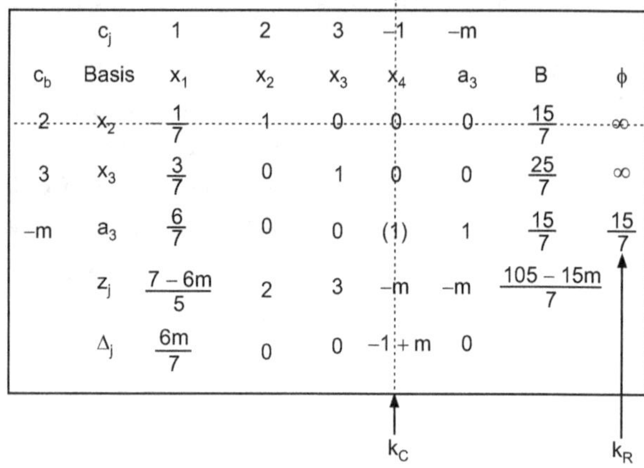

c_b	Basis	c_j 1 x_1	2 x_2	3 x_3	-1 x_4	$-m$ a_3	B	ϕ
2	x_2	$\dfrac{1}{7}$	1	0	0	0	$\dfrac{15}{7}$	∞
3	x_3	$\dfrac{3}{7}$	0	1	0	0	$\dfrac{25}{7}$	∞
$-m$	a_3	$\dfrac{6}{7}$	0	0	(1)	1	$\dfrac{15}{7}$	$\dfrac{15}{7}$
	z_j	$\dfrac{7 - 6m}{5}$	2	3	$-m$	$-m$	$\dfrac{105 - 15m}{7}$	
	Δ_j	$\dfrac{6m}{7}$	0	0	$-1 + m$	0		

k_C k_R

Here a_1 basis column is neglected or deleted.

Table 2.20

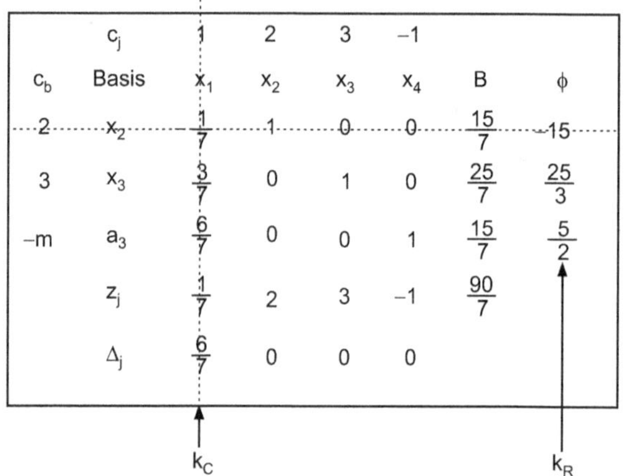

c_b	Basis	c_j 1 x_1	2 x_2	3 x_3	-1 x_4	B	ϕ
2	x_2	$\dfrac{1}{7}$	1	0	0	$\dfrac{15}{7}$	15
3	x_3	$\dfrac{3}{7}$	0	1	0	$\dfrac{25}{7}$	$\dfrac{25}{3}$
$-m$	a_3	$\dfrac{6}{7}$	0	0	1	$\dfrac{15}{7}$	$\dfrac{5}{2}$
	z_j	$\dfrac{1}{7}$	2	3	-1	$\dfrac{90}{7}$	
	Δ_j	$\dfrac{6}{7}$	0	0	0		

k_C k_R

Table 2.21

c_b	Basis	c_j 1 x_1	2 x_2	3 x_3	-1 x_4	B	ϕ
2	x_2	0	1	0	$\dfrac{1}{6}$	$\dfrac{5}{2}$	
3	x_3	0	0	1	$-\dfrac{1}{2}$	$\dfrac{5}{2}$	
1	x_1	1	0	0	$\dfrac{7}{6}$	$\dfrac{5}{2}$	
	z_j	1	2	3	0	15	
	Δ_j	0	0	0	-1		

Optimal solution

Since Δ_j is either zero or negative under all column the optimal solution is reached.

Therefore, $z_{max} = 15$ at

$$x_1 = \frac{5}{2} \quad a_1 = 0$$
$$x_2 = \frac{5}{2} \quad a_2 = 0$$
$$x_3 = \frac{5}{2} \quad a_3 = 0$$
$$x_4 = 0$$

UNIVERSITY QUESTIONS

1. Define the following for simplex method.
   (a) Leaving variable.
   (b) Entering variable.   **[May 17, Nov. 13, 4M]**

2. Explain big m method to the solve L.P.P.
   [Dec. 16, May/June 11, 6M]

3. Use big m method to solve the following LPP.
   Maximize $z = 3x_1 + 2x_2$
   Subject to : $2x_1 + x_2 \leq 1$
   $3x_1 + 4x_2 \geq 4$
   $x_1 \geq 0, \ x_2 \geq 0$
   [Dec 17, May 08, 9M]

4. Use big method to solve the following L.P.P
   Maximize $z = 2x_1 + x_2$
   Subject to : $3x_1 + x_2 = 3$
   $4x_1 + 3x_2 \geq 6$
   $x_1 + x_3 \leq 3$
   $x_1 \geq 0, \ x_2 \geq 0$
   [May 16, Nov./Dec. 14, 8M]

5. Use penalty method (Big m method) to solve the following L.P.P.
   Maximize
   $$z = x_1 + 20x_2$$
   Subject to :
   $5x_1 + 3x_2 \leq 30$
   $3x_1 + 6x_2 \leq 36$
   $2x_1 + 5x_2 \leq 20$
   $x_1 \geq 0, x_2 \geq 0$
   (Refer solved example 2.11)

2.6.2 Two Phase Method

This method is explained by Example 2.12.

Example 2.12 : *[Two phase method]*

[Dec. 17, May 05, May 14, 10M]

Use to two phase simple method to
Maximize z $= 5x_1 - 4x_2 + 3x_3$
Subject to : $2x_1 + x_2 - 6x_3 = 20$
$6x_1 + 5x_2 - 10x_3 \leq 76$
$8x_1 - 3x_2 + 6x_3 \leq 50$
$x_1 \geq 0, x_2 \geq 0, x_3 \geq 0$

Solution : Two phase method consists of the following steps :

Phase I :

Step 1 : Write the Problem in Standard Mathematical Form :

Let A_1 be artificial variable [This is to be added for = constraints] and s_1, s_2, s_3 be slack variable (with respect to constraints).

Here new objective function (artificial objective function) is
Minimize w = Sum of artificial variable.
Here $w = A_1$ (since only one artificial used for this problem).

Thus, for phase I, the standard mathematical form is :
Minimize $w = 0x_1 + 0x_2 + 0x_3 + 0s_2 + 0s_3 + A_1$
Subject to : $6x_1 + 5x_2 + 10x_3 + s_2 + 0s_3 + 0A_1 = 76$
$8x_1 - 3x_2 + 6x_3 + 0s_2 + s_3 + 0A_1 = 50$
$x_1 \geq 0, x_2 \geq 0, x_3 \geq 0, s_2 \geq 0, s_3 \geq 0, A_1 \geq 0$

Step 2 : Determine an Initial Basic Feasible Solution :

Let $x_1, x_2, x_3 = 0$ in above equation we have
$A_1 = 20$
$s_2 = 76$
$s_3 = 50$

Step 3 : Prepare Simplex Table and Check for Optimality :

Table 2.22

c_j		0	0	0	0	0	1		
c_b	Basis	x_1	x_2	x_3	s_2	s_3	A_1	B	ϕ
1	A_1	2	1	−6	0	0	1	20	10
0	s_2	6	5	10	1	0	0	76	$\frac{38}{3}$
0	s_3	(8)	3	6	0	1	0	50	$\frac{25}{4}$
	z_j	2	1	−6	0	0	1	20	
	Δ_j	−2	−1	6	0	0	0		

k_C (min. of Δ_j) k_R (min. ratio)

Here for optimal solution, Δ_j should be zero or positive. Here, Δ_j is negative at some variable column hence solution is not optimal.

Step 4 : Application of Simplex Criterion :

By applying simplex criterion, $\boxed{s_3 \rightarrow x_1}$

Step 5 : Preparation of next iteration for table or format :

Table 2.23

c_j		0	0	0	0	0	1		
c_b	Basis	x_1	x_2	x_3	s_2	s_3	A_1	B	ϕ
1	A_1	0	$\left(\frac{7}{4}\right)$	$-\frac{15}{2}$	0	$-\frac{1}{4}$	1	$\frac{15}{2}$	$\frac{30}{7}$
0	s_2	0	$\frac{29}{4}$	$\frac{11}{2}$	1	$-\frac{3}{4}$	0	$\frac{77}{2}$	$\frac{154}{29}$
0	s_3	1	$-\frac{3}{8}$	$\frac{3}{4}$	0	$\frac{1}{8}$	0	$\frac{25}{4}$	$-\frac{50}{3}$
	z_j	2	$\frac{7}{4}$	$-\frac{15}{2}$	0	$-\frac{1}{4}$	1	$\frac{15}{2}$	
	Δ_j	0	$-\frac{7}{4}$	$\frac{15}{2}$	0	$\frac{1}{4}$	0		

k_C (min. of Δ_j) k_R (min. ratio)

Second solution of Phase I

Step 6 : Continue Successive Iteration Till Optimal Solution is Achieved :

Table 2.24

c_b	Basis	c_j						B	ϕ
		0	0	0	0	0	1		
		x_1	x_2	x_3	s_2	s_3	A_1		
1	x_2	0	1	$-\dfrac{30}{7}$	0	$-\dfrac{1}{7}$	$\dfrac{4}{7}$	$\dfrac{30}{2}$	
0	s_2	0	0	$\dfrac{256}{7}$	1	$\dfrac{2}{7}$	$-\dfrac{29}{7}$	$\dfrac{52}{7}$	
0	x_1	1	0	$-\dfrac{6}{7}$	0	$\dfrac{1}{14}$	$\dfrac{3}{14}$	$\dfrac{55}{7}$	
	z_j	0	0	0	0	0	0		
	Δ_j	0	0	0	0	0	1		

Optimal solution of Phase I

Since Δ_j is zero or position solution is optimal.

Phase II :

Phase II of this method, determines optimal solution to original problem. Objective function for the initial table of phase II is the objective function of original problem (given problem).

The remaining part of initial table for phase II is the last table of phase I, with the only difference that z_j and Δ_j row in the last table for phase I are charged to account for the changes in the cost coefficient, Table 2.25, represents the initial table for phase II computations. Note that since basis column in table 2.24, contains no artificial variably, it is not considered in Table 2.25.

Table 2.25

c_b	Basis	c_j					B
		5	-4	3	0	0	
		x_1	x_2	x_3	s_2	s_3	
-4	x_2	0	1	$-\dfrac{30}{7}$	0	$-\dfrac{1}{7}$	$\dfrac{30}{2}$
0	s_2	0	0	$\dfrac{256}{7}$	1	$\dfrac{2}{7}$	$\dfrac{52}{7}$
5	x_1	1	0	$-\dfrac{6}{7}$	0	$\dfrac{1}{14}$	$\dfrac{55}{7}$
	z_j	5	-4	$\dfrac{90}{7}$	0	$\dfrac{13}{14}$	$\dfrac{155}{7}$
	Δ_j	0	0	$-\dfrac{69}{7}$	0	$-\dfrac{13}{14}$	

← Optimal solution

Since Δ_j is **zero or negative**, the solution is optimal solution.

$\therefore$ Optimal solution is

$$x_1 = \frac{55}{7},\ x_2 = \frac{30}{7},\ x_3 = 0,\ z_{max} = \frac{155}{7}$$

2.6.3 Dual Simplex Method

This method is explained by example 2.13.

Example 2.13 : *[Dual problem]*

[Dec. 17, May 05, 10M, Dec. 11, 10M]

Using the dual, solve the following LPP

$$z_{max} = 5x_1 - 2x_2 + 3x_3$$

$$Subjected\ to :\quad 2x_1 + 2x_2 - x_3 \geq 2$$

$$3x_1 - 4x_2 \leq 3$$

$$x_2 + 3x_3 \leq 5$$

$$x_1, x_2, x_3 \geq 0$$

Solution : The given problem is of maximization, hence all constraints should of ($\leq$) type, here for constraints equation :

$$2x_1 + 2x_2 - x_3 \quad \geq 2 \qquad \text{Multiply by } -1$$

$$-2x_1 - 2x_2 + x_3 \leq -2$$

$$\therefore \text{Primal problem becomes}$$

$$z_{max} = 5x_1 - 2x_2 + 3x_3$$

Subject to : $-2x_1 - 2x_2 + x_3 \leq -2$

$$3x_1 - 4x_2 \leq 3$$

$$x_2 + 3x_3 \leq 5$$

$$x_1, x_2, x_3 \geq 0$$

$$AX = B$$

$$\begin{bmatrix} -2 & -2 & 1 \\ 3 & -4 & 0 \\ 0 & 1 & 3 \end{bmatrix} \begin{bmatrix} x_1 \\ x_2 \\ x_3 \end{bmatrix} = \begin{bmatrix} -2 \\ 3 \\ 5 \end{bmatrix}$$

$\therefore$ The corresponding dual problem is :

Minimize $z' = -2v_1 + 3v_2 + 5v_3$

Subject to : $-2v_1 + 3v_2 \geq 5$

$$-2v_2 - 4v_2 + v_3 \geq -2$$

$$v_1 + 3v_3 \geq 3$$

$$v_1, v_2, v_3 \geq 0$$

$$A'V = C$$

$$\begin{bmatrix} -2 & 3 & 0 \\ -2 & -4 & 1 \\ 1 & 0 & 3 \end{bmatrix} \begin{bmatrix} v_1 \\ v_2 \\ v_3 \end{bmatrix} = \begin{bmatrix} 5 \\ -2 \\ 3 \end{bmatrix}$$

Using steps of simplex method :

Step 1 : Write the Problem in Standard Mathematical Form :

Multiplying second constraints by –1, it can be written as :

$$2v_1 + 4v_2 - v_3 \leq 2$$

Using Big m Method :

Let s_1, s_2, s_3 be slack variables and A_1 and A_2 be artificial variable.

$\therefore$ The standard mathematical form is as follows :

Minimize $z' = -2v_1 + 3v_2 \geq 5v_3 + 0s_1 + 0s_2 + 0s_3$
$+ mA_1 + mA_2$

Subject to : $2v_1 + 4v_2 - v_3 + 0s_1 + 0s_2 + 0s_3 = 2$

$v_1 + 0v_2 + 3v_3 + 0s_1 + 0s_2 + s_3 + A_2 = 3$

$v_1, v_2, v_3, s_1, s_2, s_3, A_1, A_2 \geq 0$

Step 2 : Determine an Initial Basic Feasible Solution :

Let $v_1 = v_2 = v_3 = s_1 = s_3 = 0$, we get

$A_1 = 5$

$s_2 = 2$

$A_2 = 3$

Step 3 : Prepare Simplex Table and Check for Optimality :

In table 2.26, Δ_j is positive under some column hence it is not optimal solution.

Table 2.26

c_b	Basis	c_j / v_1	v_2	v_3	s_1	s_2	s_3	A_1	A_2	B	ϕ
		-2	3	5	0	0	0	0	0	m	m
m	A_1	-2	3	0	-1	0	0	1	0	5	$\frac{5}{3}$
0	s_2	2	(4)	-1	0	1	0	0	0	2	$\frac{1}{2}$
m	A_2	1	0	3	0	0	-1	0	1	3	∞
	z_j	$-m$	$3m$	$3m$	$-m$	0	$-m$	m	m	20	
	Δ_j	$2+m$	$3-3m$	$5-3m$	m	0	m	0	0		

$\uparrow$
k_C (min. of Δ_j)

Step 4 : Application of Simplex Criterion :

Applying simplex criterion, $\boxed{s_2 \to v_2}$

Step 5 : Preparation of Next Iteration Table or Format :

Table 2.27

c_b	Basis	c_j / v_1	v_2	v_3	s_1	s_2	s_3	A_1	A_2	B	ϕ
		-2	3	5	0	0	0	m	m		
0	s_3	-15	0	0	-4	-3	1	4	-1	11	
3	v_2	$-\frac{2}{3}$	1	0	$-\frac{1}{3}$	0	0	$\frac{1}{3}$	0	$\frac{5}{3}$	
5	v_3	$-\frac{14}{3}$	0	1	$-\frac{4}{3}$	-1	0	$\frac{4}{3}$	0	$\frac{14}{3}$	
	z_j	$-\frac{76}{3}$	3	5	$-\frac{23}{3}$	-5	0	$\frac{23}{3}$	0	$\frac{85}{3}$	
	Δ_j	$\frac{70}{3}$	0	0	$\frac{23}{3}$	5	0	$-\frac{23}{3}+m$	m		

$\leftarrow$ Optimal solution (B column)

For primal problem

The optimal solution is $v_1 = 0$, $v_2 = \frac{5}{3}$, $v_3 = \frac{14}{3}$, $z' = \frac{85}{3}$.

For primal problem, variable under slack of dual are optimal value of x_1, x_2 and x_3 with changed sign.

$\therefore$ $x_1 = \frac{23}{3}$, $x_2 = 5$, $x_3 = 0$, $z_{max} = \frac{85}{3}$.

Example 2.14 : *[Simplex method]*

[May 16, Nov./Dec. 14, 8M]

Solve the following LPP by simplex method

Maximize $Z = 3X_1 + 2X_2$

Subject to :

$X_1 + X_2 \leq 4$

$X_1 - X_2 \leq 2$

$X_1, X_2 \geq 0$

Solution :

The solution is represented in following table

(i) Simplex Table 1

C_b	Basis	C_j / X_1	X_2	S_1	S_2	B	$\emptyset$
		3	2	0	0		
0	S_1	1	1	1	0	4	
0	S_2	1	1	0	1	2	
	Z_j	0	0	0	0		
	Δ_j	3	2	0	0		

(ii) Simplex Table 2

C_b	Basis	C_j / X_1	X_2	S_1	S_2	B	$\emptyset$
		3	2	0	0		
0	S_1	0	2	1	1	2	$2/2=1$
3	X_1	1	-1	0	1	2	$-2/1=-2$
	Z_j	3	-3	0	3	6	
	Δ_j	0	5	0	0		

(iii) Simplex Table 3

C_b	Basis	C_j / X_1	X_2	S_1	S_2	B	$\emptyset$
		3	2	0	0		
2	X_2	0	1	$1/2$	$-1/2$	1	
3	X_1	1	0	$1/2$	$3/2$	3	
	Z_j	3	2	$5/2$	$7/2$	11	
	Δ_j	0	0	$-5/2$	$-7/2$		

Solution: $+Z_{max}=11$, $X_1=3$, $X_2=1$

Example 2.15 : *Solve LPP by suitable method*

[May 18, Nov./Dec. 13, 8M, 12M]

Maximize :

$$Z = -X_1 + 2X_2$$

Subject to :

$$-X_1 + 3X_2 \le 10$$
$$X_1 + X_2 \le 6$$
$$X_1 - X_2 \le 2$$

Where, $X_1, X_2 \ge 0$

Solution : Please refer to Section 2.3 [Example 2.2].

Maximize :

$$Z = -X_1 + 2X_2$$

Subject to :

$$-X_1 + 3X_2 \le 10$$
$$X_1 + X_2 \le 6$$
$$X_1 - X_2 \le 2$$

Where, $X_1, X_2 \ge 0$

$$\text{at A} \quad X_1 = 0$$
$$X_2 = 3.333$$
$$Z_{max} = 6.67$$

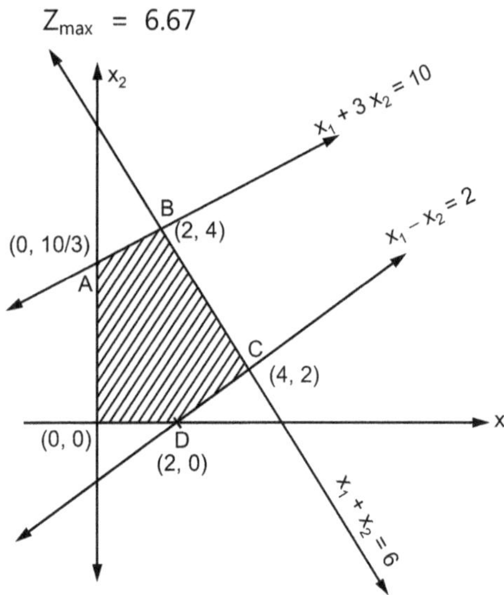

Fig. 2.4

Example 2.16 : *Old hens can be bought for ₹ 2.00 each but young ones costs ₹ 5.00 each. The old hens lay 3 eggs per week and the young ones lays 5 eggs per week. Each eggs costs ₹ 0.30 . A hens costs ₹ 1.00 per week to feed. If the financial constraints is to spend ₹ 80.00 per week for hens and the capacity constraints is that total number of hens cannot exceed 20 hens and the objectives is to earn a profit more than ₹ 6.00 per week. Formulate the LP problem.*

Solution :

Let x be the number of old hens and y be the number of young hens to be brought .

The desired LP is:

$$\text{Max Z} = 0.5y - 0.1x$$

S.T:
$$2x + 5y \le 80;$$
$$X + y \le 20;$$

> **By graphical solution : Z (0,16) = ₹ 8.00**

2.6.4 Sensitivity Analysis

[Dec. 17, May 08, 09, May 12, Nov. 13]

Sensitivity analysis deals with analysis of changes in parameters in the optimal solution. Since, in the market situation, the cost of job produced, raw material, men etc. may vary from day to day. Therefore, the profit of product changes. It is used for increasing performance of the system. The effect of parameters of linear programming problem consists of :

- Effect of right hand side of constraints (B_j).
- Effect of cost coefficient (c_j).
- Effect of coefficient of constraints (aij).
- Input of new variable.
- Input of new constraints.
- Removal of variable.
- Removal of constraints.

- **Effect of Right Hand Side of Constraints (B_j) :**

  When the right hand side constraints are changed then it changes the solution as well as objective function. For this, dual simplex method is followed to delete infeasibility and to have a feasible solution.

- **Effect of Cost Coefficient (c_j) :**

  This occurs due to change in cost or profit of either basic variable or non-basic variable. the change in coefficient of objective function, affects due to variation of cost or profit.

- **Effect of Coefficient of Constraints (a_{ij}) :**

  When coefficient of constraints changes the optimality of solution changes. It does not affect feasibility of solution.

- **Input of New Variable :**

  When new variable (new product or item) is added then it will increase profit or cost. It changes the optimal value of objective function involved.

- **Input of New Constraints :**

  When the new constraints are added then optimal solution may not changes. For this, dual simplex method is applied for finding solution.

- **Removal of Variable :**

  When non-basic variables are removed then it does not changes optimality and feasibility of solution. But when basic variable is removed then it affect optimality and feasibility of solutions. For this, usual simple method is applied.

- **Removal of Constraints :**

  When constraints (unbinding) are removed then it will not changes the optimal solution. But binding constraints changes optimal solution of the problem.

QUESTIONS FOR PRACTICE

1. Discuss concept of sensitivity. **[May 17, Dec. 14, 6M]**

2. Write short note on : Big M method

3. Discus concept of sensitivity analysis in LPP

   [May 16, Dec. 14, 6M]

4. Define the following

   (i) Slack variable

   (ii) Surplus variable

   (iii) Artificial variable

5. Define following terms of LPP.

   (i) Basic solution

   (ii) Feasible solution

6. Write dual of the following problem : **[May 18, 5M]**

   Minimize $z = 25 \times 1 + 10 \times 2$

   Subjected to condition :

   $$x_1 + x_2 \geq 50, \quad x_1 \geq 20$$
   $$x_2 \leq 40, \quad x_1, x_2 \geq 0$$

7. Write short note on formulation of linear programming problem. **[May 18, 3M]**

8. Solve LPP by Suitable Method. **[Dec 2011, 10M]**

Maximize: $Z = 10x_1 + 20x_2$

Subject to $5x_1 + 2x_2 \leq 30$

$$3x_1 + 6x_2 < 36$$
$$2x_1 + 5x_2 \leq 20$$

where $x_1 > 0, x_2 > 0$ (Refer Section 2.3 [Example 2.2].

[Ans. $Z = 90.5$, $X_1 = 5.2$, $X_2 = 1.9$] [For graphical method error of 6% is allowed]

9. A firm manufactures two products P_1P_2 on which the profits earned are ₹ 5 and ₹ 8 respectively. Each product is prepared on two machines M_1 and M_2. The machine time required for these products on the two machines and their availability is as shown below.

	Product P_1	Product P_2	Availability of machines (mins.) per day
Machine M_1	2	1	400
Machine M_2	4	1	600

Find the number of units of products P_1 and P_2 to be manufactured per day to get maximum profits?

(Refer Section 2.2 [Example 2.1].

[Ans. $P_1 = 0$, $P_2 = 400$, $Z = 3200$

[May 15, Dec 11, 10M]

10. A firm manufacturers three products S_1, S_2 and S_3 on which the profits earned are ₹ 2, ₹ 5 and ₹ 4 respectively. Each product need two types of raw materials R_1 and R_2 which the firm can purchase upto a maximum of 500 and 400 units respectively. Design production plan so as to maximize the profit.

Raw Material	Consumption of Raw Materials per unit Product		
	S_1	S_2	S_3
R_1	0.5	1	1
R_2	2	0.5	0.5

(Refer Section 2.2 [Example 2.2].

[Dec 17, May 10, 16 M]

11. A factory has decided to diversify its activities. The data collected for the sales and production departments are summarized below:

Potential demand exists for two products A and B. Market can absorb, any quantity of A, whereas the share of B for this organization is expected to be not more than 400 units per month. Contribution per unit of product A and B is expected to be ₹ 6 and ₹ 8 respectively. These products, require three different processes and the time required per unit of product is given in the table below: **[Nov/Dec 2012 12 M]**

Process	Product A	Product B	Available Hours
1	2	3	900
2	1	2	600
3	2	2	1200

Find the product mix to optimize the contribution.

(Refer Section 2.2)   **[Dec 16, May 12, 10M]**

12. A person requires 10, 12 and 12 units of chemicals A, B and C respectively per Jar. A dry product contains 1, 2 and 4 units of A, B and C per cartoon. If the liquid product sells for ₹ 3 per Jar and dry product sell for ₹ 2 per cartoon, how many of each should be purchased to minimize the cost and meet the requirements.   **[May 2013, 12M]**

(Refer Section 2.2 [Example 2.2].)

◈ ◈ ◈

CHAPTER 3
TRANSPORTATION AND ASSIGNMENT MODEL

3.1 INTRODUCTION

[May 14, 17, Nov./Dec. 04, Nov. 11]

- A problem which deals with distribution of an item produced at a several sources (origins) to a number of different destinations (warehouses) is known as *Transportation problem*.

- It is also known as distribution model or distribution problem instead of transportation problem. The various features of linear programming problems can be observed in this problem. Here the supply as well as demand of the various centres are finite and constitutes the limited resources. It is also assumed that cost of transportation is linear but this is not often true in practical problems.

3.1.1 Applications of Transportation Model

- Transportation problem is used for the following applications :
  - ➢ Scheduling of employee
  - ➢ Assignment of resources such as men, machines etc.
  - ➢ Inventory management
  - ➢ Production and Investment analysis
  - ➢ Location of factory.
- Transportation problem deals with distribution of job from plants to destinations, destination to dealers, dealers to distributes and distributor to clients. It is particular features of linear programming that concern with shipping a commodity from factories to warehouses. The main objective of transportation problem is to minimise total shipping cost in proportional to the number of jobs shipped on a given route.

3.1.2 Assumptions in Transportation Model

The transportation problem is based on the following assumptions :

- The total amount of unit available at different origins (sources) is equal to total requirements at various destinations.
- Item or unit can be distributed easily from all origins (sources) to destinations.
- The unit transportation cost of the jobs from all sources to destination is known.

- The distribution cost on a given route is directly proportional to the number of jobs shipped on that route.
- The objective is to minimize the total distribution cost for the organisation as a whole and not for individual availability and destinations.

3.2 FORMULATION OF TRANSPORTATION MODEL

[May 16]

- The main characteristic of transportation problem is that sources (origins) and products must be expressed in term of only one kind of unit. Consider that there are 'm' origins and 'n' destinations. Let a_i be the number of supply item or unit available at source i (where, i = 1, 2, 3, ... m) and b_j be the number of requirement units at destination j (where, j = 1, 2, 3, ... n).

- Let c_{ij} represents the unit transportation cost for shipping units from origins (sources) i to destination j.

- The main objective is to minimize total transportation cost for the number of units to be distributed from origin (source) i to destination j and the supply limits at the origins (sources) and demands at the distribution must be equal.

- If x_{ij} ($x_{ij} \geq 0$) is the number of units distributed from source (origin) i to destination j then the equivalent linear programming problem will be as follows :

Determine x_{ij} (i = 1, 2, 3, ... m, j = 1, 2, 3 ... n) in order to minimize

$$Z = \sum_{j=1}^{n} \sum_{i=1}^{m} x_{ij}\, c_{ij}$$

Subject to $\sum_{i=1}^{m} x_{ij} = a_i$, i = 1, 2, 3, ... , m

$$\sum_{j=1}^{n} x_{ij} = b_j, j = 1, 2, 3, ... , n$$

where, $x_{ij} \geq 0$

The system will be in balance (The two sets of constrains will be consistent)

if $$\sum_{i=1}^{m} a_i = \sum_{j=1}^{n} b_j$$

- Generally the problem will have $(m + n - 1)$ constituents and $m \times n$ unknowns.

  For feasible solution of transportation problem the necessary as well as sufficient condition is

$$\sum_{i=1}^{m} a_i = \sum_{j=1}^{n} b_j$$

- The transportation problem which satisfy this condition is referred as **Balanced Transportation Problem** or **Standard Transportation Problem**. The transportation problem which does not satisfy the above condition is referred as **Unbalanced Transportation Problem**. The unbalanced transportation problem is converted to balanced transportation problem by adding dummy source (origin) or destination.

- All the above discussion can be expressed in form of an effectiveness matrix as shown below in Table 3.1. In table 3.1, c_{ij} is the unit transportation cost from the ith source (origin) to j^{th} destination and x_{ij} is quantity distributed from i^{th} source (origin) to j^{th} destination, a_i is the supply available at source (origin) and b_j is demand at destination j.

Table 3.1

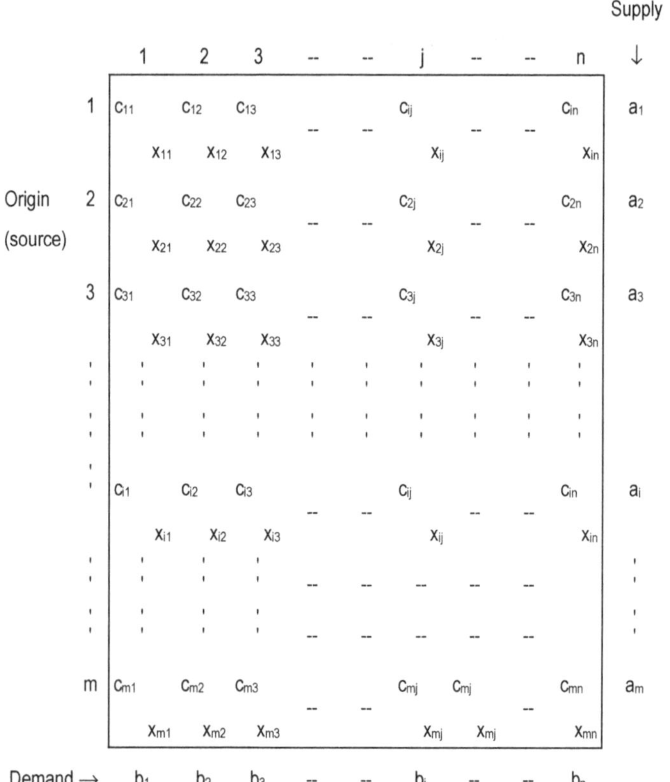

3.3 BASIC TERMINOLOGIES FOR TRANSPORTATION MODEL

Basic terms used in Transportation Problem are as follows :

[May 17, Nov./Dec. 06]

1. Basic feasible solution
2. Feasible solution
3. Optimal solution
4. Non-degenerate basic feasible solution
5. Degenerate basic feasible solution.

1. Basic Feasible Solution :

A feasible solution which contains no more than $(m + n - 1)$ non-negative allocations (where m is the number of rows and n is the number of column) is known as *Basic Feasible Solution*.

2. Feasible Solution :

A solution which is a set of non-negative allocations, x_{ij} that fulfils the rim (row and column) restriction is known as *Feasible Solution*.

3. Optimal Solution :

A feasible solution that maximizes (or minimizes) the transportation profit (cost) is known as an *Optimal Solution*.

4. Non-Degenerate Basic Feasible Solution :

A basic feasible solution in which :

(i) The total number of non-negative allocations is exactly $(m + n - 1)$ and

(ii) These $(m + n - 1)$ allocations are in independent position.

is known as Non-Degenerate Basic Feasible Solution.

5. Degenerate Basic Feasible Solution :

A basic feasible solution in which the total number of non-negative allocation is less than $(m + n - 1)$ is known as degenerate Basic Feasible Solution. This is also referred as **Degeneracy** in the transportation problem.

3.4 BASIC METHOD OF SOLVING TRANSPORTATION PROBLEM OR MODEL

(May 16, 17)

(Optimization Method like UV and Stepping Stone Method)

The solution of transportation problem is achieved by using the following steps :

Step 1 : Formulation of transportation problem.

Step 2 : Determination of an initial basic feasible solution.

Step 3 : Optimality Test or Testing initial basic solution for optimality.

Step 4 : Development of next, iterative, improved solution.

Step 5 : Repeating the iteration until optimality is achieved.

Step 1 : Formulation of Transportation Problem

The sub-steps involved are as follows :

- Express supply from origin (source), requirement at destination and cost of transportation from origin (source) to destination in the form of effectiveness matrix.

- Check the type of problem (balanced problem or unbalanced problem).

- Convert unbalanced problem into balanced problem by adding dummy origin (source) or destination.

Step 2 : Determination of an Initial Basic Feasible Solution

An initial basic feasible solution is achieved by using the following methods :

- North-West Corner Method or North-West Corner Rule (NWCM)

- Row Minima Method (RMM)

- Column Minima Method (CMM)

- Lowest Cost Entry Method (LCEM) or Matrix Minima Method

- Vogel's Approximation Method (VAM) or Unit Cost Penalty Method.

3.4.1 North-West Corner Method (NWCM) or North West Corner Rule

The steps used in North-West Corner Method are as follows :

- Start with North-West Corner (upper left) of the requirement matrix i.e. The effectiveness matrix formed in step (1) of transportation problem and compare the supply of source (origin) 1 (call it as m_1) with the requirement of destination 1 (call it as n_1).

  - ➤ If $n_1 < m_1$ i.e. if the quantity required at n_1 is less than the number of units available at m_1 then set x_{11} equal to n_1 and proceed to cell (1, 2) i.e. [proceed horizontally →]

  - ➤ If $n_1 = m_1$, set x_{11} equal to n_1 and proceed to cell (2, 2) [i.e. proceed diagonally ⬊].

  - ➤ If $n_1 > m_1$, set x_{11} equal to m_1 and proceed to cell (2, 1) [i.e. proceed vertically ↓].

- Continue in this manner, step-by-step, away from north-west corner until finally, a value is reached in the south east corner.

[**Note :** The amounts allocated are put in parenthesis () and they expresses the corresponding decision variable. These cells are known as allocated cells or loaded cells. Cells without allocation are known as *Non-Basic Cell* or *Vacant Cell* or *Empty Cell*.]

This method can be explained with Example 3.1.

SOLVED EXAMPLES

Example 3.1 : *Determine an initial basic feasible solution for the following transportation problem in which cells contain the transportation cost in rupees using north-west corner rule.*   **[May 16, Nov./Dec. 12]**

Table 3.2

		A	B	C	D	Supply ↓
Plan t	I	2	3	11	7	6
	II	1	0	6	1	1
	III	5	8	15	9	10
Requirement →		7	5	3	2	(17)

Solution : Since, the given transportation problem is balanced transportation problem the solution (Initial Basic Feasible solution) is obtained as follows :

Step 1 : Start with north-west corner (upper left) of the requirement matrix i.e. the effectiveness matrix formed in step (1) of transportation problem and compare the supply of source (origin) 1 (call it as m_1) with the requirement of destination 1 (call it as n_1).

Here $n_1 = 7$ and $m_1 = 6$.

i.e. $n_1 = 7$ and $m_1 = 6$

∴ $n_1 > m_1$

∴ Set x_{11} equal to m_1 and proceed to cell (2, 1) (i.e. proceed vertically ↓)

∴ $x_{11} = 6$

Table 3.3

		A	B	C	D	Supply ↓
Plant	I	2 (6)	3	11	7	6/0
	II	1	0	6	1	1
	III	5	8	15	9	10
Requirement →		7/1	5	3	2	

Step 2 : Continue in this manner, step-by-step, away from north-west corner until finally, a value is reached in the south east corner (S.E. corner).

Now at cell (2, 1) $m_2 = 1$ and $n_2 = 1$

      ↓   ↓

    (say as m_1)   (say as n_1)

i.e. $m_1 = 1, n = 1$

Since, $m_1 = n_1$, set x_{11} equal to n_1 and proceed to cell (2, 2) [i.e. proceed diagonally]

Table 3.4

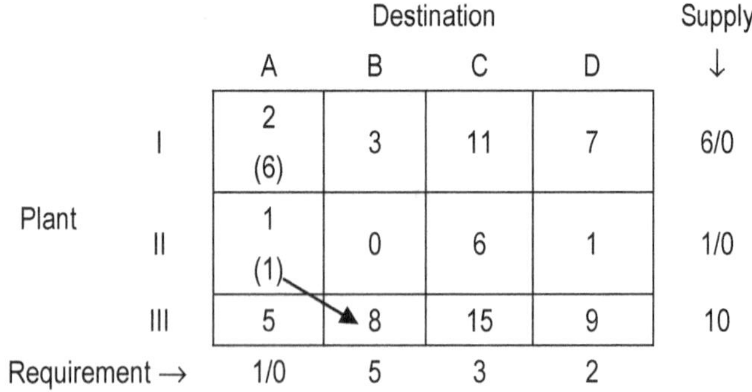

		Destination			Supply ↓
	A	B	C	D	
I	2 (6)	3	11	7	6/0
Plant II	1 (1)	0	6	1	1/0
III	5	8	15	9	10
Requirement →	1/0	5	3	2	

Similarly at cell (3, 2) $m_3 = 10$ and $n_3 = 5$

        ↓   ↓

    (say as m_1)   (say as n_1)

i.e. $n_1 < m_1$ set x_{11} equal to n_1 and proceed to cell (1, 2) [i.e. proceed horizontally →]

[Here to cell (3, 3)]

Table 3.5

		Destination			Supply ↓
	A	B	C	D	
I	2 (6)	3	11	7	6/0
Plant II	1 (1)	0	6	1	1/0
III	5	8 (5) →	15	9	10/5
Requirement →	7/1/0	5/0	3	2	

Similarly we proceed upto south east corner.

Table 3.6

		Destination			Supply ↓
	A	B	C	D	
I	2 (6)	3	11	7	6/0
Plant II	1 (1)	0	6	1	1/0
III	5	8 (5) →	15 (3) →	9(2)	10/5/2/0
Requirement →	7/1/0	5/0	3/0	2/0	

The total transportation cost is given as :

$$Z = ₹ [(2 \times 6) + 1 \times 1 + 8 \times 5 + 15 \times 3 + 9 \times 2] \times 100$$

$$= ₹ 11,600$$

3.4.2 Row Minima Method (RMM)

An initial basic feasible solution can be obtained by the following steps :

Step 1 : Allocate as much as possible in the smallest cost cell of the first row so that either the capacity of first origin (source) is exhausted or the requirement at the jth destination is fulfilled or both. In case of tie among the cost, choose arbitrarily. There are three cases involved which are as follows :

- If the capacity of first origin (source) is completely exhausted, score out the first row and process to second row.

- If the requirement at j^{th} destination is satisfied, score out the jth column and consider again the first row with the remaining capacity.

- If the capacity of first origin (source) as well as the requirement at the jth destination completely satisfied, then make a zero allocation in the second smallest cost cell of first row. Score out the row as well as j^{th} column and move down to second row.

Step 2 : Proceed further, for the resulting matrix, until all the rim condition are satisfied.

The above method can be explained by example 3.2.

Example 3.2 : *Find an initial basic feasible solution to the following transportation problem shown in Table 3.7 using row minima method (RMM).*

Table 3.7

		Destination				Supply ↓
		A	B	C	D	
	I	2	3	11	7	6
Plant	II	1	0	6	1	1
	III	5	8	15	9	10
Requirement →		7	5	3	2	(17)

Solution :

Step 1 : Allocate as much as possible in the smallest cost cell of the first row so that either the capacity of first origin (source) is exhausted or the requirement at the jth destination is fulfilled or both. In case of the tie among the cost, choose arbitrarily. Here in row 1, cell (1, 1) has smallest cost, hence allocate 6 units to this cell, since capacity of first row completely exhausted, score out first row and proceed to second row.

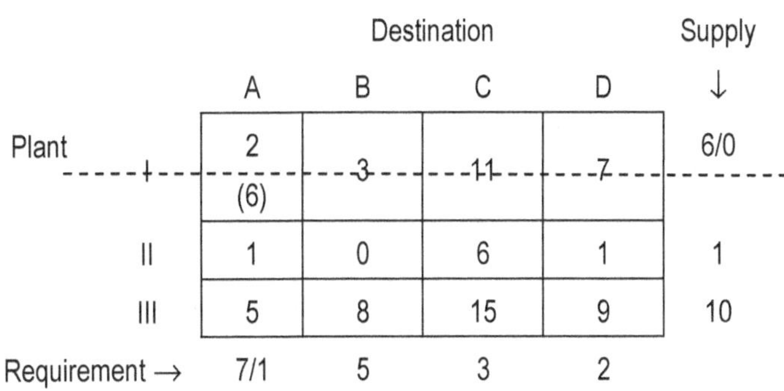

Table 3.8

	Destination A	B	C	D	Supply ↓
Plant I	2 (6)	3	11	7	6/0
II	1	0	6	1	1
III	5	8	15	9	10
Requirement →	7/1	5	3	2	

Step 2 : Proceed further, for the resulting matrix until all the rim condition are fulfilled.

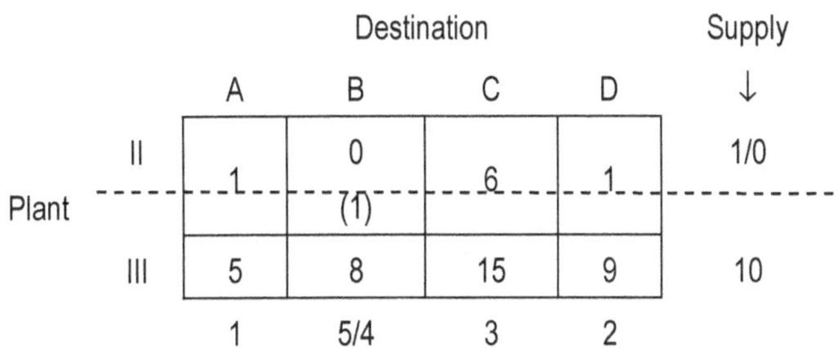

Table 3.9

	Destination A	B	C	D	Supply ↓
Plant II	1	0 (1)	6	1	1/0
III	5	8	15	9	10
	1	5/4	3	2	

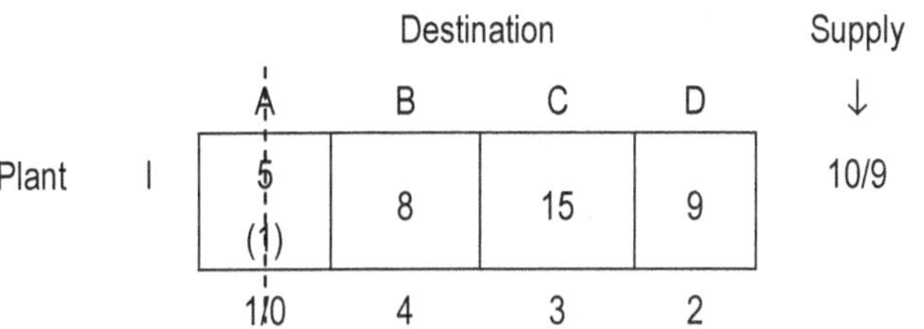

Table 3.10

	Destination A	B	C	D	Supply ↓
Plant I	5 (1)	8	15	9	10/9
	1/0	4	3	2	

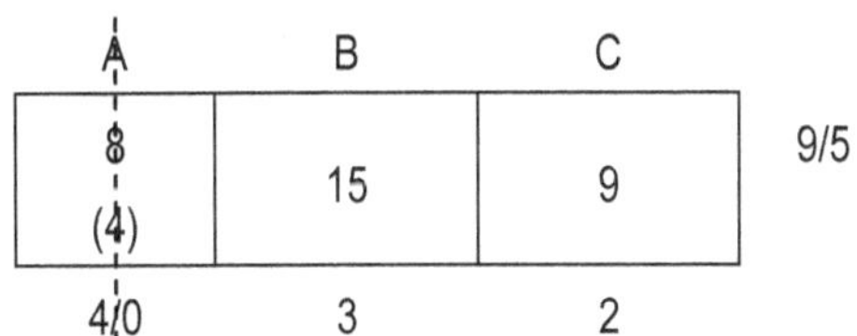

Table 3.11

A	B	C	
8 (4)	15	9	9/5
4/0	3	2	

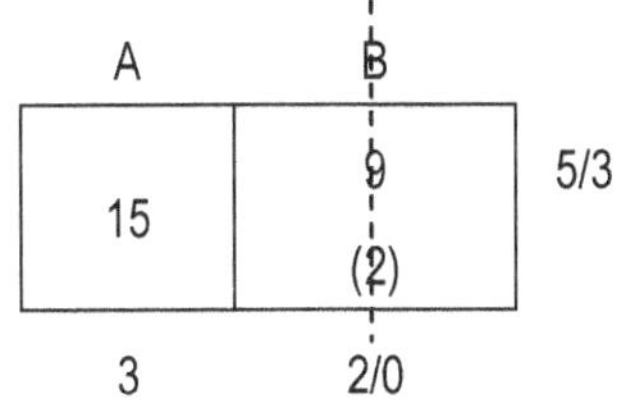

Table 3.12

A	B	
15	9 (2)	5/3
3	2/0	

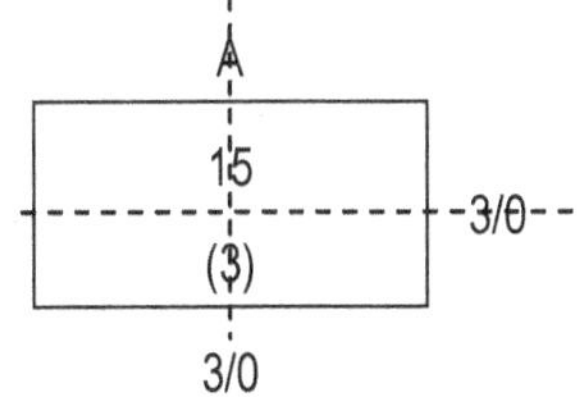

Table 3.13

A	
15 (3)	3/0
3/0	

∴   The transportation cost of the given problem is

$$Z = ₹ \,[2 \times 6 + 0 \times 1 + 5 \times 1 + 8 \times 4 + 9 \times 2 + 15 \times 3] \times 100$$

$$= ₹ \,11,200$$

3.4.3   Column Minima Method [CMM]   [Dec. 16]

An initial basic feasible solution can be obtained by the following steps :

Step 1 : Allocate as much as possible in the smallest cost cell of the first column so that either the demand of first destination is satisfied or the capacity of ith source (origin) is exhausted. For the tie among the cost cells, in the column, choose arbitrarily. There are three cases involved which are as follows :

- If the requirement of the first destination is fulfilled, score out the first column and proceed to the right of second column.

- If the capacity of ith source (origin) is satisfied, score out i^{th} row and consider again the first column with remaining requirement.

- If the requirement of first destination as well as capacity of ith source (origin) are completely satisfied, make a zero allocation in the second smallest cost cell of the first column, score out the column as well as i^{th} row and move to second column.

Step 2 : Proceed further, for resulting matrix until all the rim condition are satisfied.

The above method is explained by example 3.3.

Example 3.3 : *Find an initial basic feasible solution for the following transportation problem shown in Table 3.14 using column minima method.*

Table 3.14

		Destination A	B	C	D	Supply ↓
Plant	I	2	3	11	7	6
	II	1	0	6	1	1
	III	5	8	15	9	10
Requirement →		7	5	3	2	

Solution :

Step 1 : Allocate as much as possible in the smallest cost cell all of the first column so that either the demand of first destination is satisfied or the capacity of ith source (origin) is exhausted or both. Here cell (2, 1) has smallest cost, hence allocate 1 unit to this cell. Since, the capacity of 2nd row is fulfilled, score out this row and consider again the first column with remaining requirement.

Table 3.15

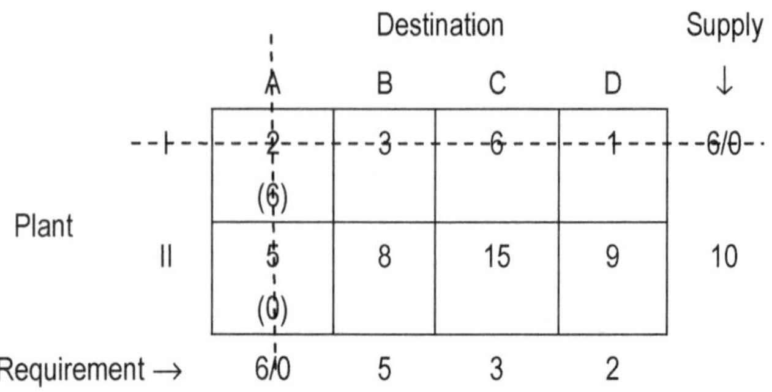

	Destination				Supply
---	A	B	C	D	↓
I	2	3	11	7	6
II	1 (1)	0	6	1	1/0
III	5	8	15	9	10
Requirement →	7/6	5	3	2	

(Plant)

Step 2 : Proceed further, for the resulting matrix, until all the rim condition are satisfied.

Table 3.16

	Destination				Supply
---	A	B	C	D	↓
I	2 (6)	3	6	1	6/0
II	5 (0)	8	15	9	10
Requirement →	6/0	5	3	2	

(Plant)

Since, the requirement of first destination as well as the capacity of first origin (source) are completely satisfied, make a zero allocation in a second smallest cost cell of first column, score out the column as well as row and move right to second column, this is as shown in Table 3.16.

Table 3.17

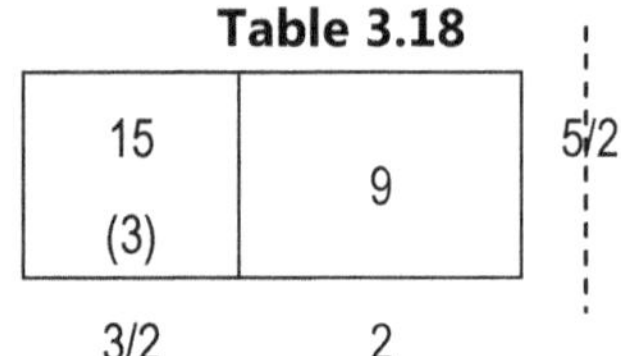

8 (5)	15	9	10/5
5/0	3	2	

Table 3.18

15 (3)	9	5/2
3/2	2	

Table 3.19

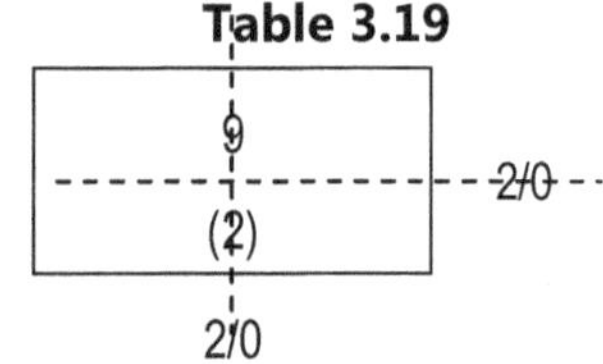

9 (2)	2/0
2/0	

∴   The transportation cost of the given problem is

$$Z = ₹ [1 \times 1 + 2 \times 6 + 5 \times 0 + 8 \times 5 + 15 \times 3 + 9 \times 2] \times 100$$

$$= ₹ 11,600$$

3.4.4 Lowest Cost Entry Method (LCEM) or Matrix Minima Method   [Dec. 16]

An initial basic feasible solution can be obtained by the following steps :

Step 1 : Select the smallest cost cell and allocate as much as possible within the restriction of rim condition to the smallest cost cell.

Step 2 : Score out row or column completely satisfied by allocation just made.

Step 3 : Consider again matrix (resulting matrix) consists of row and column where allocation have not yet been made and repeat steps (i) and (ii) until all assignment have been made.

The above method can be explained by example 3.4.

Example 3.4 : *Determine an initial basic feasible solution to the following transportation problem shown in table 3.20 (a), using lowest cost entry method.*   ***[Dec. 17, May 14]***

Table 3.20 (a)

		Warehouses						Supply
		W₁	W₂	W₃	W₄	W₅	W₆	↓
	A	2	1	3	3	2	5	50
Origin	B	3	2	2	4	3	4	40
	C	3	5	4	2	4	1	60
	D	4	2	2	1	2	2	30
Demand →		30	50	20	40	30	10	

Step 1 : Select the smallest cost cell and allocate as much as possible within the restriction of rim condition to the smallest cost cell.

Here, smallest cost cell is (1, 2), (3, 6) and (4, 4). i.e. there is tie, in this case, choose cell where allocation of more number of unit can be made.

Hence, consider cell (1, 2)

Table 3.20 (b)

		Warehouses						Supply
		W₁	W₂	W₃	W₄	W₅	W₆	↓
	A	2	1 (50)	3	3	2	5	50/0
Origin	B	3	2	2	4	3	4	40
	C	3	5	4	2	4	1	60
	D	4	2	2	1	2	2	30
Demand →		30	50/0	20	40	30	10	

Step 2 : Score out row or column completely satisfied by allocation just made. This is as shown in table 3.20 (b).

Step 3 : Consider again resulting matrix consist of row and column where allocation have not yet been made and repeat steps (i) and (ii) until all assignment have been made.

Table 3.21

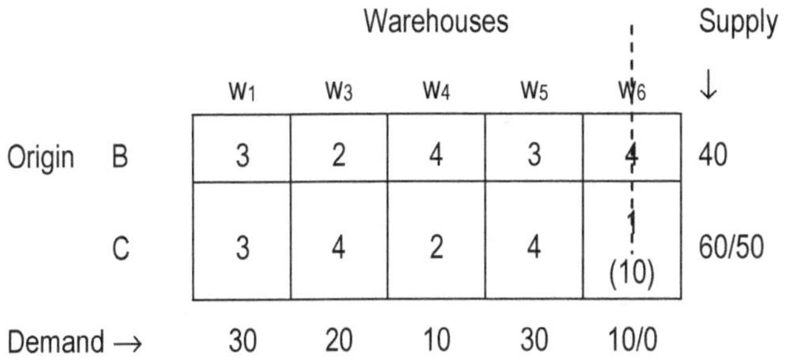

Table 3.21 (Warehouses / Supply):

	W₁	W₃	W₄	W₅	W₆	Supply ↓
Origin B	3	2	4	3	4	40
C	3	4	2	4	1	60
D	4	2	1	2	2	30/0
			(30)			
Demand →	30	20	40/10	30	10	

Table 3.22

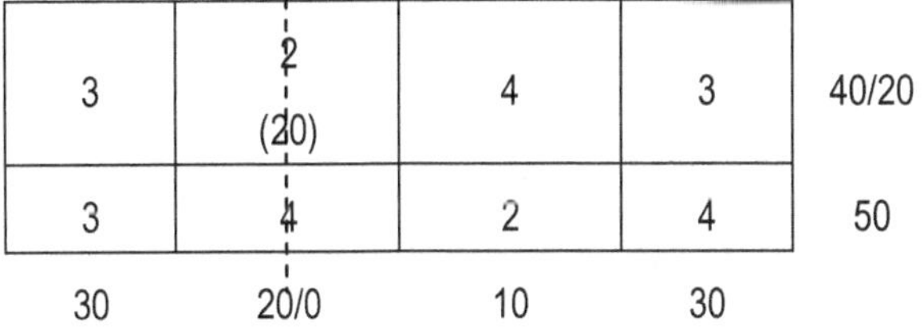

	W₁	W₃	W₄	W₅	W₆	Supply ↓
Origin B	3	2	4	3	4	40
C	3	4	2	4	1 (10)	60/50
Demand →	30	20	10	30	10/0	

Table 3.23

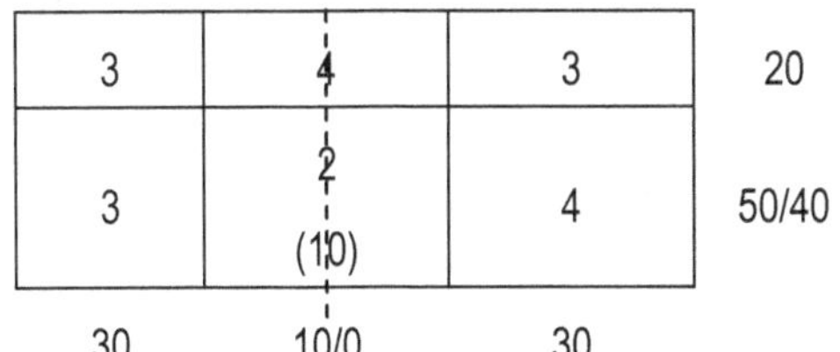

3	2 (20)	4	3	40/20
3	4	2	4	50
30	20/0	10	30	

Table 3.24

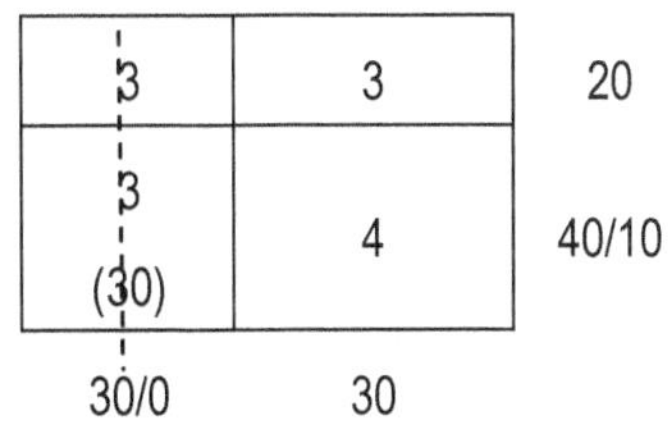

3	4	3	20
3	2 (10)	4	50/40
30	10/0	30	

Table 3.25

3	3	20
3 (30)	4	40/10
30/0	30	

Table 3.26

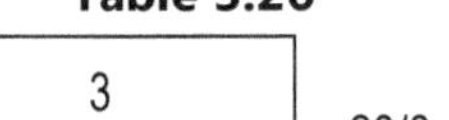

3 (20)	20/0
4	10

3/10

Table 3.27

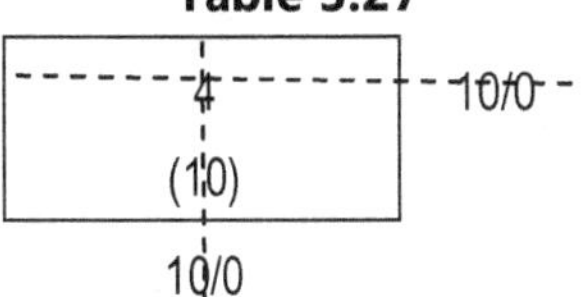

4 (10)	10/0

10/0

∴ The solution of above problem is given as :

$$Z = ₹ [1 \times 50 + 1 \times 30 + 1 \times 10 + 2 \times 20$$
$$+ 2 \times 10 + 3 \times 30 + 3 \times 20 + 4 \times 10]$$
$$= ₹ 340$$

3.4.5 Vogel's Approximation Method (VAM) or Unit Cost Penalty Method

[May 17, Nov. /Dec. 14, 8M]

This method, makes an effective use of cost information and provides better an initial basic feasible solution that achieved by the other methods.

An initial basic feasible solution can be obtained by the following steps :

Step 1 : Put the difference between smallest and second smallest element in each column below the corresponding column and difference between smallest and second smallest element in each row to the right of row. Enter these number in brackets [ ] as shown.

Step 2 : Select the row or column with highest difference (penalty) and allocate as much as possible within the restriction of the rim condition to the smallest cost cell in the row or column selected.

[**Note :** In case of tie among the highest difference (penalties) select, the row or column having smallest cost. In case of tie in smallest cost also, choose the cell which can have maximum allocation. If there is tie among maximum allocation cell also, then, choose cell arbitrarily for allocation.]

Step 3 : Cross-out the row or column completely fulfilled by the allocation just made.

Step 4 : Consider again resulting matrix consists of row and column where allocation have not yet been made and repeat steps (i) and (iii) until all assignment have been made.

The above method can be explained by example 3.5.

Example 3.5 : *A company has plants I, II and III which supply warehouses at IV, V, VI and VII. Monthly plants capacities are 250, 300 and 400 units respectively for regular production. If overtime production is utilized, plants I and II can produces 50 and 75 additional units respectively at overtime incremental costs of ₹ 4 and ₹ 5 respectively. The current warehouses demands are 200, 175, 275 and 300 units respectively. Unit transportation cost in rupees from plants to warehouses are given in table 3.28 below.*

Determine an initial basic feasible solution by VAM.

[Dec. 16, May / June 05, 16M, May 14]

Table 3.28

Warehouses

		IV	V	VI	VII
	I	11	13	17	14
Plant	II	16	18	14	10
	III	21	24	13	10

Solution : The given transportation problem can be expressed as :

Table 3.29

Warehouses

		IV	V	VI	VII	Capacity
	I	11	13	17	14	250
Plant	II	16	18	14	10	300
	III	21	24	13	10	400
Demand →		200	175	275	300	

Step 1 : Put the difference between smallest and second smallest element in each column below the corresponding column and the difference between smallest and second smallest element in each row to the right of row. Enter this number in Bracket [ ].

Table 3.30

Warehouses

		IV	V	VI	VII	Capacity
	I	11	13	17	14	250 [2]
Plant	II	16	18	14	10	300 [4]
	III	21	24	13	10	400 [3]
Demand →		200	175	275	300	
		[5]	[5]	[1]	[0]	

Step 2 : Select the row or column with highest difference and allocate as much as possible within the restriction of the rim condition to the smallest cost cell in the row and column selected.

Table 3.31

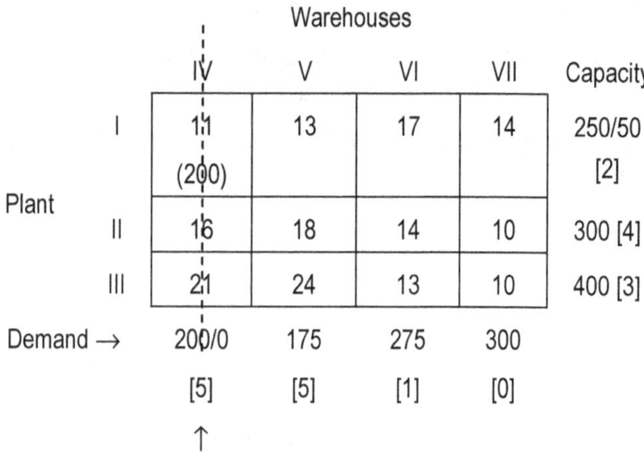

Warehouses

		IV	V	VI	VII	Capacity
	I	11 (200)	13	17	14	250/50 [2]
Plant	II	16	18	14	10	300 [4]
	III	21	24	13	10	400 [3]
Demand →		200/0	175	275	300	
		[5]	[5]	[1]	[0]	
		↑				

Step 3 : Cross-out the row or column completely fulfilled by the allocation just made. This is as shown in table 3.31.

Step 4 : Consider again resulting matrix consists of row and column where allocation have not yet been made and repeat steps (i) and (iii) until all assignment have been made.

Table 3.32

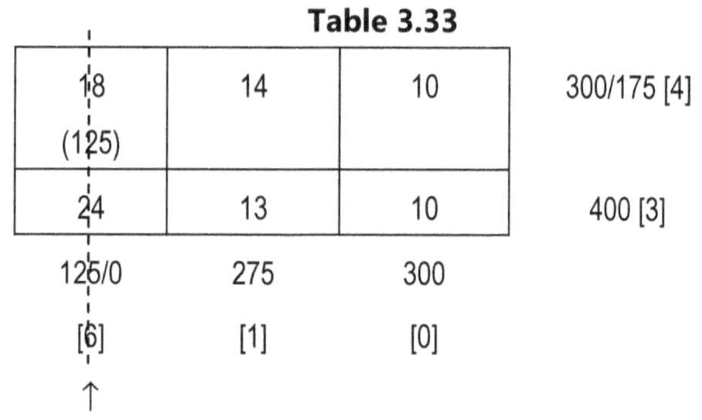

~~13~~	~~17~~	~~14~~	~~50/0 [1]~~
(50)			
18	14	10	300 [4]
24	13	10	400 [3]
175/125	275	300	
[5]	[1]	[0]	
↑			

Table 3.33

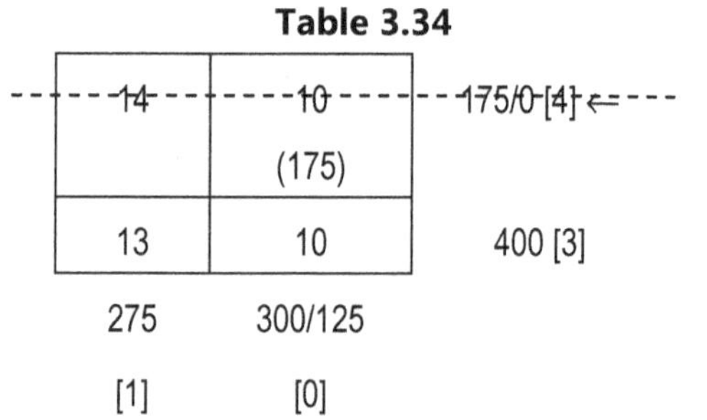

18 (125)	14	10	300/175 [4]
24	13	10	400 [3]
125/0	275	300	
[6]	[1]	[0]	
↑			

Table 3.34

~~14~~	~~10~~	~~175/0 [4]~~ ⇐
	(175)	
13	10	400 [3]
275	300/125	
[1]	[0]	

Table 3.35

13	10
275	125/0

400/275 [3] ←

[1]   [0]

Table 3.36

13 — — — — 275/0

(275)

275/0

$\therefore$ The initial basic feasible solution is given as :

$$Z = ₹\,[11 \times 200 + 13 \times 50 + 18 \times 125 + 10 \times 175 + 10 \times 125 + 13 \times 275]$$

$$= \text{Rs. } 11675$$

UNIVERSITY QUESTIONS

1. What is transportation problem?

   [May 17, Nov./Dec. 04, Nov. / Dec. 11, 4M]

2. Explain mathematical formulation of transportation problem. **[May 10]**

3. Differentiate between balanced and unbalanced transportation problem. **[Nov. /Dec. 12, 4M]**

4. Explain the following transportation problem solution method.

   (i) Vogel's approximation method.

   (ii) North-West corner method.

   (iii) Lowest cost entry method.

   (iv) Row minima method. **[May 16]**

5. Find an initial basic feasible solution to the following transportation problem by Vogel's approximation method.

   [May 16, Nov./Dec. 08, May 14, Aug. 15, 10M]

   Table 3.37

		Warehouses				Supply
		W_1	W_2	W_3	W_4	↓
	1	19	30	50	10	7
Factory	2	70	30	40	60	9
	3	40	8	70	20	18
Requirement →		5	8	7	14	

[Ans. : $x_{11} = 5$, $x_{14} = 2$, $x_{23} = 7$, $x_{24} = 2$, $x_{32} = 8$, $x_{34} = 10$]

$[19 \times 5 + 10 \times 2 + 40 \times 7 + 60 \times 2 + 8 \times 8 + 20 \times 10 = 779]$

6. Determine an initial basic feasible solution to the following transportation problem using

   (i) Vogel's approximation method.

   (ii) North-West corner method.

   State which of the method is better.

   (Solved Example 3.5 and 3.1) [Dec. 16, May 13, 8M]

   Table 3.38

		Destination				Supply
		D_1	D_2	D_3	D_4	↓
	I	2	3	11	7	6
Factory	II	1	0	6	1	1
	III	5	8	15	9	10
Requirement →		7	5	3	2	

[Ans. : 1. By VAM $\rightarrow$ Z = ₹ 10,200, 2. By NWCM $\rightarrow$ ₹ 11,600] (VAM is better since cost is low)

7. Determine an initial basic solution for the following transportation problem. **[May 15, Nov. 14]**

   Table 3.39

		Warehouses					Available
		W_1	W_2	W_3	W_4	W_5	↓
	1	68	35	4	74	15	18
	2	57	88	91	3	8	17
Factory	3	91	60	75	45	60	19
	4	52	53	24	7	82	13
	5	51	18	82	13	7	15
Requirement →		16	18	20	14	14	

[Ans. : $Z_{min} = 2202$]

8. Find an initial basic feasible solution for the following transportation problem by NWCM method

   (Solved Example 3.1) [May 16, Nov./Dec. 07, May 13]

   Table 3.40

		Warehouses				Supply
		W_1	W_2	W_3	W_4	↓
	A	14	25	45	5	6
Factory	B	65	25	35	55	8
	C	35	3	65	15	18
Requirement →		4	7	6	13	

[Ans. : $x_{11} = 4$, $x_{12} = 2$, $x_{22} = 5$, $x_{23} = 3$, $x_{33} = 3$, $x_{34} = 13$]

9. Determine an initial basic feasible solution to following transportation problem.

By least cost method

(Solved Example 3.4)   **[Dec. 16, May 14]**

Table 3.41

		Warehouses			Supply ↓
		W₁	W₂	W₃	
Plant	1	2	7	4	5
	2	3	3	1	8
	3	5	4	7	7
	4	1	6	2	14
Demand →		7	9	18	

[Ans. : ₹ 8300]

9. Solve the following Transportation problem involving three sources and four destinations. The cell entries represent the cost of transportation per unit. Obtain solution by VAM Method ?

[Dec. 16, Nov./Dec. 11, 8M]

		1	2	3	4	Supply
	1	3	1	7	4	300
Sources	2	2	6	5	9	400
	3	8	3	3	2	500
Demand →		250	350	400	200	1200

[Ans. ₹ 2850]

10. Solve the following cost-minimizing transportation problem.   **[Nov. / Dec. 13, 12M]**

	I	II	III	IV	V	VI	Available
1	2	1	3	3	2	5	50
2	3	2	2	4	3	4	40
3	3	5	4	2	4	1	60
4	4	2	2	1	2	2	30
Required	30	50	20	40	30	10	180

11. Solve the following Transportation problem involving three sources and four destinations. The cell entries represent the cost of transportation per unit. Obtain solution by VAM method. Find optimum solution by using Steeping Stone Method.

[May 17, Nov./Dec. 13, 10M]

Destinations

		1	2	3	4	Supply
	1	3	1	7	4	300
Sources	2	2	6	5	9	400
	3	8	3	3	2	500
Demand		250	350	400	200	

(Step 3 of Section 3.4)

(Optimization Method like U-V Method and Stepping Stone Method)

This test is used to determine whether the obtained an initial basic feasible solution is optimal or not. Optimality test can be applied only on that initial basic feasible solution in which :

*   The number of allocation is (m + n − 1) where m is number of row and n is number of column.

*   This (m + n − 1) allocation should be in independent position.

Optimization Methods Which are Used for Optimal Solution are as Follows :

1. Stepping stone method

2. U-V method or modified distribution method or MODI method.

3.5.1 Stepping Stone Method   **[May 16]**

Since, optimality includes checking of each empty cell (vacant cells) to determine whether or not making an allocation in it minimizes the total cost of transportation, in this method, procedure start with empty cell (vacant cell). The steps involved are as follows :

Step 1 : Choose the empty cell (vacant cell) (i.e. cell without allocation) arbitrarily.

Step 2 : Draw a path in the matrix consisting of a series of alternate horizontal and vertical line such that path begin and ends in the chosen empty cell (vacant cell) and cell corner of path lie in the cell for which allocation have been made. The path may skip over any number of allocated or vacant cell.

Step 3 : Consider the corner of path in the chosen vacant cell as positive and other corner of path alternately negative, positive and so on.

Step 4 : Allocate 1 unit to chosen vacant cell, subtract and add 1 unit from cell at the corner of path, without disturbing the row and column requirement.

Step 5 : Evaluate the chosen vacant cell (i.e. the net change in cost of transportation).

[**Note :** This method is inefficient for large transportation problem since (m − 1) (n − 1) cell evaluation should be made.]

The above method can be explained with example 3.6.

Example 3.6 : *Solve the following transportation problem and use stepping stone method to test optimality of the solution.*   **[May 16, Nov. / Dec. 12, 10M]**

Table 3.42

		Destination			
		D_1	D_2	D_3	D_4	Supply
Plant	I	2	3	11	7	6
	II	1	0	6	1	1
	III	5	8	15	9	10
Requirement		7	5	3	2	

Solution : Using VAM, an initial basic feasible solution is as follows :

Table 3.43 (a)

		Destination				
		D_1	D_2	D_3	D_4	Supply
Plant	I	2 (1)	3 (5)	11	7	6
	II	1	0	6	1 (1)	1
	III	5 (6)	8	15 (3)	9 (1)	10
Requirement		7	5	3	2	

The above solution can be checked for optimality by stepping stone method.

Step 1 : Choose the empty cell (vacant cell) (i.e. cell without allocation) arbitrarily.

Let us consider cell (3, 2) (empty cell).

Step 2 : Draw a path in matrix consisting of a series of alternate horizontal and vertical line such that path begins and ends in the chosen empty cell (vacant cell) and cell corner of path lie in the cell for which allocation have been made. The path may skip over any number of allocated or vacant cells.

Table 3.43 (b)

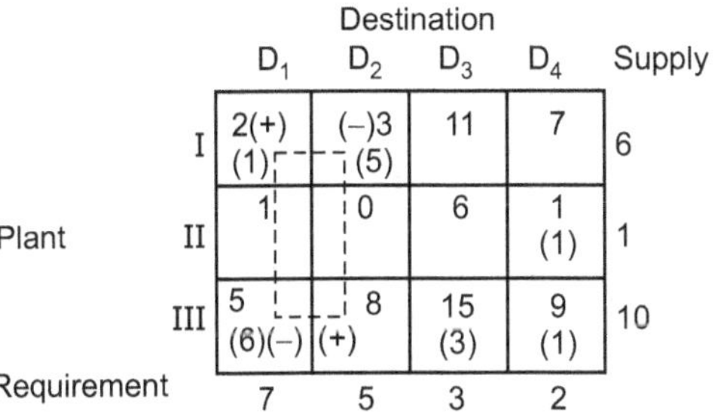

		Destination				
		D_1	D_2	D_3	D_4	Supply
Plant	I	2(+) (1)	(−)3 (5)	11	7	6
	II	1	0	6	1 (1)	1
	III	5 (6)(−)	8 (+)	15 (3)	9 (1)	10
Requirement		7	5	3	2	

Step 3 : Consider the corner of path in the chosen vacant cell as positive and other corner of path alternatively negative, positive, negative and so on.

Table 3.43 (c)

		Destination				
		D_1	D_2	D_3	D_4	Supply
Plant	I	2 (+1) (1)	(−1) 3 (5)	11	7	6
	II	1	0	6	1	1
	III	5 (−1) (6)	(+1) 8	15 (3)	9 (1)	10
Requirement		7	5	3	2	

Step 4 : Allocate 1 unit to chosen vacant cell, subtract and add 1 unit from cell at the corner of path without disturbing the row and column requirement.

Table 3.43 (d)

		Destination				
		D_1	D_2	D_3	D_4	Supply
Plant	I	2 (+1) (1)	(−1) 3 (5)	11	7	6
	II	1	0	6	1 (1)	1
	III	5 (−1) (6)	(+1) 8	15 (3)	9 (1)	10
Requirement		7	5	3	2	

Step 5 : Evaluate the chosen vacant cell [i.e. the net change in cost of transportation]

$\therefore$ Evaluation of cell (3, 2)

= [Product of '1' unit allocation and corresponding cost of cell at corner]

= ₹ $[8 \times 1 - 5 \times 1 + 2 \times 1 - 3 \times 1] \times 100$

= ₹ 200

Using same procedure, for remaining empty cells (vacant cell), we have

For cell (1, 3) = $C_{13} - C_{33} + C_{31} - C_{11}$
= $11 - 15 + 5 - 2 = -1$

For cell (1, 4) = $C_{31} - C_{14} + C_{34} - C_{11}$
= $5 + 7 - 9 - 2 = +1$

For cell (2, 1) = $C_{21} - C_{24} + C_{34} - C_{31}$
= $1 - 1 + 9 - 5 = +4$

For cell (2, 2) = $C_{22} - C_{24} + C_{34} - C_{31} - C_{12}$
= $0 - 1 + 9 - 5 + 2 - 3 = +2$

For cell (2, 3) = $C_{23} - C_{24} + C_{34} - C_{33}$
= $6 - 1 + 9 - 15 = -1$

For cell (3, 2) = $C_{32} - C_{31} + C_{11} - C_{12}$
= $8 - 5 + 2 - 3 = +2$

[**Note :** When any cell evaluation is negative, the transportation cost is minimized so that the initial basic solution can be improved i.e. it is not optimal. Likewise, if all cell evaluation are positive or zero, the solution will be optimal.]

Here, evaluations of cell (1, 3) and cell (2, 3) are negative, therefore an initial basic solution is not optimal.

3.5.2 U-V Method or (Modified Distribution Method) or MODI Method   [May 17]

U-V method or MODI method saves time compared to stepping stone method. It is also applied for checking optimality. The steps involved are as follows :

Step 1 : Consider an effectiveness matrix having cost concerned with cell for which allocation have been made.

Step 2 : Consider dual variable (U_i, V_j) with respect to supply and demand constraints.

Step 3 : Write the empty cells with addition of U_i and V_j.

Step 4 : From the actual cost matrix, subtract the cell value of matrix of step (3).

Step 5 : The optimality of solution depends on signs of values in cell evaluation matrix. The significance of signs are as follows :

(i) A negative value in vacant cell indicates that a better solution can be obtained by allocation units to the cell.

(ii) A positive value in vacant cells shows that a poorer solution will result by allocating units of cell.

(iii) A zero value in vacant cells indicate that another solution of same total value can be obtained by allocation units to this cell.

3.5.3 Development of Next, Iterative, Improved Allocation

(This is step 4 of Section 3.4)

For development of next, iterative, improved allocation, the steps involved are as follows :

Step 1 : Determination of non-basic cell (Identified cell) : Identify the cell with most negative cell evaluation from the cell evaluation matrix. It is the rate by which the total cost of transportation can be minimized, if one unit is allocated to this cell. If more units are allocated, the cost of transportation will decreases proportionately.

Therefore, as many units as possible (as per rim condition) will be allocated to this cell to reduce the cost by highest amount. If there is tie in the cell evaluation select the cell with maximum allocation. This cell is referred as identified cell (non-basic cell) (with reference to simplex method of LPP). This non-basic cell has been decided to be made basic by making allocation on it.

Step 2 : Consider again the initial basic feasible solution which is to be improved. Tick (✓) mark the identified cell (non-basic cell). For making non-basic cell to basic, change its present to zero. This is achieved by step (3) as below.

Step 3 : Draw a closed path in the matrix. The features of this closed path are as follows :

(i) It starts and ends in the identified cell (non-basic cell).

(ii) It has series of alternate horizontal and vertical lines only (no diagonal).

(iii) It can be drawn clockwise or counterclockwise.

(iv) All the remaining corner of path lie in the occupied cells (allocated cells) only.

(v) The closed path may skip over any number of occupied cells (allocated cells) or empty cells (vacant cells).

(vi) There may always be one and only one closed path which may drawn. The closed path has even numbers of corner (4, 6, 8, ...) and any occupied cell (allocated cell) can be considered only once. The path may or may not be rectangular or square in geometry, it may have a peculiar structure and lines may even cross-over.

Step 4 : Consider the identified cell (non-basic cell) as positive and each occupied cell (allocated cell) at the corners of path alternately, negative, positive and negative and so on.

Step 5 : Put a new allocation in the identified cell (non-basic cell) by entering lowest allocation on the path that has been assigned a negative sign. Subtract and add this new allocation from the cells at the corner of path, keeping the row and column requirements. This gives one basic cell to become zero and remaining cell as non-negative. The basic cell whose allocation have been made zero, leave the solution. The above method can be explained by example 3.7.

Example 3.7 : *Use U-V method, to test optimality of the following transportation problem.*   **[Dec. 16, May 14, 12M]**

Table 3.44

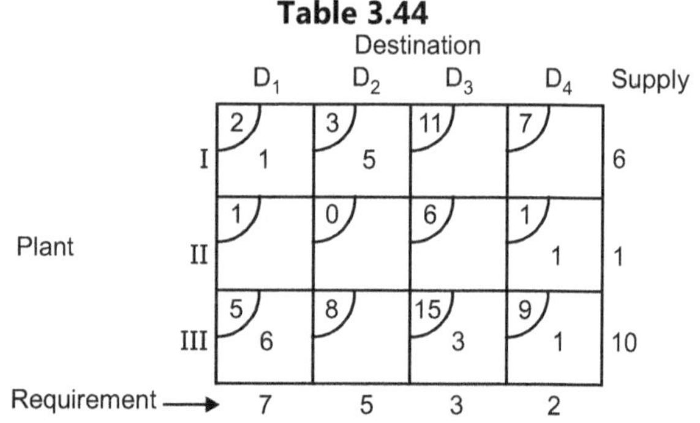

Plant	D$_1$	D$_2$	D$_3$	D$_4$	Supply
I	2 / 1	3 / 5	11	7	6
II	1	0	6	1 / 1	1
III	5 / 6	8	15 / 3	9 / 1	10
Requirement	7	5	3	2	

Solution : The optimal solution can be obtained by U-V method as follows :

Step 1 : Consider an effectiveness matrix having cost concerned with cell for which allocation have been made. Here table 3.45 is matrix having allocation with given cost.

Table 3.45

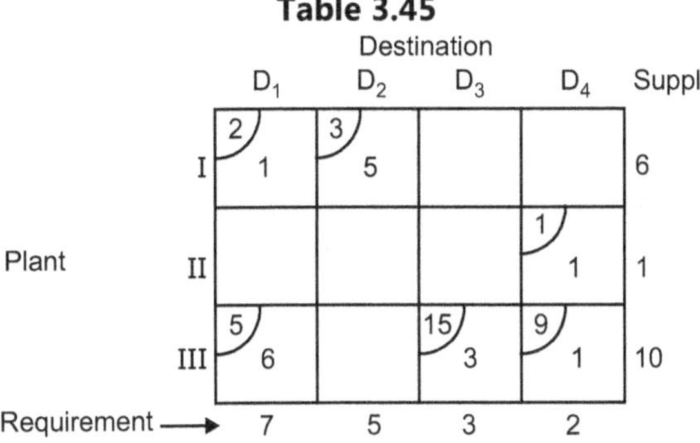

Step 2 : Consider dual variable (U_i, V_j) with respect to supply and demand constraints.

If there are 'm' source and 'n' destination, then there will be (m + n) dual variables.

Let U_i (i = 1, 2, 3, ... m) and

V_j (j = 1, 2, 3, ... n) be the dual variable with respect to supply and demand constrains.

Variables U_i and V_j are satisfying the equation.

$$\boxed{U_i + V_j = C_{ij}}$$ for all occupied cells (allocated cells)

There, put a set of numbers U_i (i = 1, 2, 3) (since there are 3 sources I, II, III) along the left of matrix and V_j (j = 1, 2, 3, 4) (since there are 4 destinations D_1, D_2, D_3 and D_4) across the top of matrix so that there addition equal the cost entered in step (1) (table 3.46).

Thus,

$$\left.\begin{array}{l} U_1 + V_1 = 2 \\ U_1 + V_2 = 3 \\ U_2 + V_4 = 1 \\ U_3 + V_1 = 5 \\ U_3 + V_3 = 15 \\ a_3 + b_4 = 9 \end{array}\right\}$$

Since, here the number of variables are [(m + n) = (3 + 4) = 7] in the given example and total allocations [(m + n − 1) = (3 + 4 −1 = 6)] one dual variable is assumed arbitrarily to zero.

Table 3.46   **Table 3.47**

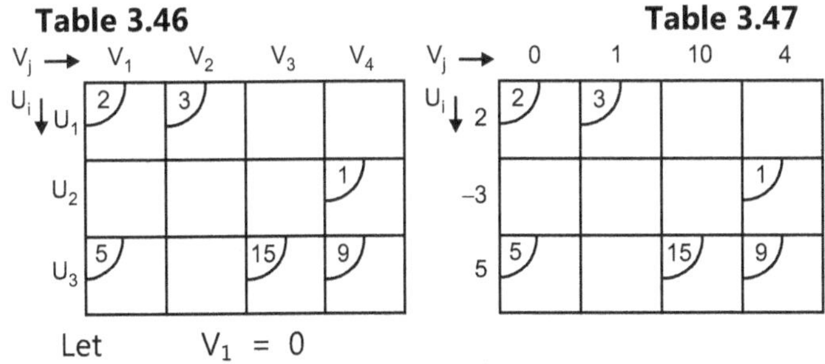

Let $V_1 = 0$

∴ Now put this in above equation, we have

$U_1 = 2$, $V_2 = 1$, $U_3 = 5$, $V_3 = 10$, $V_4 = 4$, $V_2 = -3$

The value of these variables obeys the complementary slackness theorem [which states that if primal constraints are equation, the dual variable are unrestricted in sign] i.e. [from duality theorem]. Therefore, matrix is given in table 3.47.

Now for any unoccupied cell (empty cell).

The implicit and actual cost are as follows :

Implicit cost $= U_i + V_j$

Actual cost $= C_{ij}$

These two costs are compared and $C_{ij} - (U_i + V_j)$ are estimated for each empty cell.

If all $C_{ij} - (U_i + V_j) \geq 0$ then using complementary slackness theorem it can be seen that the related solution is optimum.

If any $C_{ij} - (U_i + V_j) < 0$ then solution is not optimal.

Here the evaluate of cell $= C_{ij} (U_i + V_j)$.

Step 3 : Write the empty cells with addition of U_i and V_j.

Table 3.48

$V_j \rightarrow$	0	1	10	4
$U_i \downarrow$ 2	X	X	12 ⊗	6
−3	−3	−2	7	X
5	X	6	X	X

[**Note :** X → occupied cell or allocated cell

⊗ → 12 = 2 + 10 (addition of U_i and V_j) unoccupied cell or empty cell.]

Similar addition at other empty cells.

Step 4 : From the actual cost matrix, subtract the cell value of matrix of step (3).

Table 3.49

$V_j \rightarrow$	0	1	10	4
$U_i \downarrow$ 2	X	X	11 − 12 ⊗	7 − 6
−3	1 + 3	0 + 2	6 − 7	X
5	X	8 − 6	X	X

Table 3.50

$V_j \rightarrow$	0	1	10	4
$U_i \downarrow$ 2	X	X	− 1	1
−3	4	2	1	X
5	X	2	X	X

[**Note :** ⊗ at cell (1, 3)   →Actual cost − Cell value

→ 11 − 12

→ − 1]

Similarly at other empty cell, value are written as shown in table 3.49.

The table 3.50 is referred as cell evaluation matrix.

Step 5 : The optimality of solution depends on signs of value in cell evaluation matrix. The significance of signs are as follows :

- A negative value in vacant cell indicates that a better solution can be obtained by allocation units to the cell.

- The positive value in vacant cell shows that a poorer solution will result by allocating units of cell.

- A zero value in vacant cell – indicates that another solution of same total value can be obtained by allocations unit to this cells.

Development of Next, Iterative, Improved Allocation :

For development of next, iterative, improved allocation, the steps involved are as follows :

Step 1 : Determination of Non-Basic Cell (Identified Cell) : Identify the cell with most negative cell evaluation from the cell evaluation matrix. It is the rate by which the total cost of transportation can be minimized, if one unit is allocated to this cell. If more units are allocated, the cost of transportation will decreases proportionately. Therefore, as many units as possible (as per rim condition) will be allocated to this cell to reduce the cost by highest amount. If there is tie in the cell evaluation, select the cell with maximum allocation. The cell is referred as Identified cell (non-basic cell) (with reference to simplex method of LPP). This non-basic cell has been decided to be made basic by making allocation on it.

In the given example, in table 3.51, both tied cells will have the same highest allocation of '1' unit. Hence choose (1, 3) cell arbitrarily.

Step 2 : Consider again the initial basic feasible solution which is to be improved. Tick (✓) mark the identified cell (non-basic cell) for making non-basic cell to basic, change its present to zero. This is achieved by step (3) as follows :

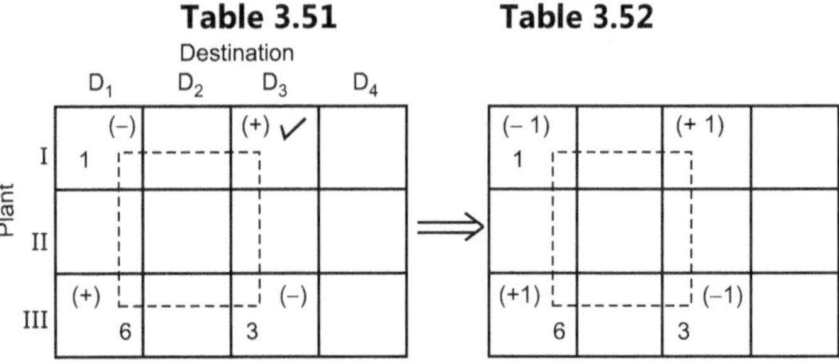

Table 3.51   **Table 3.52**

Step 3 : Draw a closed path in the matrix. The features of this closed path are as follows :

   (i) It starts and ends in the identified cell (non-basic cell).

   (ii) It has series of alternate horizontal and vertical lines only (no diagonal).

   (iii) It can be drawn clockwise or counterclockwise.

   (iv) All the remaining corner of path lie in the occupied cell (allocated cell) only.

   (v) The closed path may skip over any number of occupied cell (allocated cells) or empty cells (vacant cells).

   (vi) There may always be one and only one closed path which may be drawn. The closed path has even number of corners (4, 6, 8 …) and any occupied cell (allocated cell) can be considered only once. The path may or may not be rectangular or square in geometry. It may have a peculiar structure and lines may even cross-over.

Step 4 : Consider the identified cell (non-basic cell) as positive and each occupied cell (allocated cell) at the corner of path alternately, negative, positive, negative and so on.

Step 5 : Put a new allocation in the identified cell (non-basic cell) by entering lowest allocation on the path that has been assigned a negative sign. Subtract and add this new allocation from the cell at the corner of path, keeping the row and column requirement. This gives one basic cell to become zero and remaining cell as non-negative. The basic cell whose allocation have been made zero, leave, the solution.

Table 3.53 (a)   Table 3.53 (b)

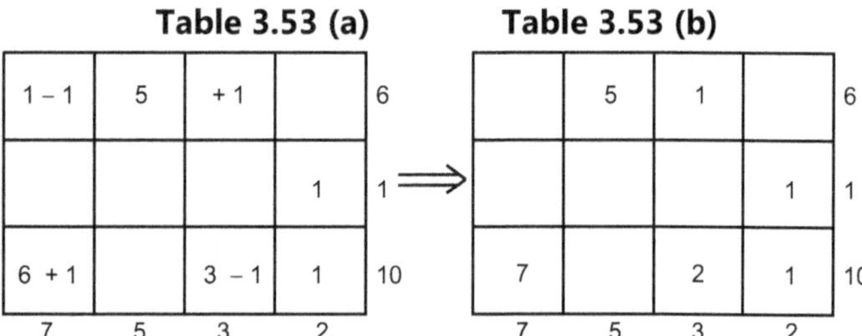

Here, the total cost of transportation should decreases by ₹ (100 × 1) = ₹ 100. Since, cell evaluation (in hundred of rupees) is −1 and 1 unit has been reallocated. This can be checked by actually estimating the cost of transportation from table 3.53.

Therefore, the total cost for IInd feasible solution is

$$= ₹\,[3 \times 5 + 11 \times 1 + 1 \times 1 + 7 \times 5 + 2 \times 15 + 1 \times 9] \times 100$$

$$= ₹\,10,100$$

This value is lower then the first (starting) feasible solution by ₹ 100.

For optimal solution, repeating the iteration, (i.e. repeat steps (3) and (4) of U-V method.

In second feasible solution :

   (i) The number of allocation is $(m + n - 1)$ i.e. $3 + 4 - 1 = 6$.

   (ii) Then $(m + n - 1)$ allocation should be in independent positions.

Here above two conditions are fulfilled, hence perform an optimality test as follows (use U-V method).

Step 1 : Consider an effectiveness matrix, having cost concerned with cell for which allocation have been made (i.e. second feasible solution of the given problem.)

Table 3.54

	D_1	D_2	D_3	D_4	
I		5	1		6
Plant II				1	1
III	7		2	1	10
	7	5	3	2	

Table 3.55 : Matrix having Allocation with given Cost

	3	11	
			1
5		15	9

Step 2 : Consider dual variable (U_i, V_j) with respect to supply and demand constraints.

Table 3.56

$V_j \rightarrow$	0	2	10	4
$U_i \downarrow$ 1		3	11	
-3				1
5	5		15	9

Thus, $u_1 + v_2 = 3$

$u_1 + v_3 = 11$

$u_2 + v_4 = 1$

$u_3 + v_1 = 5$

$u_3 + v_3 = 15$

$u_3 + v_4 = 9$

Let $V_1 = 0$

$\therefore$ We have,

$$U_3 = 5$$
$$V_3 = 10$$
$$U_2 = -3$$
$$U_1 = 1$$
$$V_4 = 4$$
$$V_2 = 2$$

These values are expressed in table 3.56.

Step 3 : Write the empty cell with addition of U_i and V_j.

Table 3.57

$V_j \rightarrow$	0	2	10	4
$U_i \downarrow$ 1	1	X	X	5
-3	-3	-1	7	X
5	X	7	X	X

Step 4 : From the actual cost matrix, subtract the cell value of matrix of step 3.

Table 3.58

$V_j \rightarrow$	6	2	10	4
$U_i \downarrow$ 1	$2-1$	X	X	$7-5$
-3	$1+3$	$0+1$	$6-7$	X
5	X	$8-7$	X	X

Table 3.59

$V_j \rightarrow$	0	2	10	4
$U_i \downarrow$ 1	1	X	X	2
-3	4	1	-1	X
5	X	1	X	X

Step 5 : The optimality of solution depends on signs of value in cell evaluation matrix.

Here, one cell (2, 3) value is negative, hence the second feasible solution is not optimal.

Development of Next, Iteration, Improved Solution :

(Refer steps from section 3.5.3)

Next improved solution can be obtained by the following steps given in section 3.5.3.

Therefore,

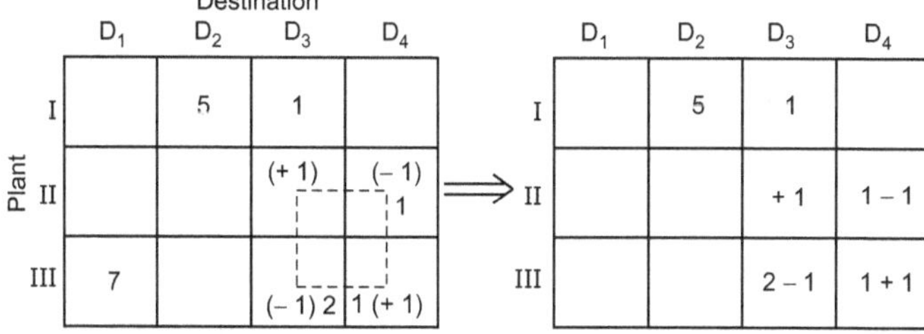

Table 3.60

	Destination			
	D_1	D_2	D_3	D_4
I		5	1	
Plant II			(+ 1)	(− 1) 1
III	7		(− 1) 2	1 (+ 1)

Table 3.61

	D_1	D_2	D_3	D_4
I		5	1	
II			+ 1	$1-1$
III			$2-1$	$1+1$

Table 3.62

	5	1	
		1	0
7		1	2

Table 3.63

	5	1	
		1	
7		1	2

For this solution, the cost of transportation is as follows :

$$= ₹ [5 \times 3 + 1 \times 11 + 1 \times 6 + 1 \times 15 + 2 \times 9 + 7 \times 5] \times 100$$

$$= ₹ 10,000$$

Now check optimality of IIIrd basic feasible solution.

If all cell value (of cell evaluation matrix) are positive then basic feasible solution is optimal.

For IIIrd feasible solution, repeat iterative until optimality is achieved.

In IIIrd feasible solution :

- The number of allocation is (m + n − 1) i.e. 3 + 4 − 1 = 6.

- (m + n − 1) allocate should be in independent position.

Here above two conditions are satisfied, hence perform, an optimality test as follows : (i.e. use U–V method).

Using steps given in Section 3.5.2, we have

Table 3.64 : Matrix having Allocation with given Cost

$V_j \rightarrow$	D_1	D_2	D_3	D_4
I		3	11	
Plant II			6	
III	5		15	9

Thus, $u_1 + v_2 = 3$

$u_1 + v_3 = 11$

$u_2 + v_3 = 6$

$u_3 + v_1 = 5$

$u_3 + v_3 = 15$

$u_3 + v_4 = 9$

Let $v_1 = 0$

∴   From above equation

$U_3 = 5$

$V_3 = 10$

$U_4 = 4$

$U_2 = -4$

$V_1 = 1$

$V_2 = 2$

These values are expressed in table 3.65.

Table 3.65

$V_j \rightarrow$	0	2	10	4
$U_i \downarrow$ 1		3	11	
−4			6	
5	5		15	9

Table 3.66 : Empty cell with addition of U_i and V_j

$V_j \rightarrow$	0	2	10	4
$U_i \downarrow$ 1	1	X	X	5
−4	−4	−2	6	0
5	X	7	X	X

Table 3.67 : From the actual cost matrix, Subtract the cost value of matrix table 3.66

Table 3.68 : Cell Evaluation Matrix

$V_j \rightarrow$				
$U_i \downarrow$ 2 − 1	X	X	7 − 5
1 + 4	0 + 2	X	1 − 0
X	8 − 7	X	X

$\Longrightarrow$

1	X	X	2
5	2	X	1
X	1	X	X

Since, all cell values (in cell evaluation matrix, table 3.68) are positive, the IIIrd basic feasible solution is optimal. Therefore, this optimal solution is given as follows :

Table 3.69

Plant	D_1	D_2	D_3	D_4	Supply
I	2	3 / 5	11 / 1	7	6
II	1	0	6 / 1	1	1
III	5 / 7	8	15 / 1	9 / 2	10
Requirement →	7	5	3	2	

Destination (column headers), Plant (row labels).

Table 3.70 : Optimal Solution

Plant (Source)	Destination	No. of Unit Transported	Transportation cost/unit (₹)	Total Cost of Transportation
I	D_2	5	300	1,500
	D_3	1	1,100	1,100
II	D_3	1	600	600
	D_1	7	500	3,500
III	D_3	1	1,500	1,500
	D_4	2	900	1,800
				Total = ₹ 10,000

UNIVERSITY QUESTIONS

1. Explain the following transportation method. **[May 17]**

    (a) U-V method or modified distribution method.

    (b) VAM

    (c) Stepping stone method

    (d) Least cost method

    (e) N-W corner rule.

2. Using VAM approximation method, obtain an initial feasible solution of the following transportation problem. **(Example 3.5)**   **[Dec. 16, May 10, 8M]**

Table 3.71

$$\begin{array}{c|cccc|}
 & D & E & F & G & \\
\hline
A & 11 & 13 & 17 & 14 & 250 \\
B & 16 & 18 & 14 & 10 & 300 \\
C & 21 & 24 & 13 & 10 & 400 \\
\hline
 & 200 & 225 & 275 & 250 &
\end{array}$$

3. Write a note a "Degeneracy in Transportation Problem. **[May 15, Nov. /Dec. 06, 4M]**

4. List out important characteristics of Transportation Problem.

   [Dec. 16, Nov./Dec. 06, 4M]

5. Using Vogel's approximation method, solve following transportation problem.

   [May 15, Nov./Dec. 06, 6M]

Table 3.72

	D_1	D_2	D_3	D_4	
01	21	16	25	13	11 (3)
02	17	18	14	23	13 (3)
03	32	27	18	41	19 (9)
Requirement	6 (04)	10 (02)	12 (04)	15 (10)	

[**Ans.** ₹ 787 (13 × 11 + 17 × 6 + 18 × 3 + 23 × 4 + 27 × 7 + 18 × 12]

6. Use MODI method to check the optimality of the following transportation problem.

   [May 18, Nov./Dec. 06, 6M]

Table 3.73

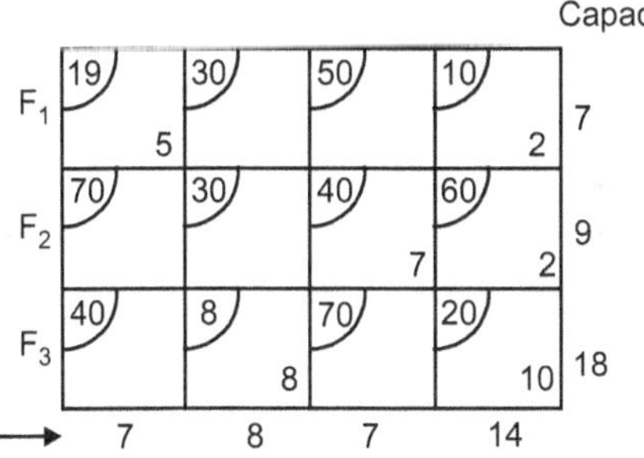

7. Prove that the minimum cost is 76 for transportation table given below. **[May 08, 9M]**

Table 3.74

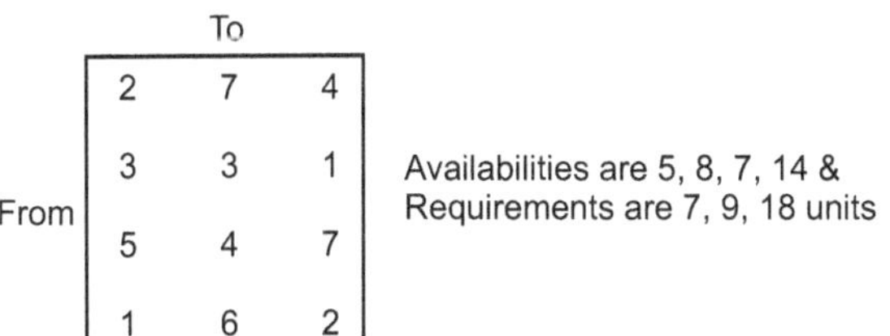

8. Obtain IBFS for 3 origin, 3 destination unbalanced transportation problem by matrix minima method. Check only, whether this IBFS is optimal or not by MODI method. Show the close path for most desirable empty cell and decide number of units to be shifted to this cell. Assume any suitable data for availabilities and requirement with total requirement is more than total availability. Assume any suitable cost along various routes.

   [Dec. 15, May/June 11, 12M]

9. Write the transportation form for assignment problem and LP form of transportation.

   [Dec. 16, May/June 11, 4M]

10. Solve the following transportation problem.

    [Dec. 15, May 09, 8M]

Table 3.75

		A	B	C	D	
				Destination		
	I	21	16	25	13	11
Source	II	17	18	14	23	13
	III	32	27	18	41	19
Requirement		6	10	12	15	43

11. Solve the following Transportation problem involving three sources and three destinations. The cell entries represent the cost of transportation per unit. Obtain the initial solution by VAM method and final optimal solution by MODI method. **[Dec. 16, May 12, 10M]**

		1	2	3	Supply
			Destinations		
	1	1	4	8	10
Sources	2	7	2	3	20
	3	5	4	2	15
Demand		23	12	10	

12. A company has three production shops supplying a product to five warehouses. The cost of production varies from shop to shop and so does the unit transportation cost from shop to warehouse. Each shop has a specific production capacity and each warehouse has certain amount of requirement. The unit transportation costs are, given below :

    [May 17, Nov./Dec. 12, 12M]

		1	2	3	4	5	Capacity
				Warehouse			
Shop	A	6	4	4	7	5	100
	B	5	6	7	4	8	125
	C	3	4	6	3	4	175
Demand		60	80	85	105	70	

The cost of manufacturing the product at different production shops is Variable

		Variable Cost (₹)	Fixed Cost (₹)
Shop	A	14	7000
	B	16	4000
	C	15	5000

Find the optimum quantity to, be supplied from each shop to different warehouses at minimum total cost.

3.6 TRANS-SHIPMENT METHOD AS AN EXTENSION OF TRANSPORTATION [Dec. 16]

- Trans-shipment method deals with shipping through intermediate or transient nodes before reaching to final destination or centre.

- Trans-shipment problem is more general than that of usual transportation problem where direct shipment only are permitted between a source (origin) and a destination.

- The trans-shipment problem can be transformed to a usual transportation problem using the concept of buffer and solved similar to transportation problem. The concept of transshipment problem is explained with help of Example 3.8.

Example 3.8 : *Two plants P_1 and P_2 are linked to three distributors D_1, D_2 and D_3 by way of two centres C_1 and C_2 according to network shown in Fig. 3.1. The supply amount of plants P_1 and P_2 are 1000 and 1200 vehicles and demand amount at distributor D_1, D_2 and D_3 are 800, 900 and 500 vehicles. The shipping cost per vehicle (in thousands of rupees) between pairs of node is expressed on connecting links or (arc) of network. Solve the trans-shipment problem.*

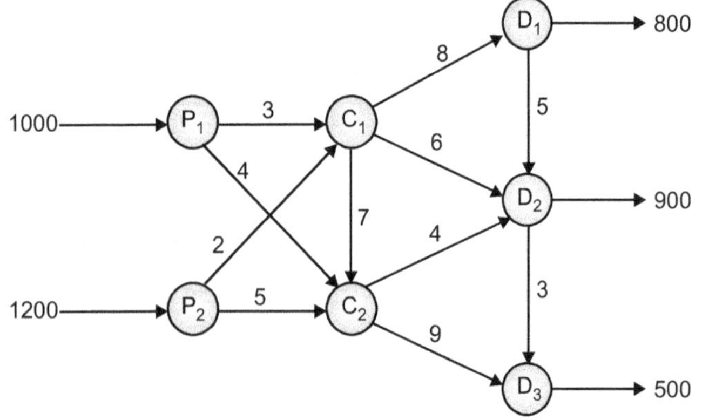

Fig. 3.1 : Trans-shipment network between plants and distributors

Solution : In Fig. 3.1, trans-shipment occurs in the network because the total supply amount of 2200 (1000 + 1200) vehicles from nodes P_1 and P_2 could potentially path through any node of network before ultimately reaching their destinations at nodes D_1, D_2 and D_3. In this situation, the nodes of network with both start and end arc (C_1, C_2, D_1 and D_2) act as both sources (origins) and destination and these are treated as **Trans-Shipment Nodes.** The remaining nodes are either power supply nodes (P_1) or pure requirement nodes (D_3). The trans-shipment problem can be transformed into usual transportation problem with 6 sources (origins) (P_1, P_2, C_1, C_2, D_1 and D_2) and five destinations (C_1, C_2, D_1, D_2 and D_3). The unit of supply and requirement of different nodes are estimated as :

Supply at a pure supply node   = Actual supply

Requirement at a pure demand node = Actual demand or requirement

Supply at trans-shipment node   = Actual supply + Buffer unit

Demand at trans-shipment node = Actual requirement + Buffer unit

The buffer unit should be sufficiently large to allow the total actual supply (or demand) amounts to pass through any of trans-shipment nodes.

Let B_u be the expected buffer unit, then

$$B_u = \text{Total supply (or demand)}$$
$$= 1000 + 1200 \text{ (or } 800 + 900 + 500)$$
$$\therefore \quad B_u = 2200 \text{ vehicles}$$

Using the buffer B_u and the unit shipping cost provided in the network, the equivalent usual transportation problem is developed as in table 3.76, the solution of this problem is expressed in Fig. 3.2.

Table 3.76

	C_1	C_2	D_1	D_2	D_3	
P_1	3	4	M	M	M	1000
P_2	2	5	M	M	M	1200
C_1	0	7	8	6	M	Bu
C_2	M	0	M	4	9	Bu
D_1	M	M	0	5	M	Bu
D_2	M	M	M	0	3	
	Bu	Bu	800 + Bu	900 + Bu	500	

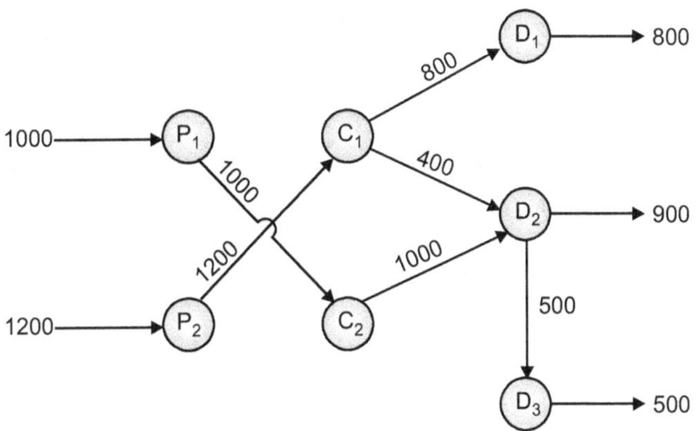

Fig. 3.2 : Solution of trans-shipment problem

The Effect of Trans-Shipment : Distributor receives 1400 vehicles, keeps 900 vehicles to satisfy its demand and sends the remaining 500 vehicles to distributor D_3.

UNIVERSITY QUESTIONS

1. Write a short note on trans-shipment **[Dec. 16]**

2. A company product is to be transported from sources A, B, C to destinations D_1, D_2, D_3 and D_4. The supply at the sources, the requirement at the destination and time of shipment is shown in table 3.77. Work out a transportation plan so that the time requirement of shipment is the lowest. **[May 15]**

Table 3.77

		Destination				
		D_1	D_2	D_3	D_4	Supply
	A	10	22	0	22	8
Source	B	15	20	12	8	13
	C	20	22	10	15	11
Requirement ⟶		5	11	8	6	32/32

3. The network in Fig. 3.3 gives the shipping paths from nodes 1 and 2 to nodes 5 and 6 by way of nodes 3 and 4. The unit shipping costs are represented on respective arc :

   (i) Develop the corresponding trans-shipment problem.

   (ii) Solve the problem and show how the shipment are routed from the origins to destination.

   [May 13]

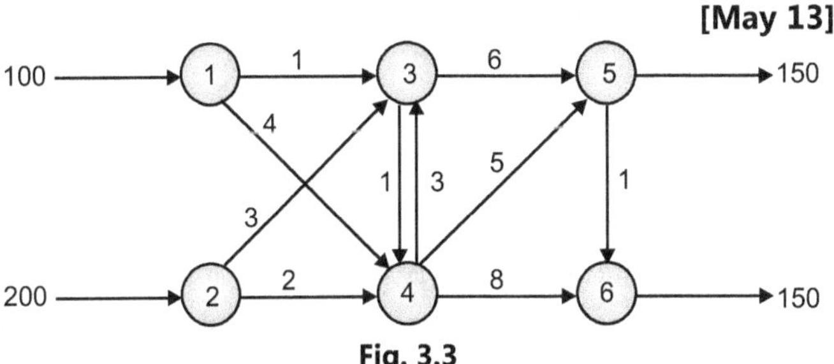

Fig. 3.3

4. Determine the shortest route between nodes a and g of the network in Fig. 3.4 by constructing the problem as trans-shipment model. The distance between the different nodes is represented in network. Assume that node a has a net supply of 1 unit of 'g' has c net demand also of 1 unit. **[May 15]**

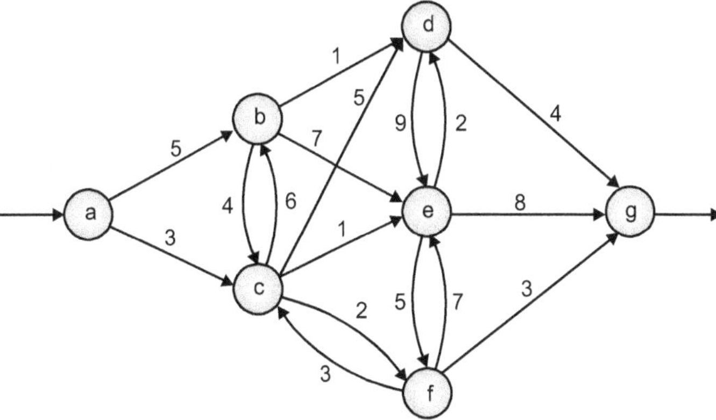

Fig. 3.4

5. In the water pipeline network shown in Fig. 3.5, the various nodes indicates pumping and receiving stations. Distances in miles between stations are shown on the network. The transportation cost per gallon between two nodes is directly proportional to the length of pipeline. Construct the associated trans-shipment problem and determine optimum solution.

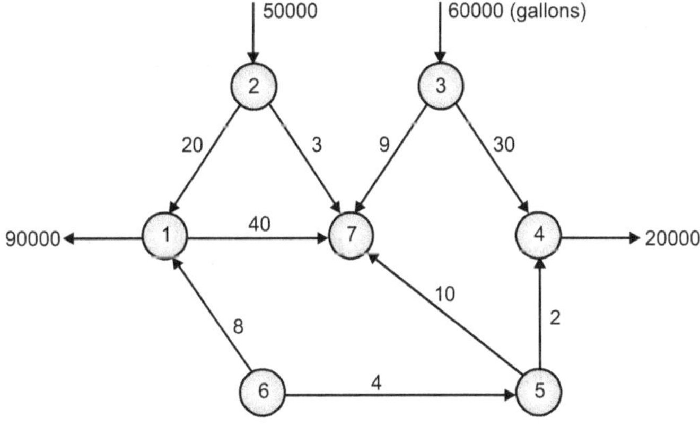

Fig. 3.5

3.7 ASSIGNMENT MODEL　　　　　　　　**[May 17]**

The main aim of assignment problem is to allocate a number of resources (men, material, machines etc.) to various locations at minimum cost or maximum profit.

3.7.1 Applications of Assignment Problem

The assignment problem is used for the following field :

- For assignment of items to factories.
- For assignment of workmen to machines.
- For assignment of salesman to various centres.
- Cars to various routes.
- Innovation activities to group of people.

It is particular case of transportation problem. A computational method which gives optimise solution to specific problem is known as Assignment Problem or Assignment Model.

The general assignment problem with 'n' workmen and 'n' destination (item) is represented in Table 3.78.

Table 3.78

Destination (Item)

	1	2	3 -------------------- n	Supply (ai)
1	C_{11}	C_{12}	C_{13}-------------------- C_{1n}	1
2	C_{21}	C_{22}	C_{23}-------------------- C_{2n}	1
3	C_{31}	C_{32}	C_{33}-------------------- C_{3n}	1
n	C_{n1}	C_{n2}	C_{n3} -------------------- C_{nn}	1

Workmen (rows 1 to n on the left)

Requirement (Bj)   1   1   1-------------------- 1

Table 3.78 represents the assignment of n workmen to n destination (item), C_{ij} is the cost of assigning i^{th} workmen to j^{th} destination (item), x_{ij} represents the assignment of ith worken to j^{th} destination (item). If i^{th} workmen can be assigned to j^{th} destination (item) $x_{ij} = 1$, otherwise zero.

This shows that an assignment problem can be expressed by $n \times n$ matrix, which constitutes n! possible ways of doing assignments.

3.7.2 Mathematical Formulation of Assignment Problem

Let x_{ij} expresses the assignment of ith workmen to j^{th} destination (item).

$$x_{ij} = \begin{bmatrix} 1, \text{ if the } i^{th} \text{ workmen is assigned to } j^{th} \text{ destination (item)} \\ 0, \text{ if the } i^{th} \text{ workmen is not assigned to } j^{th} \text{ destination (item)} \end{bmatrix}$$

This problem is given by

$$\text{Minimize } z = \sum_{i=1}^{n} \sum_{j=1}^{n} C_{ij} x_{ij}$$

Subject to constraints :

$$\sum_{j=1}^{n} x_{ij} = 1 \text{ for } i = 1, 2, 3 \ldots n \text{ (one destination}$$

(item) is assigned to i^{th} workmen)

$$\sum_{i=1}^{n} x_{ij} = 1, \text{ for } j = 1, 2, 3 \ldots n \text{ (one workmen is}$$

assigned to j^{th} destination (item))

and $\qquad x_{ij} = 0$ or $1 \left(\text{or } x_{ij} = x_{ij}^{2} \right)$

From above, we observe that if the last condition as replaced by $x_{ij} \geq 0$, we have transportation problem with all requirement and available resources equal to 1.

3.7.3 Hungarian Method to Solve Assignment Model   [Dec. 17]

The assignment problem is solved by the Hungarian method suggested by Mr. Kenig of Hungary. It is also referred as reduced matrix method or the floods technique. This method is very efficient and saves substantial time over the other techniques or methods.

The Steps of Hungarian Method are as Follows :

Step 1 : Prepare cost matrix from given problem statement. The cost matrix (n × n) should be square. For m × n matrix (m ≠ n), add a dummy row or column with zero cost as per case to have a square matrix.

Step 2 : Subtract the smallest element of each row of the effectiveness matrix from all the elements of respective row.

Step 3 : Subtract the smallest element of each column of the resulting matrix achieved in step 2 from all the element of respective column.

Step 4 : Make an optimal assignment in resulting matrix of step 3.

(a) Check rows successively until a row with exactly one unmarked zero is achieved. Make an assignment to this single zeros by making square (□) around it.

(b) Cross (×) all other zero in the column as they will not considered for making any additional assignment in that column. Proceed in this way until all rows have been checked.

(c) Now, check, successively until a column with exactly one unmarked zero is found. Make an assignment thereby a square (□) around it.

(d) Cross (×) any other zeros in the same row. In case, when there is no row or column having unmarked zero (they have more than one unmarked zero) then mark square (□) around any unmarked zero arbitrarily and cross (×) all other zero in its row and column. Proceed in the way till there is no unmarked zero left the cost matrix (resulting matrix obtained in step 3). Repeat sub-steps (a) to (d) till one of the following two things are available :

(i)   There is one assignment in each row and in each column, in this case, the optimal assignment can be made in the current solution (current feasible solution is an optimal solution). The least number of lines crossing all zero is 'n' the order of matrix given in problem.

(ii) There is some column or row without assignment. In this case, an optimal assignment cannot be made in the current solution. The least number of lines crossing all zeros have to be obtained in this situation by the following steps.

Step 5 : Draw the least (minimum) number of lines crossing all zeros. This is obtained by the following sub-steps :

(a) Mark (✓) the row which do not have assignment.

(b) Mark (✓) the column (not already marked) which have zeros in marked rows.

(c) Mark (✓) the rows (not already marked) which have assignment in marked columns.

(d) Repeat sub-steps (b) and (c) till no more rows or columns can be marked.

(e) Draw lines through all unmarked rows and marked columns. This gives least number of lines crossing all zeros. If this is equal to the order of matrix (n) then it is an optimal solution otherwise proceed to next step (6).

Step 6 : Check the uncovered elements. Choose the smallest element and subtract it from all the uncovered elements. Add this element to every element that lies at the intersection of the straight lines. Keep the remaining elements of the matrix as such. This gives second feasible solution.

Step 7 : Repeat steps (4) through (6) successively until the number of straight lines crossing all zeros becomes equal to order of the matrix i.e. each row or column will have one assignment. This shows that an optimal solution has been obtained.

Example 3.9 : (Minimization Problem) :

Five workmen of a factory are to be assigned to five jobs which can be done by any of them. Because of different number of year with the firm, the workers get different wages per hour. These are ₹ 5 per hour for A, B and C each and ₹ 3 per hour for D and E each. The amount of time taken by each employee to do a job is given in table 3.79. Obtain the assignment pattern that :

   (i) Minimizes the total time taken and

   (ii) Minimizes the total cost of getting five units of workdone.   **[May 17, Nov./Dec. 05, 12M]**

Table 3.79

| | | Workmen | | | | |
		A	B	C	D	E
	I	7	9	3	3	2
	II	6	1	6	6	5
Job	III	3	4	9	10	7
	IV	1	5	2	2	4
	V	6	6	9	4	2

Solution : Using steps given by Hungarian method, to above problem.

Step 1 : Prepare cost matrix from given problem statement.

Table 3.80 (a)

| | | Workmen | | | | |
		A	B	C	D	E
	I	7	9	3	3	2
	II	6	1	6	6	5
Job	III	3	4	9	10	7
	IV	1	5	2	2	4
	V	6	6	9	4	2

Step 2 : Subtract the smallest element of each row of the effectiveness matrix from all the elements of respective row.

Table 3.80 (b)

| | | Workmen | | | | |
		A	B	C	D	E
	I	5	7	1	1	0
	II	5	0	5	5	4
Job	III	0	1	6	7	4
	IV	0	4	1	1	3
	V	4	4	7	2	0

Step 3 : Subtract the smallest element of each column of the resulting matrix achieved in step (2) from the element of respective column.

Table 3.80 (c)

| | | Workmen | | | | |
		A	B	C	D	E
	I	5	7	0	0	0
	II	5	0	4	4	4
Job	III	0	1	5	6	4
	IV	0	4	0	0	3
	V	4	4	6	1	0

Step 4 : Make an optimal assignment of resulting matrix of step (3).

(a) Check rows successively until a row with exactly one unmarked zero is achieved. Make an assignment to this zeros by making square (□) around it.

(b) Cross (×) all other zeros in the column as they will not considered for making any additional assignment in that column. Proceed in this way until all rows have been checked.

(c) Now, check, successively until a column with exactly one unmarked zero is found. Make an assignment thereby a square (□) around it.

(d) Cross (×) any other zeros in the same row.

Table 3.81

Workmen

		A	B	C	D	E
	I	5	7	⓪	✗	✗
	II	5	⓪	4	4	4
Job	III	⓪	1	5	6	4
	IV	✗	4	✗	⓪	3
	V	4	4	6	1	⓪

Since, each row and column have assignment, this is solution of problem.

Job		Workmen		Cost
I	→	C	→	$3 \times 5 = ₹\,15$
II	→	B	→	$1 \times 5 = ₹\,5$
III	→	A	→	$3 \times 5 = ₹\,15$
IV	→	D	→	$2 \times 3 = ₹\,6$
V	→	E	→	$2 \times 3 = ₹\,6$
				Total = ₹ 47

3.7.4 Assignment Problem Variations OR Types of Assignment Problem   [Dec. 16]

The types or variations of the assignment problem are as follows :

1. Unbalanced assignment problem (non-square matrix)
2. Maximization problem
3. Restriction on assignments
4. Alternate solution (optimal solution).

1. Unbalanced Assignment Problem (Non-Square Matrix) :

When in a given cost matrix of assignment problem, the number of rows are not equal to number of columns then it is known as unbalanced assignment problem. In this situation, a dummy rows or columns are added with zero cost in the matrix to make it square matrix. Again the Hungarian method or procedure is used to it to determine optimal solution.

2. Maximization Problem :

When in assignment of employee to machine or destination the main objective is to maximize profit or sale then it is known as maximization problem. This maximization assignment problem is transferred to minimization problem and solved by the Hungarian method. This transformation may be achieved in either the following two ways :

- By subtracting all the element from largest element of the matrix given in problem.
- By multiplying the matrix element by – 1.

3. Restriction on Assignment :

In some situations, due to certain prohibitions such as place, legal, technical, particular resource (machine, material, men etc.) cannot be assigned to perform particular task, in this case, a very large cost (infinite cost, ∞) is assigned to the respective cell and such tasks will then be automatically neglected from further consideration (making assignment).

4. Alternate Solution (Optimal Solution) :

In some situations in assignment problem, it is positive to have two or more ways to strike-off all zero elements in reduced matrix, in this case, there will be more than one solution, which depends on decision person's choice. Since, he can select the one which is most appropriate to his demand.

Example 3.10 : *(Unbalanced Assignment Problem, Non-square Matrix)*

A company has one surplus car in each of the districts 1, 2, 3, 4 and 5 and one deficit car in each of districts I, II, III, IV, V and VI. The distance between district in kilometres is given in table 3.82. Determine the assignment of cars from districts in surplus to districts in deficit so that total distance covered by car is minimum.   ***[May 16]***

Table 3.82

	I	II	III	IV	V	VI
1	12	10	15	22	18	8
2	10	18	25	15	16	12
3	11	10	3	8	5	9
4	6	14	10	13	13	12
5	8	12	11	7	13	10

Solution :

Step 1 : Prepare cost matrix from given problem statement. The cost matrix (n × n) should be square for m × n matrix (m ± n), add dummy row or column with zero cost as per case to have a square matrix. Here, to make cost matrix square, add dummy row with zero as cost.

i.e. Non-square matrix → Square matrix

Table 3.83

Destination with deficit

		I	II	III	IV	V	VI
	1	12	10	15	22	18	8
	2	10	18	25	15	16	12
District	3	11	10	3	8	5	9
surplus	4	6	14	10	13	13	12
	5	8	12	11	7	13	10
	6	0	0	0	0	0	0

Dummy row

Step 2 : Subtract the smallest element of each row of the effectiveness matrix from all the elements of respective row.

Table 3.84

	I	II	III	IV	V	VI
1	4	2	7	14	10	0
2	0	8	15	5	6	2
3	8	7	0	5	2	6
4	0	8	4	7	7	6
5	1	5	4	0	6	3
6	0	0	0	0	0	0

Step 3 : Subtract the smallest element of each column of the resulting matrix obtained in step (2) from all the element of respective column. Here we get some matrix as last row has zero cost.

Table 3.85

	I	II	III	IV	V	VI
1	4	2	7	14	10	0
2	0	8	15	5	6	2
3	8	7	0	5	2	6
4	0	8	4	7	7	6
5	1	5	4	0	6	3
6	0	0	0	0	0	0

Step 4 : Make an optimal assignment in resulting matrix of step (3) as per substeps (a) to (d) given in Hungarian method.

Table 3.86

	I	II	III	IV	V	VI
1	4	2	7	14	10	[0]
2	[0]	8	15	5	6	2
3	8	7	[0]	5	2	6
4	✗	8	4	7	7	6
5	1	5	4	[0]	6	3
6	✗	[0]	✗	✗	✗	✗

Step 5 : Draw the least number of lines crossing all zeros. This is obtained as follows :

(i)   Mark (✓) the row which do not have assignment.

(ii) Mark (✓) the column (not already marked) which have zeros in marked rows.

(iii) Mark (✓) the row (not already marked) which have assignment in marked columns.

(iv) Repeat sub-step (b) and (c) till no more row or column can be marked.

(v)   Draw lines through all unmarked rows and marked columns. This gives least number of lines crossing all zeros. If this equal to the order of matrix (n) then it is an optimal solution otherwise proceed to next step (7).

Table 3.87

	I	II	III	IV	V	VI
1	4	2	7	14	10	[0]
2	[0]	8	15	5	6	2
3	8	7	[0]	5	2	6
4	[0]	8	4	7	7	6
5	1	5	4	[0]	6	3
6	✗	[0]	✗	✗	✗	✗

Step 6 : Check the uncovered elements. Choose the smallest element and substract it from all the uncovered elements. Add this element to every element that lies at the intersection of the straight lines. Keep the remaining elements of the matrix as such.

Table 3.88

	I	II	III	IV	V	VI
1	6	2	7	14	10	0
2	0	6	13	3	4	0
3	10	7	0	5	2	6
4	0	6	2	5	5	4
5	3	5	4	0	6	3
6	2	0	0	0	0	0

Step 7 : Repeat step (4) through (6) successively until each row and column will have one assignment (i.e. an optimal solution).

Table 3.89

	I	II	III	IV	V	VI
1	6	2	7	14	10	[0]
2	[0]	6	13	3	4	✗
3	10	7	[0]	5	2	6
4	✗	6	2	5	5	4
5	3	5	4	[0]	6	3
6	2	[0]	✗	✗	✗	✗

Table 3.90

	I	II	III	IV	V	VI
1	6	[0]	5	12	8	✗
2	✗	4	11	1	2	[0]
3	12	7	[0]	5	2	8
4	[0]	4	✗	3	3	4
5	5	5	4	[0]	6	5
6	4	✗	✗	✗	[0]	2

Thus, the optimal solution is given as :

District		District		Distanced Travelled (km)
1	→	II	→	10
2	→	VI	→	12
3	→	III	→	3
4	→	I	→	6
5	→	IV	→	7
				Total = 38 km

Example 3.11 : *(Maximization Problem)*

Five different machines can do any of five required components with different profit resulting from each assignment as shown in table 3.91. Find out maximum profit possible through optimum assignment.

[Dec. 16, May/June 07, 12M]

Table 3.91

Machine

		1	2	3	4	5
	A	30	37	40	28	40
	B	40	24	27	21	36
Component	C	40	32	33	30	35
	D	25	38	40	36	36
	E	29	62	41	34	39

Solution :

Since, the problem is of maximization, transfer it to minimization.

Maximization problem $\rightarrow$ Minimization problem

Here largest element is 62, subtract all elements from 62, we have

Table 3.92

Machine

		1	2	3	4	5
	A	32	25	22	34	22
	B	22	38	35	41	26
Component	C	22	30	29	32	27
	D	37	24	22	26	26
	E	33	0	21	28	23

Now, above problem can be solved as minimization problem as per steps given by Hungarian method :

Step 1 : Prepare cost matrix from given problem statement.

Table 3.93

Machine

	1	2	3	4	5
A	32	25	22	34	22
B	22	38	35	41	26
C	22	30	29	32	27
D	37	24	22	26	26
E	33	0	21	28	23

Step 2 : Subtract the smallest element of each row of the effectiveness matrix from all the elements of respective row.

Table 3.94

	1	2	3	4	5
A	10	3	0	12	0
B	0	16	13	19	4
C	0	8	7	10	5
D	15	2	0	4	4
E	33	0	21	28	23

Step 3 : Subtract the smallest element of each row of effectiveness matrix obtained in step (2) from all the elements of respective column.

Table 3.95

	1	2	3	4	5
A	10	3	0	8	0
B	0	16	13	15	4
C	0	8	7	6	5
D	15	2	0	0	4
E	33	0	21	24	23

Step 4 : Make an optimal assignment in resulting matrix of step (3) using sub-steps (a) to (d) given in Hungarian method.

Table 3.96

	1	2	3	4	5
A	10	3	[0]	8	⊗
B	[0]	16	13	15	4
C	⊗	8	7	6	5
D	15	2	⊗	[0]	4
E	33	[0]	21	24	23

Step 5 : Draw the least numbers of lines crossing all zeros by following sub-steps (a) to (e) as given in section 3.7.3. This is as shown in table 3.96.

Step 6 : Check the uncovered elements. Choose the smallest element and subtract it from all the uncovered elements. Add this element to every element that lies at the intersection of the straight lines keep the remaining the elements of matrix as such.

Table 3.97

	1	2	3	4	5
A	14	3	0	8	0
B	0	12	9	11	0
C	0	4	3	2	1
D	19	2	0	0	4
E	37	0	21	24	23

Step 7 : Repeat steps (4) through (6) successively until each row and column will have one assignment (i.e. an optimal solution).

Table 3.98

	1	2	3	4	5
A	14	3	[0]	8	⊗
B	⊗	12	9	11	[0]
C	[0]	4	3	2	1
D	19	2	⊗	[0]	4
E	37	[0]	21	24	23

Therefore, the optimal solution is given as :

Component		Machine		Cost
A	→	3	→	40
B	→	5	→	36
C	→	1	→	40
D	→	4	→	36
E	→	2	→	62
				₹ 214

Example 3.12 : *(Restriction on Assignment)*

Four new machines A, B, C and D are to be installed in a machine shop. There are five empty spaces. Because of limited space problem B cannot be placed at P_3 and C cannot be placed at P_1. Installation cost of machine to various places are given in table 3.99 below. Find the solution for this assignment problem for minimization.

[Dec. 17]

Table 3.99

	P_1	P_2	P_3	P_4	P_5
A	4	6	10	5	6
B	7	4	–	5	4
C	–	6	9	6	2
D	9	9	7	2	3

Solution :

Step 1 : Prepare cost matrix from given problem statement.

Table 3.100

			Place		
	P_1	P_2	P_3	P_4	P_5
A	4	6	10	5	6
B	7	4	∞	5	4
Machine C	∞	6	9	6	2
D	9	9	7	2	3
E	0	0	0	0	0

Dummy machine

Since, the given matrix is non-square matrix, convert it into square by adding dummy row with zero as cost. Due to restriction, we assign infinite cost (∞) in cell (C, P_1) and (B, P_3) as in table 3.100. Neglect the cell in further consideration.

Step 2 : Subtract the smallest element of each row of the effectiveness matrix from all the elements of respective row.

Table 3.101

	P_1	P_2	P_3	P_4	P_5
A	0	2	6	1	2
B	3	0	∞	1	0
Machine C	∞	4	7	4	0
D	7	7	5	0	1
E	0	0	0	0	0

Step 3 : Subtract the smallest element of each column of resulting matrix obtained in step (2) from the element of respective column.

Table 3.102

	P_1	P_2	P_3	P_4	P_5
A	0	2	6	1	2
B	3	0	∞	1	0
C	∞	4	7	4	0
D	7	7	5	0	1
E	0	0	0	0	0

Step 4 : Make an optimal assignment in resulting matrix of step (3).

Table 3.103

	P_1	P_2	P_3	P_4	P_5
A	[4]	6	10	5	6
B	7	[4]	–	5	4
C	–	6	9	6	[2]
D	9	9	7	[2]	3

As there is no row or no column without assignment, this is optimal solution. Hence, need not to proceed to steps (5), (6) and (7). The optimal solution is given as :

Table 3.104

	P_1	P_2	P_3	P_4	P_5
A	[4]	6	10	5	6
B	7	[4]	–	5	4
C	–	6	9	6	[2]
D	9	9	7	[2]	3

Machine		Place		Cost
A	→	P_1	→	2
B	→	P_2	→	2
C	→	P_3	→	2
D	→	P_4	→	2
				₹ 12

Example 3.13 : *(Alternative Solution)*

Determine optimal solution for the minimal assignment problem as given in table 3.105.   **[Dec. 16]**

Table 3.105

Place

Workmen		A	B	C	D
	I	2	3	4	5
	II	4	5	6	7
	III	7	8	9	8
	IV	3	5	8	4

Solution :

Step 1 : Prepare cost matrix from given problem statement.

Table 3.106

Place

Workmen		A	B	C	D
	I	2	3	4	5
	II	4	5	6	7
	III	7	8	9	8
	IV	3	5	8	4

Step 2 : Subtract the smallest element of each row of the effectiveness matrix from all the elements of respective row.

Table 3.107

	A	B	C	D
I	0	1	2	3
II	0	1	2	3
III	0	1	2	1
IV	0	2	5	1

Step 3 : Subtract the smallest element of each column of the resulting matrix achieved in step (2) from all the elements of respective column.

Table 3.108

	A	B	C	D
I	0	0	0	2
II	0	0	0	2
III	0	0	0	0
IV	0	1	3	0

Step 4 : Make an optimal assignment in resulting matrix of step (3).

Table 3.109

	A	B	C	D
I	[0]	✖	✖	2
II	✖	[0]	✖	2
III	✖	✖	[0]	✖
IV	✖	1	3	[0]

As there is no row or column without assignment, there is one optimal solution which is follows :

Machine		Place		Cost
I	→	A	→	2
II	→	B	→	5
III	→	C	→	9
IV	→	D	→	4

Total = ₹ 20

Similarly, we have an **alternate solution (optimal solution)** as shown in table 3.110, likewise many more alternative solution can be achieved.

Table 3.110

	A	B	C	D
I	✖	[0]	✖	2
II	✖	✖	[0]	2
III	✖	✖	✖	[0]
IV	[0]	1	3	✖

Optimal Solution (Alternate Solution) :

Workmen		Place		Cost
I	→	B	→	3
II	→	C	→	6
III	→	D	→	8
IV	→	A	→	3

Total = ₹ 20

3.7.5 Travelling Salesman as an Extension of Assignment Problem (Branch and Bound Method)   [May 15]

A salesperson must visit number of cities as per product to be sale or work to be assigned. The time or distance or cost between every pair of cities is known. He starts from his home place (city) passes through each city once and only once and come back to his home city. The problem is to determine the path shortest in time or cost or distance.

This Problem may be Classified as Symmetrical and Asymmetrical Problem.

(a) Symmetrical Problem :

When the time (or cost or distance) between every pair of cities is independent of direction of travel then the problem is said to symmetrical problem.

(b) Asymmetrical Problem :

When for one or more of pair of cities the time (or cost or distance) changes with directive then problem is said to asymmetrical problem.

For Example :

(i)   Flight of an aeroplane.

(ii) Scheduling of school bus.

The number of possible paths will be $(n - 1)!$ for n cities to be visited. Since, the salesman has to go to all 'n' cities, the shortest route will be independent of selective of starting city.

3.7.6 Formulation of Travelling Salesman Problem

Let, C_{ij} be the cost of going from city i to city j.

x_{ij} be the '1' or '0' when salesman goes directly from city i to j.

This problem may be stated as follows :

$$\text{Minimize } z = \sum_{i=1}^{n} \sum_{j=1}^{n} C_{ij}\, x_{ij}$$

Subject to constraints :

$$\sum_{j=1}^{n} x_{ij} = 1$$

$$\sum_{i=1}^{n} x_{ij} = 1$$

and $x_{ij} = 1$ or 0, $i = 1, 2, 3, \ldots , n$

$$j = 1, 2, 3, \ldots , n$$

There are two additional constraints that no city is to be visited twice before the tour of all the cities is completed and that going from city 'i' directly to 'i' is not permitted. This mean (i) = ∞.

Therefore, the travelling salesman problem can be written as assignment problem as given in table 3.111.

Table 3.111

		To city			
	1	2	3 ·· ·· ··	1	
1	∞	C_{12}	C_{13} ·· ··	··	C_{1n}
2	C_{21}	∞	C_{23}		C_{2n}
3	C_{31}	C_{32}	∞		C_{3n}
From city	:	:	: : : : :		
	:	:	: : : : :		
	:	:	: : : : :		
n	C_{n1}	C_{n2}	C_{n3}	∞	

The above problem is solved as previous (i.e. for small assignment problem). But for large assignment problem the method developed by J.D.C. Little is applied. This method is also referred as Little method or Branch and Bound method.

3.7.7 Steps for Branch and Bound Method

[May 15]

The steps involved for this method are explained considering the following assignment problem :

Example 3.14 :

Table 3.112

		To city			
	1	2	3	4	5
1	0	2	5	7	1
2	6	0	3	8	2
From city 3	8	7	0	4	7
4	12	4	6	0	5
5	1	3	2	8	0

Solution :

Step 1 : (i) Prepare a cost matrix from given problem statement, since going from $1 \rightarrow 1$, $2 \rightarrow 2$ etc. is not allowed, assign a large penalty (cost of travels) for these cells in the cost matrix (say set S) as shown in table 3.113.

Table 3.113

		To city			
	1	2	3	4	5
1	∞	2	5	7	1
2	6	∞	3	8	2
From city 3	8	7	∞	4	7
4	12	4	6	∞	5
5	1	3	2	8	∞

(ii) Subtract the smallest element of each row of the effectiveness matrix from all the elements of respective row.

Table 3.114

		To city			
	1	2	3	4	5
1	∞	1	4	6	0
2	4	∞	1	6	0
From city 3	4	3	∞	0	3
4	8	0	2	∞	1
5	0	2	1	7	∞

(iii) Subtract the smallest element of each column (if needed) of resulting matrix achieved in steps (i), (ii) from all the elements of respective column, till there is a zero in every row or column. The total reduction 'r' is the sum of element subtracted, we call resulting matrix as $[C'_{ij}]$.

Table 3.115

		To city			
	1	2	3	4	5
1	∞	1	3	6	0
2	4	∞	0	6	0
From city 3	4	3	∞	0	3
4	8	0	1	∞	1
5	0	2	0	7	∞

$\therefore$ Total reduction = r = [1 + 2 + 4 + 4 + 1] + 1 = 12 + 1
$$= 13.$$

Step 2 : Determine the penalty of not using each zero cell in $[C_{ij}']$.

We argue that when we do not use link (h, k), we must use other element in row h and some element in column k.

Therefore, the cost of not using (h, k) is atleast equal to addition of smallest element in row h and smallest element in column k. Record these penalties in the top left corners of zero cell in the table 3.116.

Table 3.116

	1	2	3	4	5
1	∞	1	3	6	[1] 0
2	4	∞	[0] 0	6	[0] 0
3	4	3	∞	[9] ⊛ 0	3
4	[2] 8	0	1	∞	1
5	[4] ✓ 0	2	[0] 0	7	∞

For example, consider zero in cell (1, 5). The sum of smallest element in row 1 and column 5 [excluding zero in cell (1, 5)] is 1 + 0 = 1.

For cell (2, 3), the sum is 0 + 0 = 0.

For cell (5, 1), the sum is 0 + 4 = 4 ($\checkmark$).

For cell (2, 5), the sum is 0 + 0 = 0.

For cell (3, 5), the sum is 3 + 6 = 9 and so on.

Step 3 : Highest penalty

Let (h, k) be the zero entry with the highest penalty.

In case of tie, choose arbitrarily.

We now split the given set 'S' into two subsets : S (h, k) which contains the link (h, k) and S(h, $\overline{k}$), which does not contain the link (h, k), after this estimate lower bounds on the costs of all the routes in each subset. We know that if (h, k) is not used in addition to the reduction r, there will be cost of atleast $P_{h, k}$. Thus, the lower bound θ (h, $\overline{k}$) is given as :

$$\theta \ (h, \ \overline{k}) = r + P_{h, k}$$

In the given example, r = 13 and $P_{3, 4}$ = 9.

$\therefore$　　$\theta(3, \overline{4})$ = 13 + 9 = 22

$\therefore$　The total cost of the route not containing highest penalty link = $\theta \ (3, \overline{4})$ = 22.

Step 4 : Determining the total cost of route having the highest penalty link i.e. lower bound for S (h, k). This is achieved as follows :

We see that if we use link (h, k), we cannot use link (k, h); for; if we use (h, k) and (k, h) we would go from h to k and back to 'h' without visiting other cities.

We ignore the use of (h, k) by and a very heavy cost i.e. C_{ij} = ∞.

Moreover, if we use (h, k), we will not use any other link in row h and column k. This is confirmed by deleting row h and column k. In the remaining matrix, we select our element from each row and column. So that the cost will be at least the amount by which the remaining matrix can be reduced.

Let this be $r_{h, k}$, then the power bound θ(h, k) for S(h, k) is expressed as :

$$\boxed{\theta(h, \ k) = r + r_{h, k}}$$

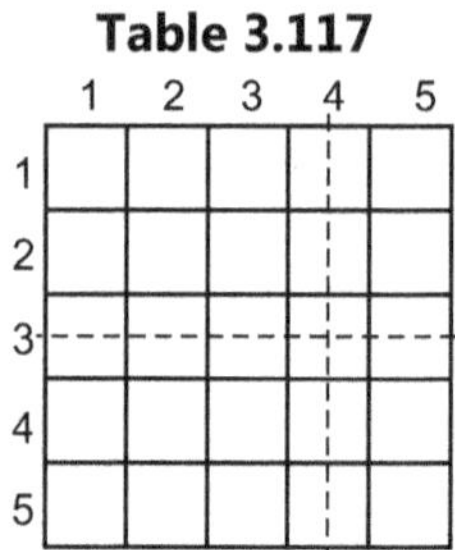

Fig. 3.6

In the example $C_{3, 4}' = \infty$ and we delete row 3 and column 4. The resulting matrix (table 3.118) can be reduced by only zero giving $\theta(3, 4)$ = 13 + 0 = 13. The results achieved in steps (3) and (4) are given in Fig. 3.6.

Table 3.117

	1	2	3	4	5
1					
2					
3					
4					
5					

Table 3.118

	1	2	3	5
1	∞	1	3	0
2	4	∞	0	0
4	8	0	∞	1
5	0	2	0	∞

Step 5 : Determining $\theta(3, \overline{4})$ and $\theta(3, 4)$, the one which is lower for further partitioning the subsets (h, k) when $\theta(3, 4)$ is lower, then return to step (2) and repeat it on the cost matrix obtained in   step (4) (table 3.118).

When $\theta(3, \bar{4})$ is selected then return to the original reduced cost matrix [or step (i)] put cost in cell (3, 4) = ∞ and repeat the step (2) onwards.

In the example, $\theta(3, 4)$ is chosen as it is smaller. Using step (2), we get table 3.119. in which penalties have been recorded in the top left corners of cell with zero entries.

Table 3.119

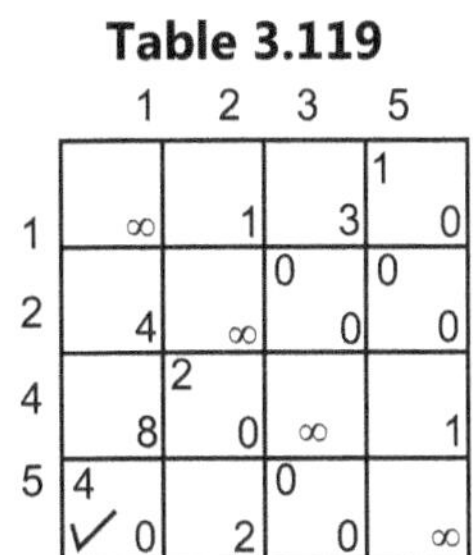

Step 6 : Cell (5, 1) has the highest penalty of 4, we now split the subset (h, k) into two parts, one containing cell (5, 1) and other not containing cell (5, 1) and estimate lower bounds on the costs of all routes in every part total cost of route not having the highest penalty link = $\theta(5, \bar{1})$ = 13 + 4 = 17.

Step 7 : Determining the total cost route containing the highest penalty link.

This is obtained as follows :

When we can use the link (5, 1), then we can not use link (1, 5).

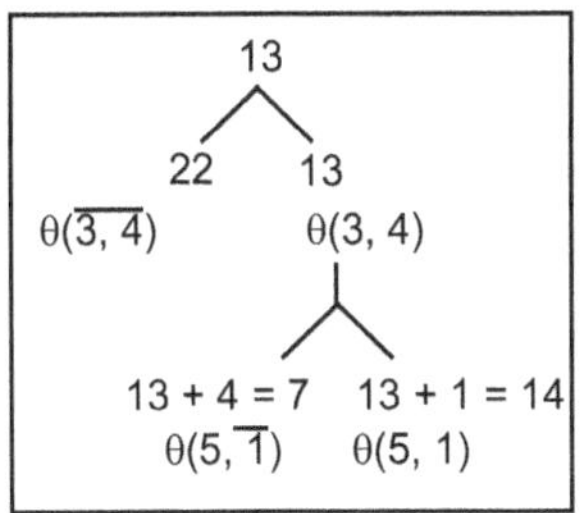

Fig. 3.7

Therefore, we put $C'_{1, 5}$ = ∞ and delete row 5 and column 1.

In the remaining matrix, we choose one element from each row and column so that cost will be atleast the amount by which the remaining matrix can be reduced. In the given example, lower bound for the matrix of table 3.119 i.e. total cost of route containing the highest penalty link (5, 1) is θ (5., 1) = 13 + 1 = 14, as the resulting matrix (table 3.121) can be reduced by 1 in row 1. The result achieved in the steps (6) and (7) are noted in Fig. 3.7 and table 3.120 and 3.121.

Step 8 : Select out of θ (J, T) and θ(5, 1), the one which is lower.

Here θ (5, 1) is lower, thus, we return to step (2) and repeat it on the cost matrix given in step (7) (Table 3.121). After reducing it so that every row and column contain zero entry (table 3.122) cell. Penalties have been recorded in top left corner of cells with zero entries in table 3.123.

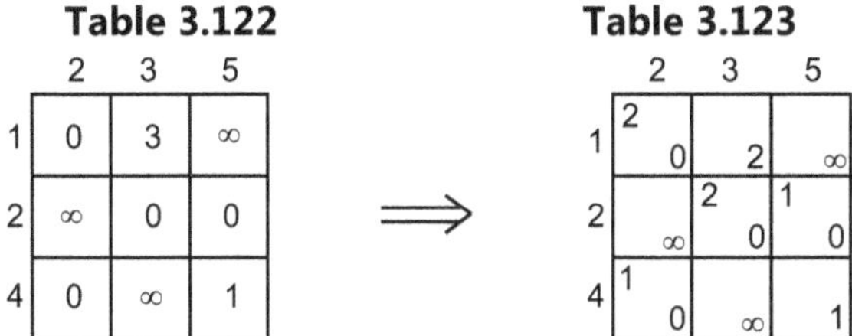

Step 9 : Cell (1, 2) has the highest penalty of 2 [Cell (2, 3)] also has same penalty of 2] out of these two cells, we choose cell (1, 2) arbitrarily.

∴   Cost of route not containing cell (1, 2) :

$$\theta(1, \bar{2}) = 14 + 2 = 16$$

Step 10 : When we use link (1, 2) then we have the chain (5, 1), (1, 2). Hence, we must execute the link (2, 5). Therefore, we put C'_{25} = ∞. We delete row 1 and column 2. the resulting matrix (table 3.125) can be reduced by 1.

∴   Cost of route containing the cell (1, 2) :

$$\theta(1, 2) = 14 + 1 = 15$$

The results obtained in steps (9) and (10) shown in Fig. 3.8 and Tables 3.124 and 3.125.

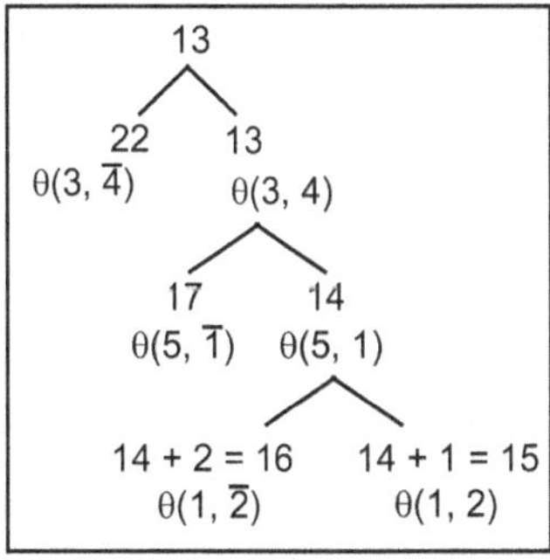

Fig. 3.8

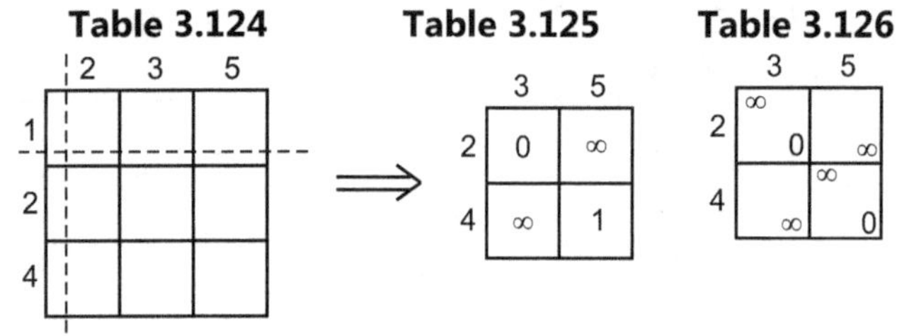

Table 3.124   **Table 3.125**   **Table 3.126**

Step 11 : Reducing the resulting matrix of step (10). So that every row and every column has atleast one zero entry cell. This is given in table 3.126.

Penalties are noted in the top left corner of cell containing zero entries in table 3.126.

Step 12 : Let us choose cell (2, 3) having a penalty of ∞.

∴   Cost of route not containing cell (2, 3)

i.e. $\theta(2, \overline{3}) = 15 + \infty = \infty$

Step 13 : Delete row 2 and column 3 we have table 3.127.

Table 3.127

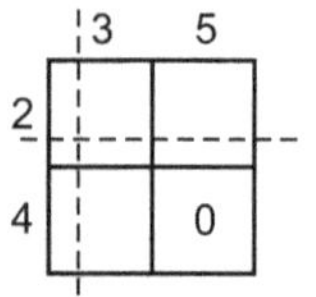

Cost of route containing cell (2, 3)

i.e. $\theta(2, 3) = 15 + 0 = 15$

∴   The least cost route is $3 \to 4 \to 5 \to 1 \to 2 \to 3$

∴

Total cost of travel = cell (2, 4) + cell (4, 5) + cell (5, 1)
        + cell (1, 2) + cell (2, 3)

        = ₹ [4 + 5 + 1 + 2 + 3]

        = ₹ 15

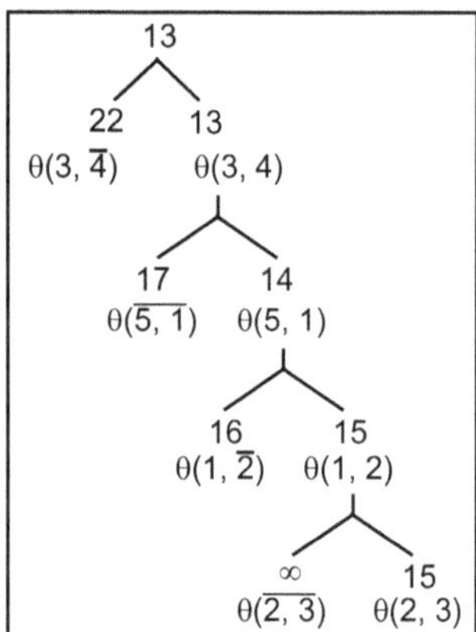

Fig. 3.9

Example 3.15 : *Obtain Initial basic feasible solution (IBFS) for following transportation problem using VAM and use MODI method to test optimality of transportation problem.*

[May 17, Nov. / Dec. 14, 10M]

		D1	D2	D3	D4	Supply
Plant		2	3	11	7	6
		1	0	6	1	1
		5	8	15	9	10
Demand		7	5	3	2	

Solution :

By VAM

		D1	D2	D3	D4	Supply
Plant		2 (1)	3 (5)	11	7	6
		1	0	6	1 (1)	1
		5 (6)	8	15 (3)	9 (1)	10
Demand		7	5	3	2	

By MODI Method

		D1	D2	D3	D4	Supply
Plant		2	3 (5)	11 (1)	7	6
		1	0	6 (1)	1	1
		5 (7)	8	15 (1)	9 (2)	10
Demand		7	5	3	2	

Example 3.16 : *Five jobs are to be assigned to five machines to minimize the total time required to process the jobs on machines .The times in hours for processing each job on each machine are given in the matrix below. By using assignment algorithm make the assignment for minimizing total time of processing.(06)*

[May 16, Nov. / Dec. 12, 8M]

	m/c A	m/c B	m/c C	m/c D	m/c E
Job 1	2	4	3	5	4
Job 2	7	4	6	8	4
Job 3	2	9	8	10	4
Job 4	8	6	12	7	4
Job 5	2	8	5	8	8

Solution : Using Hungarian method the solution is given as

Solution I :

M/c	Job	Time
A	3	3
B	2	4
C	5	4
D	4	7
E	1	2
		20 hr

Solution II (Alternative Solution)

M/c	Job	Time
A	4	5
B	2	4
C	1	2
D	5	4
E	3	5
		20 hr

Example 3.17 : Solve the following Transportation problem involving three sources and four destinations. The cell entries represent the cost of transportation per unit. Obtain solution by VAM method. Find optimum solution by using Steeping Stone Method.   **[May 15, Nov. /Dec. 13, 10M]**

Destinations

		1	2	3	4	Supply
	1	3	1	7	4	300
Sources	2	2	6	5	9	400
	3	8	3	3	2	500
	Demand	250	350	400	200	

OR

Solution : Please refer to Sections 3.4 and 3.5.1 [Example 3.6].

Destinations

		1	2	3	4	Supply
	1	3	1	/	4	300
Sources	2	2	6	5	9	400
	3	8	3	3	2	500
	Demand	250	350	400	200	

Destinations

		1	2	3	4	Supply
	1	3	1 (300)	7	4	300
Sources	2	2 (250)	6	5 (150)	9	400
	3	8	3 (50)	3 (250)	2 (200)	500
	Demand	250	350	400	200	

Optimum transportation cost

$$= 1 \times 300 + 2 \times 250 + 5 \times 150 + 3 \times 50 + 3 \times 250 + 2 \times 200$$

$$= \boxed{₹ \ 2850}$$

Example 3.18 : Four factories A,B ,C and D produces sugar and capacity of each factory is given below :

Factory A produces 10 tons of sugar and B produces 8 tons of sugar,C produces 5 tons of sugar and that of D is 6 tons of sugar.The sugar has demand in three markets X,Y and Z .The demand of markets X is 7 tons, that of markets Y is 12 tons and the demand of market Z is 4 tons .The following matrix gives the transportation cost of 1 ton of sugar from each factory to the destinations .Find the initial basic feasible solution to the following transportation problem by VAM   **[May 16, Oct. 15, 8M]**

Factories	Cost in ₹ per ton (x 100) Markets			Availability in Tons
	X	Y	Z	
A	4	3	2	10
B	5	6	1	8
C	6	4	3	5
D	3	5	4	6
Requirements in tons	7	12	4	Σb=29, Σd=23

Solution :

From	To	Load	Cost in ₹
A	X	3	3 x 4=12
A	Y	7	7 x 3=21
B	X	3	3 x 5=15
B	Z	5	5 x 1=5
C	Y	5	5 x 4=20
D	x	1	1 x 3=3
D	dummy	5	5 x 0=0
		Total ₹	76

Example 3.19 : Five jobs are to be assigned to 5 machines to minimizes the total time required to process the jobs on machines. The times in hours for processing each job on each machine are given in the matrix below. By using assignment algorithm make the assignment for minimizing the time of processing   **[May 15, 17, Oct. 15, 8M]**

Jobs	V	W	X	Y	Z
A	2	4	3	5	4
B	7	4	6	8	4
C	2	9	8	10	4
D	8	6	12	7	4
E	2	8	5	8	8

Solution :

Using Hangerian Method :

Solution I :

Jobs	Machines	Time in hrs
A	X	3
B	W	4
C	Z	4
D	Y	7
E	V	2
		Total 20 hrs

Solution II

Jobs	Machines	Time in hrs
A	Y	5
B	W	4
C	V	2
D	Z	4
E	X	5
		Total 20 hrs

QUESTIONS FOR PRACTICE

1. Write short note on travelling salesman problem with branch of bound method.

[May 17, Nov./Dec. 05]

2. Explain the role of branch and bound method to solve travelling salesman problem.

3. Discuss the similarities and differences between the transportation and, assignment problem

[Dec. 16, May 12, 6M]

4. Solve the following salesman problem by branch and bound method. **[Dec. 15, May/June 04, 16M]**

Table 3.128

	1	2	3	4	5
1	0	3	6	2	3
2	3	0	5	2	3
3	6	5	0	6	4
4	2	2	6	0	6
5	3	3	4	6	0

Ans. : $1 \to 4 \to 2 \to 3 \to 5$ or

$1 \to 5 \to 3 \to 2 \to 4 \to 1$

Minimum cost = ₹ 16.

5. A workman processes five type of jobs on his machine each week and must choose a sequence for them. The set up cost per change depends on the job. Presently on the machine and set-up to be made according to table 3.129, if he processes each type of job once and only once each week, how should he sequence the job on his machine.

Table 3.129

	1	2	3	4
1	∞	4	7	3
2	4	∞	6	3
3	7	6	∞	7
4	3	3	7	∞

Ans. : $1 \to 4 \to 2 \to 3$ or $1 \to 3 \to 2 \to 4$

Minimum cost = ₹ 19.

6. Solve the following travelling saleman problem given tables 3.130 and 3.131.

Table 3.130

		I	II	III	IV	V
	I	∞	10	25	25	10
	II	1	∞	10	15	2
From	III	8	9	∞	20	10
	IV	14	10	24	∞	15
	V	10	8	25	27	∞

Ans. : $I \to V \to II \to III \to IV \to I$ or

Minimum cost = ₹ 62.

Table 3.131

		A	B	C	D	E
	A	−	12	24	25	15
	B	6	−	16	18	7
From	C	10	11	−	18	12
	D	14	17	22	−	16
	E	12	13	23	25	−

Ans. : $A \to B \to C \to D \to E$ or

$A \to B \to E \to C \to D \to A$

Minimum distance = 74 km.

7. Solve the assignment problem of four sales person to four different regions as shown in table below such that total sales are maximized ?

[May 18, Nov./Dec. 11, 8M]

Sales Region

		1	2	3	4
	1	10	22	12	14
Salesman	2	16	18	22	10
	3	24	20	12	18
	4	16	14	24	20

[Ans. $1 \to 22$

$2 \to 22$

$3 \to 24$

$\underline{4 \to 20}$

Total Sale $\to 88$]

8. A college is having a degree programme for which the effective semester time available is very less and the programme requires field work. Hence, a few hours can be saved from the total number of class hours and can be utilized for the field work. Based on past experiences, the college has estimated the number of hours required to each subject by each faculty. The course in its present semester has 5 subjects and the college has considered 6 existing faculty members to teach there courses. The objective is to assign the best 5 teachers out of these 6 faculty members to teach 5 different subjects so that the total number of class hours required is minimized. The data of this problem is summarized as below. Solve the assignment problem optimally ?

[May 16, Nov. 11, 14M]

Subjects

		1	2	3	4	5
	1	30	39	31	38	40
	2	43	37	32	35	38
Faculty	3	24	41	33	41	34
	4	39	36	43	32	36
	5	32	49	35	40	37
	6	36	42	35	44	42

[**Ans.** 　1 → 3 (31)

　　　　2 → 2 (37)

　　　　3 → 1 (24)

　　　　4 → 4 (32)

　　　　5 → 5 (37)

　　　　<u>6 → 6 (0)</u>

　　　　Total → 161]

9. A company is faced with the problem of assigning six machines to five different Jobs. The costs estimated in hundreds of rupees are given in the table below. Solve the problem assuming that the objective is to minimize the total cost. **[Dec. 16, May 12, 10M]**

Jobs

		1	2	3	4	5
	1	2.5	5	1	6	2
	2	2	5	1.5	7	3
Machines	3	3	6.5	2	8	3
	4	3.5	7	2	9	4.5
	5	4	7	3	9	6
	6	6	9	5	10	6

10. A Municipal Corporation has decided to out road repairs oil four main arteries of the city. Municipal Corporation has granted ₹ 50 Lakh for this work with a condition that the repair must be done at the lowest cost and quickest time. The Municipal Corporation has floated tenders and five contractors have .sent in their bids. In order to expedite work, one road will be awarded to only one contractor.

(i) Find the best way of assigning the repairs to the contractors and associated cost.

(ii) If it is necessary to seek supplementary grant, what should be the amount sought.

(iii) Which of the five contractors will be unsuccessful in-his bid? **[May 17, Nov./Dec. 12, 12M]**

Cost of repairs on road

(₹ Lakhs)

		1	2	3	4
	C1	9	14	19	15
	C2	7	17	20	19
Contractors	C3	9	18	21	18
	C4	10	12	18	19
	C5	10	15	21	16

11. Four different jobs are to be done on four different machines. Table below indicate the cost of producing job I on machine j in rupees. Solve the problem assuming that the objective is to minimize the total cost. **[Dec. 17, May 13, 10M]**

Machines

		1	2	3	4
	1	5	7	11	6
Jobs	2	8	5	9	6
	3	4	7	10	7
	4	10	4	8	3

12. What is the unbalanced assignment problem? How is it solved by the Hungarian method? Explain with suitable example. **[May 17, Nov. /Dec. 13, 6M]**

13. Five jobs are to be assigned to 5 machines to minimize the total time required to process the jobs on machines. The times in hours for processing each job on each machine are given in the matrix below. By using assignment algorithm make the assignment for minimizing the time of processing.

[May 15, Nov. / Dec. 13, 10M]

Machines

		A	B	C	D	E
	1	2	4	3	5	4
	2	7	4	6	8	4
Jobs	3	2	9	8	10	4
	4	8	6	12	7	4
	5	2	8	5	8	8

14. Solve the following transportation problem and use stepping stone method to test optimality of solution. **[May 18, 8M]**

	D1	D2	D3	D4	Supply
Plant I	2	3	11	7	6
Plant II	1	0	6	1	1
Plant III	5	8	15	9	10
Requirement	7	5	3	2	

15. Five different machines can do any of five required components with different profit resulting from each assignment as shown in following table. Find out maximum profit possible through optimum assignment.

[May 18, 8M]

		1	2	3	4	5
	A	30	37	40	28	40
Component	B	40	24	27	21	36
	C	40	32	33	30	35
	D	25	38	40	36	36
	E	29	62	41	34	39

◈ ◈ ◈

CHAPTER 4
WAITING LINE MODELS (QUEUING MODEL)

4.1 INTRODUCTION

- **Queuing Theory** is the mathematical study of waiting lines or *queues*. The theory enables mathematical analysis of several related processes, including arriving at the (back of the) queue, waiting in the queue (essentially a storage process), and being served at the front of the queue. The theory permits the derivation and calculation of several performance measures including the average waiting time in the queue or the system, the expected number waiting or receiving service, and the probability of encountering the system in certain states, such as empty, full, having an available server or having to wait a certain time to be served.

- Queuing theory has applications in diverse fields, including telecommunications, traffic engineering, computing and the design of factories, shops, offices and hospitals.

4.1.1 Evolution of Queuing Model

- Queuing Theory had its beginning in the research work of a Danish engineer named A. K. Erlang. In 1909, Erlang experimented with fluctuating demand in telephone traffic.

- Eight years later he published a report addressing the delays in automatic dialing equipment. At the end of World War II, Erlang's early work was extended to more general problems and to business applications of waiting lines.

Some More Examples of Waiting Lines are Given as Follows:

Queuing Examples Situation	Arriving Customers	Service Facility
1. Passage of customers through a supermarket checkout	Shoppers	Checkout counters
2. Flow of automobile traffic through a road network	Automobiles	Road network
3. Manually placed assembly line	Parts to be assembled	Assembly line
4. Inventory of items in a warehouse	Order for withdrawal	Warehouse
5. Banking Transactions	Bank Patrons	Bank tellers
6. Ships entering a port	Ships	Docks
7. Maintenance and repair of machines	Machine break-down	Repair crew
8. Scheduling of patients in a clinic	Patients	Medical care
9. Number of runways at an airport	Airplanes	Runways
10. Parking lot	Automobiles	Parking spaces
11. Capacity of a motel	Motorists	Lodging facilities
12. Arrival of automobiles at a Garage	Automobiles	Repair of automobiles
13. Transfer of electronic messages	Electronic messages	Transmission lines
14. Flow of computer programmes through a computer system	Computer programmes	Central Processing unit
15. Sale of theatre tickets	Theatre-goers	Ticket windows
16. Arrival of trucks at central market	Trucks	Loading crews
17. Registration of unemployed at an employment exchange	Unemployed personnel	Registration assistants
18. Calls at police control room	Service calls	Policemen

- In our everydays activity, we see that there is a flow of customer to avail some service from service facility. The rate of flow depends on the nature of service and the serving capacity of the station. In many situations, there is a congestion of items arriving from service because an item cannot be serviced immediately on arrival and each new arrival has to wait for some time before it is attended. This situation occurs where the total number of customers requiring service exceeds the number of facilities. So we can define a queue as **"A group of customers/items waiting at some place to receive attention/service including those receiving the service."**

- In this situation, if queue length exceeds a limit, the customers gets frustrated and leave the queue to get the service at some other service station. In this case, the organization looses the customer goodwill. Similarly some service facility waits for arrival of customers when the total capacity of system is more than the number of customers requiring service. In this case, service facility remains idle for a considerable time causing a burden of exchequer.

- So, in absence of a perfect balance between the service facility and the customers, waiting is required either by the customer or by the service facility. The imbalance between the customer and service facility, known as congestion, cannot be eliminated completely but efforts/techniques can be evolved and applied to reduce the magnitude of congestion or waiting time of a new arrival in the system or the service station. The method of reducing congestion by the expansion of servicing counter may result in an increase in idle time of the service station and may become uneconomical for the organization. Thus, both the situations namely of unreasonable long queue or expansion of servicing counters are uneconomical to individual or managers of the system.

- Therefore, if the length of the queue is longer, the waiting time of the customer will increase causing dissatisfaction of customer and to avoid the longer waiting time of customer, if the management increases the service facilities, then many a time we see that the service facilities will remain idle causing burden on the organization. To avoid this situation, the theory of waiting line will help us to reduce the waiting time of the customer and suggest the organization to install optimal number of service facilities, so that customer will be happy and the organization can run the business economically.

- The arrival pattern of the customer and the service time of the facility depend on many factors and they are not under the control of the management. Both cannot be estimated or assessed in advance and moreover their arrival pattern and service times are random in nature. The waiting line phenomenon is the direct result of randomness in the operation of service facility and random arrival pattern of the customer. The customer arrival time cannot be known in advance to schedule the service time and the time required to serve each customer depends on the magnitude of the service required by the customer. So the time required for both customers will vary. The randomness of arrival pattern and service time makes the waiting line theory more complicated and needs careful study. The theory tries to strike a balance between the costs associated with waiting and costs of preventing waiting and help us to determine the optimal number of service facilities required and optimal arrival rate of the customers of the system.

4.1.2 Elements of Queuing Model

- One thing we have to remember is that when we speak about queue, we have to deal with two elements, i.e. Arrivals and Service facility. Entire queuing system can be completely described by **(a) The input (Arrival pattern), (b) The service mechanism or service pattern, (c) The queue discipline and (d) Customer behavior.** Components of the queuing system are arrivals, the element waiting in the queue, the unit being served, the service facility and the unit leaving the queue after service. This is shown in Fig. 4.1.

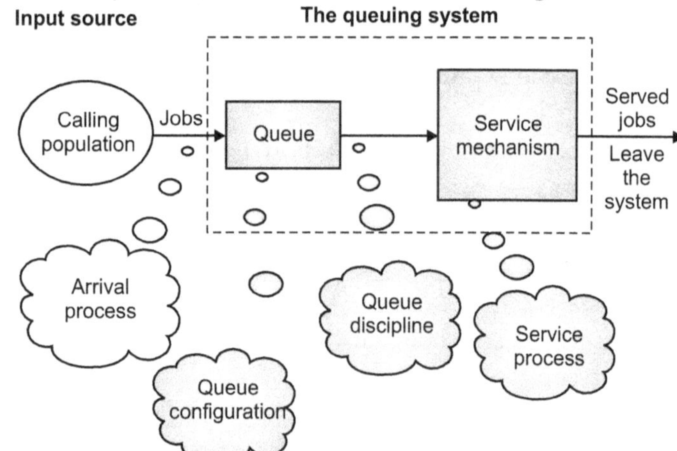

Fig. 4.1: Components of the queuing system

1. Input Process:

The input describes the way in which the customers arrive and join the system. In general, customer arrival will be in random fashion, which cannot be predicted, because the customer is an independent individual and the service organization has no control over the customer.

The input process is described mainly by the characteristics as: nature of arrivals, capacity of the system and behavior of the customers.

- **Size of Arrivals:** In our discussion in this chapter, it is dealt with those queuing system in which the customers arrive in **Poisson** or **Completely random** fashion. In fact, there are many more arrival patterns are available but for simplicity, only Poisson arrivals are considered.

- **Inter-Arrival Time:** The period between the arrival of individual customers may be constant or may be scattered in some distribution fashion. Most queuing models assume that the some inter-arrival time distraction applies for all customers throughout the period of study. In this chapter, mostly distribution of service time, which are important, are considered and they are **Negative exponential distribution** and **Erlang or Gamma distribution.**

- **Capacity of the Service System:** In queuing context, the capacity refers to the space available for the arrivals to wait before taken to service. The space available may be limited or unlimited.

- **Customer Behaviour:** The length of the queue or the waiting time of a customer or the idle time of the service facility mostly depends on the behaviour of the customer. Here the behaviour refers to the impatience of a customer during the stay in the line.

2. **Service Mechanism or Service Facility:**   [Dec. 2008]

Service facilities are arranged to serve the arriving customer or a customer in the waiting line is known as *service mechanism*. The time required to serve the customer cannot be estimated until we know the need of the customer. Many a time it is statistical variable and cannot be determined by any means such as number of customers served in a given time or time required to serve the customer, until a customer is served completely.

- **Service Facility Design:** Arriving customers may be asked to form a single line (Single queue) or multi-line (multi-queue) depending on the service need. When they stand in single line it is known as **Single channel facility** (For example, Petrol bunks, the vehicle enters the petrol station. If there is only one petrol pump, it joins the queue near the pump and when the term comes, get the fuel filled and soon after leaves the queue. Or let us say there is a single ticket counter, where the arrivals will form a queue and one by one purchases the ticket and leaves the queue) when they stand in multi-lines it is known as **multi channel facility.** (In case service is provided to customer in different stages or phases, which are in parallel, then it is known as *multi-channel multi-phase* queuing system).

- **Queue Discipline or Service Discipline :** When the customers are standing in a queue, they are called to serve depending on the nature of the customer. The order in which they are called is known as *Service discipline*. There are various ways in which the customer called to serve. They are:

  ➢ **First In First Out (FIFO) or First Come First Served (FCFS):** We are quite aware that when we are in a queue, we wish that the element which comes should be served first, so that every element has a fair chance of getting service.

  ➢ **Last In First Out (LIFO) or Last Come First Served (LCFS):** In this system, the element arrived last will have a chance of getting service first. Let us assume a bin containing some inventory. The present stock is being consumed and suppose the material ordered will arrive that is loaded into the bin. Now the old material is at the bottom of the stock whereas fresh arrived material at the top. While consuming the top material (which is arrived late) is being consumed. This is what we call Last come first served). This can also be written as First In Last Out (FILO).

  ➢ **Service In Random Order (SIRO):** In this case, the items are called for service in a random order. The element might have come first or last does not bother; the servicing facility calls the element in random order without considering the order of arrival. This may happen in some religious organizations but generally it does not followed in an industrial/business system.

Economic Life of an Asset

- Consider an investment in a machine with an initial purchase price of $1000. The yearly operating costs and salvage value of the machine depend on its age as shown in the table below. We anticipate requiring the use of the machine far into the future. Given that the salvage value is decreasing and operating costs are increasing, there must be some optimal time to replace it. The optimal replacement time is called the economic life of the machine

Year of Ownership	Operating Cost During Year	Salvage Value at Year End
1	$1000	$500
2	1200	300
3	1500	100
4	1900	0
5	2500	0

- Investment analysis recognizes that money spent or earned in the future has less value when viewed from the present. This is called the time value of money principle. We compute the present value of an amount cn received n years from now as

$$P = \frac{c_n}{(1 + i)^n}$$

- The quantity i is a percentage expressed as a decimal, and is variously called the interest rate, discount rate, or minimum acceptable rate of return. The term $1/(1 + i)n$, is the discount factor. When i is a positive quantity the discount factor is less than 1.

- The network representation of decision problems involving time value of money is not obvious. In this section, we describe the problem of determining the *economic life* of an investment. The approach is illustrative of a wide variety of problems that can be handled using network flow programming.

- The investment can be viewed as a series of cash flows in time. This is explained in detail in cash flow analysis and crashing of network in Unit 5(Project management technique)

4.2 QUEUING PROBLEMS

- The most important information required to solve a waiting line problem is the nature and probability distribution of arrivals and service pattern. The answer to any waiting line problem depending on finding:

1. **Queue Length:** The probability distribution of queue length or the number of persons in the system at any point of time. Further we can estimate the probability that there is no queue.

2. **Waiting Time:** This is probability distribution of waiting time of customers in the queue. That is, we have to find the time spent by a customer in the queue before the commencement of his service, which is called **his waiting time in the queue**.

- The total time spent in the system is the waiting time in the queue plus the service time. The waiting time depends on various factors, such as:

➢ The number of units already waiting in the system,

➢ The number of service stations in the system,

➢ The schedule in which units are selected for service,

➢ The nature and magnitude of service being given to the element being served.

3. **Service Time:** It is the time taken for serving a particular arrival.

4. **Average Idle Time or Busy Time Distribution:** The average time for which the system remains idle. We can estimate the probability distribution of busy periods. If we suppose that the server is idle initially and the customer arrives, he will be provided service immediately.

- During his service time some more customers will arrive and will be served in their turn according to the system discipline. This process will continue in this way until no customer is left unserved and the server becomes free again after serving all the customers. At this stage we can conclude, that the busy period is over.

- On the other hand, during the idle periods no customer is present in the system. A busy period and the idle period following it together constitute a busy cycle. The study of busy period is of great interest in cases where technical features of the server and its capacity for continuous operation must be taken into account.

4.3 LIST OF SYMBOLS USED IN QUEUING MODEL

A queue is designated or described as shown below: A model is expressed as:

A/B/S : (d/f) where,

A : Arrival pattern of the units, given by the probability distribution of inter-arrival time of units.

For example, Poisson distribution, Erlang distribution, and inter-arrival time is 1 minute or 10 units arrived in 30 minutes etc.

B : The probability distribution of service time of individual being actually served. For example, the service time follows negative exponential distribution and 10 units are served in 10 minutes or the service time is 3 minutes etc.

S : The number of service channels in the system. For example, the item is served at one service facility or the person will receive service at 3 facilities etc.

d : Capacity of the system. That is the maximum number of units the system can accommodate at any time. For example, the system has limited capacity of 40 units or the system has infinite capacity etc.

f : The manner or order in which the arriving units are taken into service i.e. FIFO/LIFO/SIRO/Priority.

4.3.1 Notations Used [May 2009, 2010]

X : Inter arrival time between two successive customers (arrivals).

Y : The service time required by any customer.

w : The waiting time for any customer before it is taken into service.

v : Time spent by the customer in the system.

n : Number of customers in the system, that is customers in the waiting line at any time, including the number of customers being served.

$P_n(t)$: Probability that n customers arrive in the system in time $.t.$.

$\phi(t)$: Probability that n units are served in time $.t.$.

U (T) : Probability distribution of inter arrival-time P (t ≤ T).

V (T) : Probability distribution of servicing time P (t ≤ T).

F (N) : Probability distribution of queue length at any time P (N ≤ n).

E_n : Some state of the system at a time when there are n units in the system.

λ_n : Average number of customers arriving per unit of time, when there are already 'n' units in the system.

λ : Average number of customers arriving per unit of time.

μ_n : Average number of customers being served per unit of time when there are already 'n' units in the system.

μ : Average number of customers being served per unit of time.

$1/\lambda$: Inter-arrival time between two arrivals.

$1/\mu$: Service time between two units or customers.

$\rho = \left(\dfrac{\lambda}{\mu}\right)$: System utility or traffic intensity which tells us how much time the system was utilized in a given time. For example, given time is 8 hours and if ρ = 3/8, it means to say that out of 8 hours the system is used for 3 hours and (8 − 3 = 5) 5 hours is idle.

4.4 GENERALIZED POISSON QUEUING MODEL (QUEUING DECISION MODELS)

- Most elementary queuing models assume that the *inputs/arrivals* and *outputs/departures* follow a **birth** and **death** process.

- Any queuing model is characterized by situations where both arrivals and departures take place simultaneously. Depending upon the nature of inputs and service faculties, there can be a number of queuing models as shown below:

  ➢ **Probabilistic Queuing Model:** Both arrival and service rates are some unknown random variables.

  ➢ **Deterministic Queuing Model:** Both arrival and service rates are known and fixed.

  ➢ **Mixed Queuing Model:** Either of the arrival and service rates is unknown random variable and other known and fixed.

- Earlier we have seen how to designate a queue. Arrival pattern / Service pattern / Number of channels / (Capacity / Order of servicing). (A /B/ S / (d / f).

In general,

M : is used to denote Poisson distribution (**Markovian**) of arrivals and departures.

D : is used to constant or Deterministic distribution.

Ek : is used to represent Erlangian probability distribution.

G : is used to show some general probability distribution.

- In general, queuing models are used to explain the descriptive behavior of a queuing system. These quantify the effect of decision variables on the expected waiting times and waiting lengths as well as generate waiting cost and service cost information.

- The various systems can be evaluated through these aspects and the system, which offers the minimum total cost is selected.

4.4.1 Poisson Arrival/Poisson Output/Number of Channels/Infinite Capacity/FIFO

Model I: (M/M/1): (∞/FIFO):

[Dec. 2009, 2010, May 2010]

Formulae Used:

- Average number of arrivals per unit of time = λ.

- Average number of units served per unit of time = μ.

- Traffic intensity or utility ratio = $\rho = \dfrac{\lambda}{\mu}$ the condition is $(\mu > \lambda)$.

- Probability that the system is empty = $P_0 = (1 - P)$.
- Probability that there are 'n' units in the system = $P_n = \rho^n P_0$.
- Average number of units in the system = $E(n) = \dfrac{\rho}{(1 - \rho)}$ or $\dfrac{\lambda}{(\mu - \lambda)} = L_q + \dfrac{\lambda}{\mu}$.
- Average number of units in the waiting line = $E_L = \dfrac{\rho^2}{(1 - \rho)} = \dfrac{\lambda^2}{(\mu - \lambda)}$.
- Average waiting length (mean time in the system) = $E\left(\dfrac{L}{L} \geq 0\right)$.

$$= \dfrac{1}{(\mu - \lambda)} = \dfrac{1}{(1 - \rho)}$$
$$= E(w) + \dfrac{1}{\mu} = \dfrac{L}{\lambda}$$

- Average length of waiting line with the condition that it is always greater than zero

$$= V(n) = \dfrac{\rho}{(1 - \rho)^2} = \dfrac{\lambda}{(\mu - \lambda)^2}$$
$$= \dfrac{L_q}{\lambda} = \dfrac{\lambda}{\mu(\mu - \lambda)}$$

- Average time an arrival spends in the system = $E(v)$

$$= \dfrac{1}{\mu(1 - \rho)} = \dfrac{1}{(\mu - \lambda)} = E\left(\dfrac{w}{w} > 0\right)$$

- $P(w > 0)$ = System is busy = ρ.
- Idle time = $(1 - \rho)$.
- Probability distribution of waiting time = $P(w)\, dw = \mu\rho (1 - \rho)\, e^{-\mu w(1 - \rho)}$.
- Probability that a consumer has to wait on arrival = $P(w > 0) = \rho$.
- Probability that a new arrival stays in the system = $P(v)\, dv = \mu(1 - \rho)\, e^{-\mu v(1 - \rho)}\, dv$.

SOLVED EXAMPLES

Example 4.1: *A T.V. repairman finds that the time spent on his jobs have an exponential distribution with mean of 30 minutes. If he repairs sets in the order in which they come in, and if the arrival of sets is approximately Poisson with an average rate of 10 per 8 hour day, what is repairman's expected idle time each day? How many jobs are ahead of the average set just brought in?*

Solution:

This problem is Poisson arrival/Negative exponential service/single channel/infinite capacity/FIFO type problem.

Data given: λ = 10 sets per 8 hour day = $\dfrac{10}{8}$

$$= \dfrac{5}{4} \text{ sets per hour}$$

$\dfrac{1}{\mu}$ = 30 minutes, hence μ, = $\left(\dfrac{1}{30}\right) \times 60 = 2$ sets per hour.

Hence,

Utility ratio = $\rho = \left(\dfrac{\lambda}{\mu}\right) = \dfrac{\left(\dfrac{5}{4}\right)}{2} = \dfrac{5}{8} = 0.625$. This means out of 8 hours, 5 hours the system is busy i.e. repairman is busy.

Probability that there is no queue = The system is idle = $(1 - \rho) = 1 - \left(\dfrac{5}{8}\right) = \dfrac{3}{8}$ = That is out of 8 hours the repairman will be idle for 3 hours.

Number of sets ahead of the set just entered = Average number of sets in system

$$= \dfrac{\lambda}{(\mu - \lambda)} = \dfrac{\rho}{(1 - \rho)} = \dfrac{0.625}{(1 - 0.625)}$$
$$= \dfrac{5}{3} \text{ ahead of jobs just came in.}$$

Example 4.2: *The arrivals at a telephone booth are considered to be the following Poisson law of distribution with an average time of 10 minutes between one arrival and the next. Length of the phone call is assumed to be distributed exponentially with a mean of 3 minutes.*

(i) What is the probability that a person arriving at the booth will have to wait?

(ii) What is the average length of queue that forms from time to time?

(iii) The telephone department will install a second booth when convinced that an arrival would expect to wait at least three minutes for the phone. By how much must the flow of arrivals be increased in order to justify a second booth? **[Dec. 2009]**

Solution:

Data given: Time interval between two arrivals = 10 min. = $\dfrac{1}{\lambda}$, Length of phone call = 3 min. = $\dfrac{1}{\mu}$. Hence, $\lambda = \dfrac{1}{10}$ = 0.1 – per min. and $\mu = \dfrac{1}{3}$ = 0.33 per min. and $\rho = \dfrac{\lambda}{\mu} = \dfrac{0.10}{0.33}$ = 0.3.

(i) Any person who is coming to booth has to wait when there is somebody in the queue. He need not wait when there is nobody in the queue i.e. the queue is empty. Hence, the probability of that an arrival does not wait = $P_0 = (1 - \rho)$.

Hence, the probability that an arrival has to wait = 1 – The probability that an arrival does not wait

$= (1 - P_0) = 1 (1 - \rho) = \rho = 0.3$. That means 30% of the time the fresh arrival has to wait. That means that 70% of the time is idle.

(ii) Average length of non-empty queue from time to time = (Average length of the waiting line with the condition that it is always greater than zero = $1 (1 - \rho)$

i.e. $\left(\dfrac{L}{L} > 0\right) = \dfrac{1}{(1 - 0.30)} = 1.43$ persons.

(iii) The installation of the second booth is justified if the waiting time is greater than or equal to three. If the new arrival rate is λ', then for $\mu = 0.33$, we can work out the length of the waiting line. In this case, $\rho = \dfrac{\lambda'}{\mu}$.

Length of the waiting line of λ' and $\mu = 0.33 = E(w) =$
$\left\{\dfrac{\lambda'}{\mu(\mu - \lambda')}\right\} \geq 3$ or $\lambda' = (3\mu^2 - 3\rho\mu\lambda')$ or $\lambda' = \dfrac{(3\mu^2)}{(1 + 3\mu)}$

$= \dfrac{(3 \times 0.33^2)}{(1 + 3 \times 0.33)}$ i.e. $\lambda' \geq 0.16$. That is the arrival rate must be at least 0.16 persons per minute or one arrival in every 6 minutes. This can be written as 10 arrivals per hour to justify the second booth.

Example 4.3: *In a departmental stored one cashier is there to serve the customers. And the customers pick up their needs by themselves. The arrival rate is 9 customers for every 5 minutes and the cashier can serve 10 customers in 5 minutes. Assuming Poisson arrival rate and exponential distribution for service rate. Find:*

(i) Average number of customers in the system.

(ii) Average number of customers in the queue or average queue length.

(iii) Average time a customer spends in the system.

(iv) Average time a customer waits before being served.

Solution:

Data given: Arrival rate is $\lambda = \left(\dfrac{9}{5}\right) = 1.8$ customers per minute.

Service rate $= \mu = \left(\dfrac{10}{5}\right) = 2$ customers per minute. Hence,

$\rho = \left(\dfrac{\lambda}{\mu}\right) = \left(\dfrac{1.8}{2}\right) = 0.9.$

(i) Average number of customers in the system

$= E(n) = \dfrac{\rho}{(1 - \rho)} = \dfrac{0.9}{(1 - 0.9)} = \dfrac{0.9}{0.1} = 9$ customers.

(ii) Average time a customer spends in the system

$= E(v) = \dfrac{1}{\mu(1 - \rho)} = \dfrac{1}{(\mu - \lambda)} = \dfrac{1}{(2 - 1.8)} = 5$ minutes.

(iii) Average number of customers in the queue

$= E(L) = \dfrac{\rho^2}{(1 - \rho)} = \dfrac{\lambda^2}{\mu(\mu - \lambda)} = (\rho) \times \dfrac{\lambda}{(\mu - \lambda)}$

$= 0.9 \times \dfrac{1.8}{(2 - 1.8)} = 8.1$ customers.

(iv) Average time a customer spends in the queue

$= \dfrac{\rho}{\mu(1 - \rho)} = \dfrac{\lambda}{\mu(\mu - \lambda)} = \dfrac{0.9}{2(1 - 0.9)} = \dfrac{0.9}{0.2}$

$= 4.5$ minutes.

Example 4.4: *There is a congestion of the platform of a railway station. The trains arrive at the rate of 30 trains per day. The waiting time for any train to hump is exponentially distributed with an average of 36 minutes.*
Calculate: (i) The mean queue size,
(ii) The probability that the queue size exceeds 9.

Solution:

Data given: Arrival rate 30 trains per day, service time = 36 minutes.

$\lambda = 30$ trains per day.

Hence, inter arrival time $= \dfrac{1}{\lambda} = \dfrac{(60 \times 24)}{30} = \dfrac{1}{48}$ minutes.

Given that the inter service time $= \dfrac{1}{\mu} = 36$ minutes.

Therefore, $\rho = \left(\dfrac{\lambda}{\mu}\right) = \dfrac{36}{48} = 0.75.$

(i) The mean queue size $= E(n) = \dfrac{\rho}{(1 - \rho)} = \dfrac{0.75}{(1 - 0.75)}$

$= \dfrac{0.75}{0.25} = 3$ trains.

(ii) Probability that queue size exceeds 9 = Probability of queue size $\geq 10 = 1 -$ Probability of queue size less than $10 = 1 - (P_0 - P_1 + P_2 \ldots + P_9)$

$= 1 - \left\{(1 - \rho)\left[\dfrac{(1 - \rho^{10})}{(1 - \rho)}\right]\right\} = 1 - (1 - \rho^{10}) = \rho^{10}$

$= (0.75)^{10} = 0.06$ approximately.

Example 4.5: *Arrival rate of telephone calls at a telephone booth is according to Poisson distribution, with an average time of 9 minutes between consecutive arrivals. The length of telephone call is exponentially distributed with a man of 3 minutes.*
Find:

(i) Determine the probability that a person arriving at the booth will have to wait.

(ii) Find the average queue length that forms from time to time.

(iii) The telephone company will install a second booth when conveniences that an arrival would expect to have to wait at least four minutes for the phone. Find the increase in flow of arrivals, which will justify a second booth.

(iv) What is the probability that an arrival will have to wait for more than 10 minutes before the phone is free?

(v) What is the probability that they will have to wait for more than 10 minutes before the phone is available and the call is also complete?

(vi) Find the fraction of a day that the phone will be in use.

Solution:

Data given: Arrival rate $\lambda = \dfrac{1}{9}$ per minute and service rate $\mu = \dfrac{1}{3}$ per minute.

(i) Probability that a person has to wait (person will wait when the system is busy i.e. we have to find ρ)

$$\rho = \frac{\lambda}{4} = \frac{\left(\dfrac{1}{9}\right)}{\left(\dfrac{1}{3}\right)} = \frac{3}{9} = 0.33 \text{ i.e. 33\% of the time the}$$

customer has to wait. This means that 67% of the time the customers will get the phone soon after arrival.

(ii) Average queue length that forms from time to time =

$$\text{time} = \frac{\mu}{(\mu - \lambda)} = \frac{(1/3)}{[(1/3) - (1/9)]} = \frac{(1/3)}{(2/9)} = \frac{(1/3)}{(9/2)} = 1.5$$

persons.

(iii) Average waiting of time the queue = $E(w)$ =

$$\frac{\lambda_1}{\mu(\mu - \lambda_1)} = 4.$$

$$4 = \frac{\lambda_1}{\left(\dfrac{1}{3}\right)} \times \left[\left(\frac{1}{3}\right) - \lambda_1\right] = \left(\frac{1}{9}\right) - \left(\frac{\lambda_1}{3}\right) = \left(\frac{\lambda_1}{4}\right)$$

$$\text{or } \lambda_1 \times \left(\frac{7}{2}\right) = \frac{1}{9}$$

$$\lambda_1 = \frac{12}{(7 \times 9)} = \left(\frac{4}{21}\right) \text{ arrivals per minute.}$$

Hence, increase in the flow of arrivals = $\left(\dfrac{4}{21}\right) - \left(\dfrac{1}{9}\right)$

$$= \frac{5}{63} \text{ per minute.}$$

(iv) Probability of waiting time ≥ 10 $\displaystyle\int_{10}^{\infty} (\lambda/\mu)(\mu - \lambda) \times$ $e^{-(\mu - \lambda)t}$ dt.

$$= (\lambda/\mu)(\mu - \lambda) \times \left[\frac{(e^{-(\mu - \lambda)t})}{-(\mu - \lambda)}\right]_{10}^{\infty} = \left(\frac{\lambda}{\mu}\right) \times [0 - e^{-(\mu - \lambda)10}]$$

$$= \left(\frac{\lambda}{\mu}\right) \times e^{-(\mu - \lambda)10}$$

$$= \left(\frac{1}{3}\right) \times e^{-(1/3 - 1/9) \times 10} = \left(\frac{1}{3}\right) \times e^{-(20/9)} = \frac{1}{30}$$

(v) Probability that an element spends in system ≥ 10

$$= \int_{10}^{\infty} (\mu - \lambda) \, e^{-(\mu - \lambda)10} \, dt.$$

$$= \left(\frac{\mu}{\lambda}\right) \int_{10}^{\infty} \left(\frac{\lambda}{\mu}\right)(\mu - \lambda) \, e^{-(\mu - \lambda)t} \, dt = \left(\frac{\mu}{\lambda}\right)\left(\frac{1}{30}\right)$$

$$= \left[\frac{\left(\dfrac{1}{3}\right)}{\left(\dfrac{1}{9}\right)}\right] \times \left(\frac{1}{30}\right) = \frac{1}{10} = 0.1$$

(vi) The expected fraction of a day that the phone will be in use = $\rho = \left(\dfrac{\lambda}{\mu}\right) = 0.33$.

Example 4.6: *In large maintenance department, fitters draw parts from the parts stores, which is at present staffed by one storekeeper. The maintenance foreman is concerned about the time spent by fitters in getting parts and wants to know if the employment of a stores helper would be worthwhile. On investigation it is found that:*

(i) A simple queue situation exists,

(ii) Fitters cost ₹2.50 per hour,

(iii) The storekeeper costs ₹2/- per hour and can deal on an average with 10 fitters per hour.

(iv) A labour can be employed at ₹1.75 per hour and would increase the capacity of the stores to 12 per hour.

(v) On an average 8 fitters visit the stores each hour.

Solution:

Data given: $\lambda = 8$ fitters per hour, $\mu = 10$ per hour.

Number of fitters in the system = $E(n) = \dfrac{\lambda}{(\mu - \lambda)}$ or $\dfrac{\rho}{(1 - \rho)}$

$$= \frac{8}{(10 - 8)} = 4 \text{ fitters.}$$

With stores labour $\lambda = 8$ per hour, $\mu = 12$ per hour.

Number of fitters in the system = $E(n) = \dfrac{\lambda}{(\mu - \lambda)}$

$$= \frac{8}{(12 - 8)} = 2 \text{ fitters.}$$

Cost per hour = Cost of fitter per hour + cost of labour per hour = $2 \times$ ₹ 2.50 + ₹ 1.75

= ₹ 6.75.

Since, there is a net saving of ₹ 3.25 per hour, it is recommended to employ the labourer.

4.4.2 Model II: (M /M / 1) : (FCFS/∞ / ∞) : (Birth – Death Model) Generalization of Model

- In waiting line system, each arrival can be considered to be a birth i.e. if the system is in the state E_n, i.e. there are n units in the system and there is an arrival then the state of the system changes to the state E_{n+1}. Similarly when there is a departure from the system the state of the system becomes E_{n-1}.
- Hence, E_{n+1}. Similarly, when there is a departure from the system the state of the system becomes E_{n-1}. Hence, whole system is thus viewed as a birth and death process. When λ is the arrival rate of the system, will never be fixed and dependent on the queue length 'n', then it will mean that some person interested in the joining the queue may not join due to longer queue. Similarly, if μ is also dependent on the queue length it may affect the service rate. Hence, in the case both λ and μ cannot be taken to be fixed.
- Three cases may occur, which are described below.
- In this model, arrival rate and service rate i.e. λ and μ do not remain constant during the queuing phenomenon and vary to $\lambda_1, \lambda_2 \dots \lambda_n$ and $\mu_1, \mu_2 \dots \mu_n$ respectively. Then,

$$P_1 = \left(\frac{\lambda_0}{\mu_1}\right) P_0$$

$$P_2 = \left(\frac{\lambda_0}{\mu_1}\right)\left(\frac{\lambda_1}{\mu_2}\right) P_0$$

$$\vdots$$

$$P_n = \left(\frac{\lambda_0}{\mu_1}\right)\left(\frac{\lambda_1}{\mu_2}\right) \dots \left(\frac{\lambda_{n-2}}{\mu_{n-1}}\right) \times \left(\frac{\lambda_{n-1}}{\mu_n}\right) P_0$$

But there are some special cases when:

1. $\lambda_n = \lambda$ and $\mu_n = \mu$ then,

$$P_0 = 1 - \left(\frac{\lambda}{\mu}\right), \quad P_n = \left(\frac{\lambda}{\mu}\right)^n \times \left[1 - \left(\frac{\lambda}{\mu}\right)\right]$$

2. When $\lambda_n = \dfrac{\lambda}{(n + 1)}$ and $\mu_n = \mu$

$$P_0 = e^{-\rho}$$

$$P_n = \left[\frac{\rho^n}{(n!)}\right] \times e^{-\rho}, \text{ where } \rho = \left(\frac{\lambda}{\mu}\right)$$

3. When $\lambda_n = \lambda$ and $\mu_n = n \times \mu$ then $\rho_0 = e^{-\rho}$ and

$$\rho_n = \left(\frac{\rho^n}{n!}\right) \times e^{-\rho}$$

Example 4.7: *A transport company has a single unloading berth with vehicles arriving in a Poisson fashion at an average rate of three per day. The unloading time distribution for a vehicle with 'n' unloading workers is found to be exponentially with an average unloading time (1/2) × n days. The company has a large labour supply without regular working hours, and to avoid long waiting lines, the*

company has a policy of using as many unloading group of workers in a vehicle as there are vehicles waiting in line or being unloaded.

Under these conditions find,

(i) What will be the average number of unloading group of workers working at any time?

(ii) What is the probability that more than four groups of workers are needed?

Solution:

Let us assume that there are 'n' vehicles waiting in line at any time. Now service rate is dependent on waiting length hence μ_n = 2n vehicles per day (when there are 'n' groups of workers in the system).

Now λ = 3 vehicles per day and μ = 2 vehicles per day. (With one unloading labour group)

$$\text{Hence,} \quad P_n = \left(\frac{\rho^n}{n!}\right) \times e^{-\rho} \text{ for } n \geq 0$$

Therefore, expected number of group of workers working any specified instant is

$$E(n) = \sum_{n=0}^{\infty} n \times P_n \times \frac{(\rho^n e^{-\rho})}{n!}$$

$$= \rho \times e^{-\rho} \times \sum_{n=1}^{\infty} \frac{(\rho^{n-1})}{(n-1)} = \left(\frac{\lambda}{\mu}\right)$$

$$= 1.5 \text{ labour group}$$

The probability that the vehicle entering in service will require more than four groups of workers is

$$\sum_{n=5}^{\infty} P_n = 1 - \sum_{n=0}^{4} \left(\frac{\rho^n}{n!}\right) e^{-\rho} = 0.019$$

4.4.3 Model III: Finite Queue Length Model: (M / M / 1) : FCFS / N / ∞

This model differs from the above model in the sense that the maximum number of customers in the system is limited to N. Therefore, the equations of above model is valid for this model as long as n < N and arrivals will not exceed N under any circumstances. The various equations of the model are:

1. $P_0 = \dfrac{(1 - \rho)}{(1 - \rho^{N+1})}$, where $\rho = \dfrac{\lambda}{\mu}$ and $\dfrac{\lambda}{\mu} > 1$ is allowed.

2. $P_n = \dfrac{(1 - \rho)\,\rho^n}{(1 - \rho^{N+1})}$ for all n = 0, 1, 2, ... N.

3. Average queue length,

$$E(n) = \frac{\rho[1 - (1 + N)\,\rho^N + N\rho^{N+1}]}{(1 - \rho)(1 - \rho^{N+1})}$$

$$= \left[\frac{(1 - \rho)}{(1 - \rho^{N+1})}\right] \times \sum_{n=0}^{n} n\rho^n = P_0 \times \sum_{n=0}^{N} n \times \rho^n$$

4. The average length of the waiting line = E(L)

$$= \frac{[1 - N\rho^{N+1} + (N-1)\rho^2]}{(1-\rho)(1-\rho^{N+1})}$$

5. Waiting time in the system = E(v) = E(n)/λ' where λ' = λ(1 − ρ_N).

6. Waiting time in the queue = E(w) = E(L)/λ' = [(E(n)/λ'/(1/μ)].

Example 4.8: *In a railway marshalling yard, good train arrives at the rate of 30 trains per day. Assume that the inter arrival time follows an exponential distribution and the service time is also to be assumed as exponential with a mean of 36 minutes.*

Calculate: (i) The probability that the yard is empty, (ii) The average length assuming that the line capacity of the yard is 9 trains.

Solution:

Data given: $\lambda = \frac{30}{(60 \times 24)} = \frac{1}{48}$ trains per minute.

And $\mu = \frac{1}{16}$ trains per minute.

Therefore, $\rho = \left(\frac{\lambda}{\mu}\right) = \frac{36}{48} = 0.75$.

(i) The probability that the queue is empty is given by

$= P_0 = \frac{(1-\rho)}{(1-\rho^{N+1})}$, where, N = 9.

$$\frac{(1-0.75)}{[1-(0.75)^{9+1}]} = \frac{0.25}{0.90} = 0.28 \text{ i.e. } 28\% \text{ of the time the line is empty.}$$

Average queue length $= \left[\frac{(1-\rho)}{(1-\rho^{N+1})}\right] \times \sum_{n=0}^{N} n\,\rho^n$

$$\left[\frac{(1-0.75)}{(1-0.75^{10})}\right] \times \sum_{n=0}^{9} n\,(0.75)^n = 0.28 \times 9.58 = 3 \text{ trains}$$

Example 4.9: *A car park contains 5 cars. The arrival of cars is Poisson at a mean rate of 10 per hour. The length of time each car spends in the car park is exponential distribution with a mean of 5 hours. How many cars are in the car park on an average?*

Solution:

Data given: N = 5, λ = 10/60 = 1/6, μ = 1/2 × 60 = 1/120.

Hence, $\rho = \left(\frac{\lambda}{\mu}\right) = \frac{(1/6)}{(1/120)} = 20$

$P_0 = \frac{(1-\rho)}{1-\rho^{N+1}} = \frac{(1-20)}{(1-20^6)} = 2.9692 \times 10^{-3}$

Average cars in car park = length of the system

$= E(n) = P_0 \times \sum_{n=0}^{N} n \times \rho^n$

$= (2.9692) \times 10^{-3} \sum_{n=0}^{N} n(0.9692 \times 10^{-3})^n$

= 4 Approximately

4.4.4 Model IV: Multi Channel Queuing Model: M / M / c: (∞ / FCFS)

- The above symbols indicate a system with Poisson input and Poisson output with number of channels = c, where c is > 1, the capacity of line is infinite and first come first served discipline. Here the length of waiting line depends on the number of channels engaged. In case, the number of customers in the system is less than the number of channels i.e. n < c, then there will be no problem of waiting and the rate of servicing will be nμ as only n channels are busy, each servicing at the rate m. In case n = c, all the channels will be working and when n ≥ c, then n − c elements will be in the waiting line and the rate of service will be cμ as all the c channels are busy.

- Various formulae we have to use in this type of models are :

$$P_0 = \frac{1}{\displaystyle\sum_{n=0}^{c-1}\left[\frac{\left(\frac{\lambda}{\mu}\right)^n}{n!}\right] + \left[\frac{\left(\frac{\lambda}{\mu}\right)^c}{c!}\right] \times \left[\left(\frac{c\mu}{c\mu - \lambda}\right)\right]}$$

$$OR = \frac{1}{\left[\displaystyle\sum_{n=0}^{c-1}\frac{(c\rho)^n}{n!}\right] + [(c\rho)^c/c!(1-\rho)]}$$

$P_n = \{(\lambda/\mu)^n/n!\} \times P_0$, when $1 \leq n \leq c$

$P_n = [1/(c^{n-c} \times c!)]\,(\lambda/\mu)^n \times P_0$ when $n \geq c$

Average number of units in waiting line of the system

$$= E(n) = \left[\frac{\rho P_c}{(1-\rho)^2}\right]$$

$$= \left\{\frac{[\lambda\mu\,(\lambda/\mu)^c]}{(c-1)!\,(c\mu - \lambda)^2}\right\} P_0 + \left(\frac{\lambda}{\mu}\right)$$

Average number in the queue

$$= E(L) = \left[\frac{P c\rho}{(1-\rho)^2}\right] + c\mu$$

$$= \left\{\frac{[\lambda\mu\,(\lambda/\mu)^c]}{(c-1)!\,(c\mu - \lambda)^2}\right\} P_0$$

Average queue length = Average number of units in waiting line + number of units in service.

Average waiting time of an arrival = E(w)

$$= \frac{\text{(Average number of units in waiting line)}}{\lambda}$$

$$= (P_c\rho)/\lambda(1 - \rho)^2$$

$$= \rho/\lambda(1 - \rho)^2 \times (1/c!) \times (\lambda/\mu)^c \times P_0$$

$$= \frac{E(L)}{\lambda}$$

$$= \left\{ \frac{[\mu \times (\lambda/\mu)^c]}{[(c - 1)! \, (c\mu - \lambda)^2]} \right\} \times P_0$$

Average time an arrival spends in system = E(v)

$$\frac{\text{(Average number of items in the queue)}}{\lambda} = \left[\frac{(P_c\rho)}{\lambda \, (1 - \rho)} \right]$$

$$+ \left(\frac{c\mu}{\lambda} \right)$$

$$\frac{E(n)}{\lambda} = \left\{ \frac{[\mu \times (\lambda/\mu)^c]}{[(c - 1)! \, (c\mu - \lambda)^2]} \right\} \times P_0 + (1/\mu)$$

Probability that all the channels are occupied = $\rho(n \geq c)$

$$= \left[\frac{1}{(1 - \rho)} \right] P_c$$

$$= \frac{[\mu \times (\lambda/\mu)^c] \, P_0}{[(c - 1)! \, (c\mu - \lambda)]}$$

Probability that some units has to wait = $\rho(n \geq c + 1)$

$$= [\rho P_c/(1 - \rho)]$$

$$= 1 - \rho(n \geq c)$$

$$= 1 - \frac{[\mu \times (\lambda/\mu)^c] \, P_0}{[(C - 1)! \, (c\mu - \lambda)]}$$

The average number of units which actually wait in the system

$$= \left[\sum_{n=c+1}^{\infty} (n - c) \, P_n \right] \div \sum_{n=c+1}^{\infty} P_n$$

$$= \frac{1}{(1 - \rho)}$$

Average waiting time in the queue for all arrivals

$$= \left(\frac{1}{\lambda} \right) \sum_{n=0}^{\infty} (n - c) \, P_n = \frac{P_c}{c\mu \, (1 - \rho)^2}$$

Average waiting time in queue for those who acutely wait

$$= \frac{1}{(c\mu - \lambda)}$$

Average number of items served $= \sum_{n=0}^{c-1} n \, P_n + \sum_{n=c}^{\infty} p_n$

Average number of idle channels = c − Average number of items served

Efficiency of M/M/c model

$$= \frac{\text{Average number of items served}}{\text{Total number of channels}}$$

Utilization factor $= \rho = \left(\frac{\lambda}{c\mu} \right)$

Example 4.10: *A tax-consulting firm has 3 counters in its office to receive people who have problems concerning their income, wealth and sales taxes. On the average 48 persons arrive in an 8 hours day. Each tax adviser spends 15 minutes on the average on an arrival. If the arrivals are Poisson distributed and service times are according to exponential distribution then find:*

(i) *The average number of customers in the system,*

(ii) *Average number of customers waiting to be serviced,*

(iii) *Average time a customer spends in the system,*

(iv) *Average waiting time for a customer,*

(v) *The number of hours each week a tax adviser spends performing his job,*

(vi) *The probability that a customer has to wait before he gets service,*

(iii) *The expected number of idle tax advisers at any specified time.*

Solution:

Data given: $c = 3$, $\lambda = \dfrac{48}{8} = 6$ customers per hour,

$$\mu = \left(\frac{1}{15} \right) \times 60 = 4 \text{ customers per hour,}$$

$$\left(\frac{\lambda}{\mu} \right) = \left(\frac{6}{4} \right) = \left(\frac{3}{2} \right).$$

$$P_0 = \frac{1}{\displaystyle\sum_{n=0}^{c-1} [(\lambda/\mu)^n/n!] + [(\lambda/\mu)^c/c!] \times [(c\mu/c\mu - \lambda)]}$$

$$P_0 = \frac{1}{\displaystyle\sum_{n=0}^{2} [(\lambda/\mu)^n/n!] + [(\lambda/\mu)^3/3!] \times [(3\mu/3\mu - \lambda)]}$$

$$= \frac{1}{[1 + (\lambda/\mu) + (1/2) \, (\lambda/\mu)^2] + [(\lambda/\mu)^3/6] \times (3\mu)/(3\mu - \lambda)}$$

$$= \frac{1}{[(1) + (3/2) + (9/6)] + (27/48) \times 12/(12 - 6)}$$

$$= \frac{1}{[(29/8) + (9/8)]}$$

$$= \frac{8}{38} = 0.21 = 21\%$$

(i) Average number of customers in the system = E(n)

$$= \left\{ \frac{[\lambda\mu \times (\lambda/\mu)^c]}{[(c-1)! \, (c\mu - \lambda)^2]} \right\} \times P_0 + (\lambda + \mu)$$

$$\frac{[6 \times 4 \times (3/2)^3]}{[2! \, (12-6)^2] \times 0.21 + (3/2)} = 1.74 \text{ customers i.e.}$$

approximately 2 customers.

(ii) Average number of customer waiting to be served
= E(L) = E(n) = (λ/μ)

= 1.74 + (3/2) = 0.24 customers i.e. one customer approximately.

(iii) Average time a customer spends in the system E(n)

$$= \frac{E(L)}{\lambda} = \frac{1.74}{6} = 0.29 \text{ hours}$$

= 17.4 minutes.

(iv) Average waiting time for a customer = $\dfrac{E(L)}{\lambda} = \dfrac{0.24}{6}$

= 0.04 hours = 2.4 minutes.

(v) Utilization factor = $\rho = \left(\dfrac{\lambda}{c\mu}\right) = \left(\dfrac{6}{3 \times 4}\right) = \dfrac{1}{2}$

= 50% of the time.

(vi) Probability that a customer has to wait = P(n > c)

$$= \frac{[\mu \times (\lambda/\mu)^c] \, P_0}{(c-1)! \, (c\mu - \lambda)} = \left\{ \frac{[4 \times (3/2)^3]}{2! \times (12-6)]} \right\} \times 0.1 = 0.236.$$

(vii) When the probability of no customers waiting is P_0, all the tax advisers are idle. Now we have to find probability of one tax adviser and probability of two tax advisers are idle, which are represented as P_1 and P_2 respectively. Now we know that

$$P_n = \left\{ \frac{(\lambda/\mu)^n}{n!} \right\} \times P_0 \text{ when } 1 \le n \le c.$$

Hence, $P_1 = \left\{ \dfrac{(3/2)}{1!} \right\} \times 0.21 = 0.315$

and $P_2 = \left\{ \dfrac{(3/2)}{2!} \right\} \times 0.21 = 0.236$

Therefore, expected number of idle adviser at any specific time = $3P_0 + 2P_1 + 1P_2 = 3 \times 0.21 + 2 \times 0.315 + 1 \times 0.236 = 1.496$ i.e. approximately = 1.5.

Example 4.11: *A super market has two girls ringing up sales at the counters. If the service time from each customer is exponential with a mean of 4 minutes, and if people arrive in a Poisson fashion at the rate of 10 an hour, then find:*

(i) what is the probability of having an arrival has to wait for service?

(ii) what is the expected percentage of idle time for each girl?

Solution:

Data given: Model: M/M/c model, c = 2, μ = 1/4 services per minute, $\lambda = \dfrac{1}{(60/10)} = \dfrac{1}{6}$ per minute. $\rho = (\lambda/c\mu)$

$$= \frac{(1/6)}{2 \times (1/4)} = (1/3)$$

$$p_0 = \frac{1}{\left[\displaystyle\sum_{n=0}^{c-1} (c\rho)^n/n!\right]} = \left[\frac{(c\rho)^c}{c! \, (1-\rho)}\right]$$

$$= \frac{1}{\left[\displaystyle\sum_{n=0}^{2-1} (2\rho)^n/n!\right]} = \left[\frac{(2\rho)^c}{2! \, (1-\rho)}\right]$$

$$= \frac{1}{(1+2\rho) + [4\rho^2/2(1-\rho)]}$$

$$= 1/1 + (2/3) + (1/2!) \, (2/3)^2 \times 1/[1 - (1/3)] \text{ (because } \rho = 1/3$$

$$= 1/2$$

$$P_1 = (\lambda/\mu) \times P_2 = (2/3) \times (1/2) = 1/3$$

The probability that a customer has to wait = The probability that number of customers in the system is greater than or equal to 2 = $\rho(n \ge 2) = 1 - \rho(n < 2) = 1 - P_0 - P_1 = (1/2) - (1/3) = 0.167$.

Expected percent of idle time of girls or expected number of girls who are idle.

Let X denotes number of idle girls. X = 2 when the system is empty and both girls are free. X = 1 when the system contains only one unit and one of the girls is free. Hence, X can take two values 2 or 1. Probability P_0 and P_1 respectively.

$$E(X) = X_1 \, \rho(X = X_1) + X_2\rho(X = X_2) = (2 \times P_0) + (1 \times P_1)$$
$$= (2 \times 1/2) + (1 \times 1/3) = (4/3)$$

Probability of any girl being idle

$$= \frac{(\text{Expected number of idle girls})}{(\text{Total number of girls})} = (4/3)/2$$

$$= 0.67.$$

Expected percentage of idle time of each girl is 67%.

Example 4.12: *A telephone exchange has two long distance operators. The telephone company finds that during the peak load, long distance calls arrive in a Poisson fashion at an average rate of 15 per hour. The length of service on these calls is approximately exponentially distributed with mean length 5 minutes. What is the probability that a subscriber will have to wait for his long distance call during the peak hours of the day? If subscribers wait and are serviced in turn, what is the expected waiting time.*

Solution:

Data given: $\lambda = 15$ calls per hour, $\mu = \dfrac{60}{5} = 12$ calls per hour. Therefore,

$$\rho = \frac{(15)}{(2 \times 12)}$$

$$= \frac{5}{8P_0} \left[\frac{1}{\displaystyle\sum_{n=0}^{c-1} (c\rho)^n/n!} \right] + \left[\frac{(c\rho)^c}{c!\,(1-\rho)} \right]$$

$$= \frac{1}{1 + (5/4) + (1/2) \times (25/16)} \times \left[\frac{1}{(1-5/8)} \right]$$

$$= (12/52)$$

(i) Probability that a subscriber has to wait $= \rho(n \geq 2)$

$= 1 - P_0 - P_1$

$= [1 - (12/52) - (15/32)] = 25/52 - 0.48$

   i.e. 48% of the time the subscriber has to wait.

Expected waiting time $= E(w)$

$= [\rho/\lambda(1-\rho)^2] \times (1/c!) \times (\lambda/\mu)^c \times P_c$

$= (5/8)/15\,[1 - (5/8)]^2 \times (1/2!) \times (15/12)^2 \times (12/32)$ hours

$= 3.2$ minutes.

Example 4.13: *Ships arrive at a port at the rate of one in every 4 hours with exponential distribution of inter arrival times. The time a ship occupies a berth for unloading has exponential distribution with an average of 10 hours. If the average delay of ships waiting for berths is to be kept below 14 hours, how many berths should be provided at the port?*

Solution:

Data given: $\lambda = 1/4$ ships per hour, $\mu = 1/10$ ships per hour, $\lambda/\mu = 5.2$. For multi-channel queuing system $(\lambda/c\mu) < 1$, to ensure that the queue does not explode.

Therefore, $\dfrac{(1/4)}{(1/10)}\, c < 1$ or $c = \dfrac{5}{2}$. Let us consider $c = 3$ and calculate waiting time.

$$P_0 = \frac{1}{\displaystyle\sum_{n=0}^{c-1} (\lambda/\mu)^n/n!} + \left[\frac{(\lambda/\mu)^c}{c!} \right] \times \left[\frac{c\mu}{c\mu - \lambda} \right]$$

$$= 1/[1 + (5/2) + (1/2) \times (5/2)^2]$$

$$+ (125/6 \times 8) \times (3/10) \times (20/1)$$

$$= 1/[6.625 + 15.625] = 0.045$$

Average waiting time for ship $= E(w)$

$$= \left\{ \frac{[\mu \times (\lambda/\mu)^c]}{[(c-1)! \times (c\mu - \lambda)^2]} \right\} \times P_0$$

$$= \left\{ \frac{[(1/10) \times (5/2)^3]}{2! \times [(3/10) - (1/4)]^2} \right\} \times 0.045$$

$$= 14.06 \text{ hours, this is greater than } 14 \text{ hours.}$$

Therefore, three berths are sufficient. Let us take $c = 4$, then

$$P_0 = \frac{1}{\left[\displaystyle\sum_{n=0}^{3} [(5/2)^2/n!] \right]} + [(5/2)^4/4]$$

$$\times \{(3/10)/[(3/10) - (1/4)]\}$$

$$= \frac{1}{[1 + (5/2) + (1/2) \times (5/2)^2 + (1/6)\,(5/2)^3]}$$

$$+ 625/(24 \times 16) \times (3/10) \times (20/1)$$

$$= \frac{1}{(9.23 + 9.765)} = 0.0526 = 13.7 \text{ hours.}$$

This is less than the allowable time of 14 hours. Hence, 4 berths must be provided at the port.

Example 4.14: *A bank has two tellers working on saving account. The first teller handles withdrawals only. The second teller handles deposits only. It has been found that the service time distribution for the deposits and withdrawal both are exponential with the mean service time of 3 minutes per customer. Depositors are found to arrive in Poisson fashion throughout the day with mean arrival rate of 16 per hour. Withdrawers also arrive in Poisson fashion with mean arrival rate of 16 per hour. What would be the effect on the average waiting time of the depositors and withdrawers if each teller could handle both withdrawal and deposits? What would be the effect if this could only be accomplished by increasing the service time to 3.5 minutes?*

[Dec. 2010]

Solution:

$\mu = \dfrac{1}{3}$/minute $= 20$/hour, $\lambda_1\ 16$/hour $\lambda_2 = 14$/hour

Average waiting time for depositors,

$$\omega_{q1} = \frac{1}{\mu} \cdot \frac{\lambda_1}{\mu - \lambda_1} = \frac{1}{20} \times \frac{16}{20 - 16} = \frac{1}{5} \text{ hours}$$

$$= 12 \text{ minutes}$$

Average waiting time for withdrawers,

$$\omega_{q2} = \frac{1}{\mu} \cdot \frac{\lambda_2}{\mu - \lambda_2} = \frac{1}{20} \times \frac{14}{20 - 14} = \frac{7}{60} \text{ hours}$$

$$= 7 \text{ minutes.}$$

(i) Here, $\lambda = \lambda_1 + \lambda_2 = 16 + 14 = 30$/hours

$\mu = 20$/hours $\lambda/\mu = \dfrac{3}{2}$, $c = 2$

$\therefore \qquad P_0 = \dfrac{1}{\displaystyle\sum_{n=0}^{c-1} \dfrac{(\lambda/\mu)^n}{n!} + \dfrac{(\lambda/\mu)^c}{c!} \cdot \dfrac{c\mu}{c\mu - \lambda}}$

$$= \cfrac{1}{\sum\limits_{n=0}^{1} \dfrac{(3/2)^n}{n!} + \dfrac{(3/2)^2}{2!} \cdot \dfrac{40}{40-30}}$$

$$\therefore \qquad P_0 = \frac{1}{7}$$

$\therefore$ Average waiting time,

$$\omega_q = \frac{\mu(\lambda/\mu)^c}{(c-1)!\,(c\mu-\lambda)^2} \cdot P_0$$

$$= \frac{20.(3/2)^2}{1!\,(40-30)^2} \times \frac{1}{7} = \frac{27}{7} \text{ minutes,}$$

$$= 3.86 \text{ minutes.}$$

(ii) When the service time is 3.5 minutes

$$\mu = \frac{1}{3.5} /\text{min.} = \frac{120}{7} /\text{hr.}$$

$$\lambda = 30/\text{hr.}$$

$$\lambda/\mu = 7/4, \ c = 2$$

$$\therefore \qquad P_0 = \cfrac{1}{\sum\limits_{n=0}^{c-1} \dfrac{(\lambda/\mu)^n}{n!} + \dfrac{(\lambda/\mu)^c}{c!} \cdot \dfrac{c\mu}{c\mu-\lambda}}$$

$$= \cfrac{1}{\sum\limits_{n=0}^{1} \dfrac{(7/4)^n}{n!} + \dfrac{(7/4)^2}{2!} \cdot \dfrac{240/7}{240/7-30}}$$

$$= \cfrac{1}{\left(1+\dfrac{7}{4}\right) + \dfrac{49}{2\times16}\times8} = \frac{1}{15}$$

$\therefore$ Average waiting time,

$$\omega_q = \frac{\mu(\lambda/\mu)^c}{(c-1)!\,(c\mu-\lambda)^2} \cdot P_0$$

$$= \frac{120/7\,(7/4)^2}{1!\,(240/7-30)^2} \times \frac{1}{15}$$

$$\therefore \qquad \omega_q = \frac{343}{1800} \text{ hours} = 11.43 \text{ minutes.}$$

Example 4.15: In a railway marshalling yard, goods train arrive at a rate of 30 trains per day. Assuming that the inter-arrival time follows in exponential distribution and the service time (the time taken to jump a train) distribution is also exponential with an average of 36 minutes. Calculate, (i) expected queue length, (ii) probability that the queue size exceeds 10.

If the input of the train increases to an average of 33 per day, then what will be change in (i) and (ii)?   **[Dec. 2009]**

Solution:

Case 1: $\quad \lambda = 30/\text{day} = 1.25/\text{hr.}$

$$\mu = 36/\text{minutes} = 1.66/\text{hr.}$$

$$\frac{\lambda}{\mu} = 0.753$$

$$L_q = \frac{\lambda}{\mu}\frac{\lambda}{\mu-\lambda} = 0.753 \times \frac{30}{36-30}$$

$$L_q = 2.295 \approx 3$$

Probability of queue being greater than K = 10

$$\rho\,(> K) = \left(\frac{\lambda}{\mu}\right)^{K+1} = (0.753)^{10+1}$$

$$\therefore \qquad \rho\,(> K) = 0.044$$

Case 2: $\quad \lambda = 33/\text{day} = 1.375/\text{hr.}$

$$\mu = 36/\text{min.} = 1.66/\text{hr.}$$

$$\frac{\lambda}{\mu} = 0.828$$

$$\therefore \qquad L_q = \frac{\lambda}{\mu} \cdot \frac{\lambda}{\mu-\lambda} = 0.828 \times \frac{33}{36-33}$$

$$= 3.99 \approx 4$$

Probability of queue being greater than K = 10

$$\rho\,(> K) = \left(\frac{\lambda}{\mu}\right)^{K+1} = (0.828)^{10+1}$$

$$\therefore \qquad \rho\,(> K) = 0.125$$

$$= \frac{1}{\mu-\lambda} - \frac{1}{\mu} = \frac{\mu-\mu+\lambda}{(\mu-\lambda)\,\mu}$$

$$\omega_q = \frac{\lambda}{(\mu-\lambda)\,\mu}$$

Define Concept of Busy Period in Queuing Theory: Busy period of a server is a time during which he remains busy during servicing. Thus, time between start of time between first unit till the end of queue is called busy *period in queuing.*

Example 4.16: In a maintenance shop the inter-arrival time and tool creep are exponential with average time of 10 min. length of service time assume to be exponential distributed with mean 6 min.

Find: (i) Probability that person arriving at booth will have to wait.

(ii) Average length of queue that forms and average time that an operator spend in queue system.

(iii) Estimate fraction of data that operator is idle.

(iv) Probability that there will be 6 or more operators waiting for service.

(v) Manager of shop will installed second booth when arrival would have wait 10 min. or more for service; by how much must rate of arrival will increase in order to justify second booth.

Solution:

$$\lambda = 10 \text{ min.} = 6/\text{hr.} \frac{1}{10} \text{ hr.} \frac{1}{\text{min.}}$$

$$\mu = 6 \text{ min.} = \frac{1}{6} \text{ hr.} \frac{1}{\text{min.}}$$

(i) $$\delta = \frac{\lambda}{\mu} = \frac{\frac{1}{10}}{\frac{1}{6}} = 0.6$$

(ii) Average length of queue that form, L_q.

$$\therefore \quad L_q = \frac{\lambda^2}{\mu(\mu - \lambda)} = \frac{\left(\frac{1}{10}\right)^2}{\frac{1}{6}\left(\frac{1}{6} - \frac{1}{10}\right)}$$

(iii) ω_q = Average time that no operator spend in queue

$$= \frac{\lambda}{\mu}\left(\frac{1}{\mu - \lambda}\right)$$

$$= 0.6\left(\frac{1}{\frac{1}{6} - \frac{1}{10}}\right) = 9 \text{ min.}$$

(iv) $$= 1 - \frac{\lambda}{\mu} = 1 - 0.6$$

$$P_0 = 0.4$$

(v) $$P_n = \left(\frac{\lambda}{\mu}\right)^n \left(1 - \frac{\lambda}{\mu}\right)$$

$$= (0.6)^5 \, [(1 - 0.6) \, x]$$

$$p_0 = 0.4$$
$$p_1 = 0.24$$
$$p_2 = 0.144$$
$$p_3 = 0.0864$$
$$p_4 = 0.05184$$
$$p_5 = 0.0311$$
$$p_6 = 0$$
$$p_n = 1 - [p_0 + p_1 + p_2 + p_3 + p_4 + p_5]$$
$$= 0.047$$

Let, λ' = be new arrival rate

Let, ω_s = 10

$$10 = \frac{1}{\mu - \lambda'}$$

$$= \frac{1}{\frac{1}{6} - \lambda'}$$

$$\frac{1}{6} - \lambda' = 0.1$$

$$\lambda' = \frac{1}{15} \frac{1}{\text{min.}}$$

Increase in flow rate of arrival $= \frac{1}{15} - \frac{1}{10} = -0.033$

Let, ω_q = 10

$$10 = \frac{\lambda'}{\mu}\left(\frac{1}{\mu - \lambda'}\right)$$

$$= \frac{\lambda'}{\frac{1}{6}}\left[\frac{1}{\frac{1}{6} - \lambda'}\right]$$

$$1.667 = \frac{\lambda'}{\frac{1}{6} - \lambda'}$$

$$\frac{\lambda'}{\frac{1}{6} - \lambda'} = 1.667$$

$$\frac{1}{6\lambda'} - 1 = 1.667$$

$$\frac{1}{6\lambda'} = 2.667$$

$$\lambda' = \frac{1}{16}$$

Increase in flow rate of arrival $= \frac{1}{16} - \frac{1}{10} = -0.037$

$\therefore$ Addition of second booth is not necessary.

Example 4.17: *At what average rate must a clerk in a superportel work to ensure a probability of 0.90 that a customer will not have to wait longer than 12 min. customer arrive at counter as Poisson fashion with rate of 13/hr. service time exponentially distributed.*

$$\lambda = \frac{13}{60} \, 13/\text{hr.}$$

Solution:

Probability (waiting time > 12 min.) that should $(1 - 0.90)$ i.e. will not have to wait = 0.10.

$$\therefore \quad 0.10 = \int_{12}^{\infty} (\mu - \lambda)\frac{\lambda}{\mu} e^{-(\mu-\lambda)t} \, dt$$

$$= (\mu - \lambda)\frac{\lambda}{\mu}\left[\frac{e^{-(\mu-\lambda)t}}{-(\mu - \lambda)}\right]_{12}^{\infty}$$

$$= (\mu - \lambda)\frac{\lambda}{\mu}\left[\frac{e^{-(\mu-13/60)\,12}}{(\mu - 13/60)}\right]$$

$$= (\mu - 13)\frac{13}{\mu}\left[\frac{e^{-(\mu-13)\,12}}{(\mu - 13)}\right]$$

$$= \frac{13}{\mu}\,[e^{-(\mu-13)12}]$$

$$\frac{0.10}{13} \times \mu = e^{-(\mu-13)12} = e^{-\mu 12 + 13 \times 12}$$

$$= 0.00769 \times \mu = e^{-(\mu-13)12}$$

$$\log(0.00769) + \log(\mu) = -(\mu - 13)\, 12 \log_e$$

$$-2.114 + \log \mu = -(\mu - 13) \times 5.211$$

$$\mu = 1/16$$

Example 4.18: *A customer arrive at railway counter of rate 12/hr. There is a clerk servicing of rate of 40/hr.*

(i) *What is probability, that there is no customer on counter i.e., system idle?*

(ii) *What is probability, that there are 2 customers in counter?*

(iii) *What is probability that no customer waiting to be serve?*

(iv) *What is probability that customer being serve and nobody is waiting.*

Solution:

$$\lambda = 12/hr.$$

$$\mu = 40/hr.$$

$$\delta = 0.3$$

(i) Probability $= 1 - P_0$

$$= \lambda/\mu = 0.3$$

(ii) $\quad P_1 = 0.09$

$$P_2 = 0.027$$

$$P_n = 0.583$$

$$P_n = P_0 + P_1 \qquad \text{(2 counters)}$$

$$= \text{there is almost one customer in service}$$

$$= P_0 + P_1 = 0.3 + 0.09$$

$$P_n = 0.39$$

(iii) $\quad P_n = p_0 \times \lambda/\mu \qquad \text{(both should be busy)}$

$$P_n = 0.09.$$

Example 4.19: *Passengers arrive at first class reservation ticket counter at rate of 15 passengers/hr. and clerk can handle four and provided assistance using computer terminal 20 person/hr. Find:*

(i) *What is average time for customer before getting service?*

(ii) *What is probability that number of customers waiting to serve?*

(iii) *What is probability that more than two customers waiting are one counter m = 2?*

(iv) *What is the probability that customer is served and that nobody is waiting?*

Solution: $\quad \lambda = 15,\ \mu = 20$

$$L_s = \frac{\lambda}{\mu - \lambda} = \frac{15}{20 - 15} = 3$$

Average waiting time in queue.

(i) $\quad \omega_q = \frac{\lambda}{\mu}\left(\frac{1}{\mu - \lambda}\right) = \frac{15}{20}\left(\frac{1}{20 - 15}\right) = 0.15$

Now $\quad \delta = \frac{\lambda}{\mu} = \frac{15}{20} = 0.75$

∴ Average waiting time in system,

$$\omega_s = \frac{1}{\mu - \lambda} = \frac{1}{20 - 15} = \frac{1}{5} = 0.2$$

(ii) $\quad P_0 = 1 - \delta = 1 - 0.75 = 0.25$

(iii) Probability that more than two customers are waiting.

$$P_0 = \left(\frac{\lambda}{\mu}\right)^2 = (\delta)^2 = (0.75)^2 = 0.14$$

Probability that no one is waiting.

(iv) $\quad P = \left(\frac{\lambda}{\mu}\right)' \times \frac{\lambda}{\mu} = 0.250$

Example 4.20: *Mean arrival rate of planes at airport during peak period is 20/hr. as per Poisson's distribution. During connection, the planes are forced to flyover field in stack waiting the land of other planes that had a arrived earlier. Find:*

(i) *How many planes would be flying in this stuck during good and bad weather?*

(ii) *How many planes will be in stuck in process of landing in good and bad weather?*

(iii) *How many stuck and landing time to allow. So that priority of landing out of control would have to be requested only once in 20 times, assume service rate 60/hr. and 30/planes/hr. in bad weather.*

Solution:

$$\lambda = 20/hr.,\ \ \mu = 60$$

For Good Weather:

(i) L_q = Average planes stuck to land in good weather

$$L_q = \frac{\lambda^2}{\mu(\mu - \lambda)} = \frac{20^2}{60(60 - 20)} = 0.166$$

Now mean time spend by planes in queue is

$$\omega_q = \frac{\lambda}{\mu} = \left(\frac{1}{\mu - \lambda}\right) = \frac{20}{60}\left(\frac{1}{60 - 20}\right) = \frac{1}{120}$$

(ii) Average waiting time for planes in system is

$$\omega_s = \frac{1}{\mu - \lambda} = \frac{1}{60 - 20} = \frac{1}{40} = 1.5 \text{ min.}$$

(iii) Probability,

$$P = \int_0^\infty (\mu - \lambda)\frac{\lambda}{\mu} e^{-(\mu-\lambda)t}\, dt$$

$$= \frac{\lambda}{\mu}[0 - e^{-(\mu-\lambda)t}]$$

$$= \frac{20}{60} e^{-(60-20)\left(1 - \frac{1}{20}\right)} = 0.333$$

∴ Actual time for landing = 1 − 0.95 = 0.05

For Bad Weather:

$\lambda = 20/hr.$ $\mu = 30$

(i) ∴ $L_q = \dfrac{\lambda^2}{\mu(\mu - \lambda)} = \dfrac{20^2}{30(30 - 20)} = 1.333$

$\omega_q = \dfrac{\lambda}{\mu}\left(\dfrac{1}{\mu - \lambda}\right) = \dfrac{20}{30}\left(\dfrac{1}{30 - 20}\right) = \dfrac{1}{15}$

(ii) $\omega_s = \dfrac{\lambda}{\mu - \lambda} = \dfrac{1}{30 - 20} = \dfrac{1}{10} = 6$ min.

(iii) Probability time = $1 - \dfrac{1}{20} = 0.95$

$$P = \int_0^\infty (\mu - \lambda)\frac{\lambda}{\mu} e^{-(\mu-\lambda)t}\, dt$$

$$= \frac{\lambda}{\mu}[0 - e^{-(\mu-\lambda)t}]$$

$$= \frac{20}{30}\left[e^{(30-20)\times 0.95}\right] = 0.666$$

Actual time for landing = 1 − 095 = 0.05.

Example 4.21: *A customer arrives at a bank counter, managed by single cashier according to Poisson's distribution with mean arrival rate of 6 customers/hr. the cashier attains customer on first come first serve basis at an average of 10 customers/hr. with service time exponential distribution. Find probability of arrivals (0 − 5) during:*

(i) 15 min. interval and second 30 min. interval probability that queuing system is idle.

(ii) Probability as associated with no customers (0 − 5) in the queuing system.

(iii) The time customer should expect in the queue, the time customer spends between leaving bank at counter.

$\lambda = 6/hr.$ more $\mu = 10/hr.$

For $t = 15$ min. $= (1/4)/hr.$

$\lambda t = 3/2$ customers.

Probability of 'n' customers arriving in time t is

$$= (\lambda t)^n \times \frac{e^{-\lambda t}}{n!}$$

∴   Probability will be given as follows:

Table 4.1

n	Probability for 15 min.
0	0.223
1	0.335
2	0.251
3	0.1255
4	0.047
5	0.014

For $t = 30 t = 1/2$ hr.

∴ $\lambda t = 3$ customers

∴   Probability of 'n' customers arriving in 30 min. time

$$= 3^n \frac{e^{-3}}{n!}$$

Table 4.2

n	Probability for 30 min.
0	0.0497
1	0.149
2	0.224
3	0.224
4	0.168
5	0.10

Time spent by customer in the queue is given by

$$\omega_q = \frac{\lambda}{\mu}\left(\frac{\lambda}{\mu - \lambda}\right) = \frac{6}{10}\left(\frac{1}{10^{-6}}\right) = 0.15$$

Time that customer least the counter is

$$\omega_s = \frac{1}{\mu - \lambda} = \frac{1}{4} = 0.25$$

Example 4.22: *Cargo Transport Ltd. has 1 typist at its head office due to variation in length of letter to the typing rate is distributed approximating quotient distribution with mean service rate of a 8 per hr. Work load for typist varies with an average of 5 letters per hr. If cost of typing is valued at ₹ 12/hr. Determine:*

(i) Percentage time typist is busy $\left(i.e.,\ \rho\dfrac{\lambda}{\mu}\right)$.

(ii) Average number of letters lying with typist at any point of time (L_s).

(iii) Average time the letter lies with typist (ω_s).

(iv) Probability of 6 letters with the typist at any time.

(v) Average cost and typing letters per day.

Solution: Data Given:

Arrival rate = λ = 5/hr.

Service rate = μ = 8/hr.

(i) Therefore, percentage of time typist is busy

$$\rho = \frac{\lambda}{\mu} = 0.625$$

(ii) Average no. of letters lying with typist at any point of time,

$$L_s = \frac{\lambda}{\mu - \lambda} = \frac{5}{8 - 5} = 1.67$$

(iii) Average time the letter lies with typist,

$$\omega_s = \frac{1}{\mu - \lambda}$$

$$= \frac{1}{8 - 5} = 0.333 \text{ min.}$$

(iv) Probability of 6 letters

$$P(\omega > 0) = 1 - P_0$$

where, P_0 = Probability of idle person

$$= 1 - 0.625 = 0.375$$

$$P(N = 6) = (1 - P_0)\left(\frac{\lambda}{\mu}\right)^N$$

$$= 0.375\left(\frac{5}{8}\right)^6 = 0.0224$$

(v) Average cost of typing letters/day.

Assume 8 hrs. working day

Total no. of letters typed = 64

Average cost = Total letters × Average time for letter types per hr. × Cost × Average load

$$= 64 \times \frac{1}{3} \times 5 \times 12 = ₹ 1280.$$

Example 4.23: *There are 15 counters at reservations. Average time per customer by each clerk is 15 min. The average arrivals per hr. during 3 types of activities period have been calculated. Similar customers have been survey to determine how long they are willing to get during each type of period.*

Table 4.3

Time of Period	Arrivals per hr.	Accepted Cost in min.
Peak	110	15
Normal	60	10
Low	30	05

Write down assumptions you make on this process and hence determine how many counters open during each type of period (consider a system of 'x' different single server quick system). where, x = number of capital composed counters kept open.

Solution: Assumptions:

1. Arrival flow question distribution.

2. Server time follows exponential distribution.

3. Mean service rate μ is greater than mean arrival rate.

Arrival wait for service neither fail to join the line or leave it. System is composed at 'x' different single servers queuing system.

Arrival rate $\lambda = \dfrac{110}{x}, \dfrac{60}{x}, \dfrac{30}{x}$ during peak, normal and low period respectively.

Service rate μ = 5 min. = $\dfrac{60}{5}$ = 12/hr.

During peak, average W_T customer willing to wait is 15 min. i.e., 0.25 hrs.

Peak:

$$\omega_q = \frac{\lambda}{\mu} \times \left(\frac{1}{\mu - \lambda}\right) = \frac{\frac{110}{x}}{12}\left(\frac{1}{12 - \frac{110}{x}}\right)$$

$$= \frac{110}{12x}\left(\frac{\frac{1}{12x - 110}}{x}\right)$$

$$0.25 = \frac{110}{12x}\left(\frac{x}{12x - 100}\right)$$

$\therefore$ x peak = 13 counters.

Normal: $0.166 = \dfrac{\frac{60}{x}}{12}\left(\dfrac{1}{12 - \frac{60}{x}}\right)$

$\therefore$ x = 7.51 ≈ 8 counters.

Low: $0.25 \times 12 (12x - 30) = 30$

$$36x = 390$$

$$x = 11 \text{ counters}$$

Example 4.24: *Worker come to a tool store room to receive special tool for doing a particular work assign to him. The average time between 2 arrivals is 50s and arrivals are assumed to be Poisson's distribution. Average service time is 40s determine average length of queue, average length of non empty queue, average number of workers in the system including worker being attended. Mean waiting time of arrival, average waiting time of arrival who wait (workers) type of policy is established in other words whether to go for additional tool room which will be combined cost attendant idle time and cost of worker waiting time.*

Assume charges of skilled worker ₹ 4/hr. and that of tool room attendant 0.75/hr.

Solution:

Arrival time $= \lambda = \dfrac{1}{50}$

Service time $= \mu = \dfrac{1}{40}$

$\therefore \quad \delta = \dfrac{\lambda}{\mu} = \dfrac{\frac{1}{50}}{\frac{1}{40}} = \dfrac{40}{50} = 0.8 = 0.8\left(\dfrac{\frac{1}{50}}{\frac{1}{40} - \frac{1}{50}}\right)$

$\qquad = 3.2$ workers

$L_n = \dfrac{\mu}{\mu - \lambda} = \dfrac{\frac{1}{40}}{\frac{1}{40} - \frac{1}{50}} = 5$ workers

Average no. of workers in system,

$L_s = \dfrac{\lambda}{\mu - x} = \dfrac{\frac{1}{50}}{\frac{1}{40} - \frac{1}{50}} = 4$ workers

Average time spends by worker in queue,

$\omega_q = \dfrac{\lambda}{\mu}\left(\dfrac{1}{\mu - \lambda}\right) = 0.8\left(\dfrac{1}{\frac{1}{40} - \frac{1}{50}}\right)$

$\qquad = 0.004/\text{hr.} = 3.3$ min.

$\omega_n = \dfrac{1}{\mu - \lambda} = \dfrac{1}{\frac{1}{40} - \frac{1}{50}} = 0.555/\text{hr.}$

$\qquad = 2.64$ min.

Probability that tool room attendant is idle,

$P_0 = 1 - \dfrac{\lambda}{\mu} = 1 - \dfrac{\frac{1}{50}}{\frac{1}{40}} = 1 - 0.8 = 0.2$

Idle cost for 1 attendant,

$0.2 \times 0.75 \times 8 = 1.2/\text{day}$

Waiting time for workers = Mean waiting time of arrival × No. of workers arriving × Cost of workers

$\qquad = 0.055 \times 72 \times 8 \times 4$

$\qquad = 126.22 ₹/\text{day}$

Waiting time cost of worker > Idle time of attendant

It justifies to adopt additional tool room attendant.

Example 4.25: *Typist of an office of company receives on an 25 letters/day for typing. Typist work 8 hrs. a day and it takes an average 20 min. to type a letter.*

Cost of letter waiting to be made opportunity cost is 80 paise/hr. and cost of equipment and salary of typists = ₹

50/day. What is typist utilization rate? What average now of letters waiting to be typed?

What average waiting time needed to have a letter typed? What is total daily cost of letters to be made?

In order to improve typist service a company has a choice to take lease of 1 or 2 models of an automated type writer daily cost and resulting increasing efficiency of typist are given below:

Table 4.4

Model	Addition Cost/Day	Increasing Typist Efficiency
1	20	50%
2	25	70%

What action should company to take to minimize total waiting cost of letters to be made?

Solution:

Arrival rate = 25/day

(i) Typist utilization rate:

$\qquad \lambda = 25 \quad \mu = 3 \times 8 = 24$

$\therefore \qquad \rho = \dfrac{\lambda}{\mu} = \dfrac{25}{24} = 1.04$

(ii) Average no. of letters waiting to be typed:

$\qquad \omega = \dfrac{\lambda}{\mu}\left[\dfrac{\lambda}{\lambda - \mu}\right] = \dfrac{25}{24}\left[\dfrac{25}{1}\right] = 26.04/\text{day}$

(iii) Average waiting time for a letter to be typed.

$\qquad \omega_s = \dfrac{1}{\mu - \lambda} = \dfrac{1}{25 - 24} = 1$

(iv) Total daily cost of waiting letters:

Average no. of letters in system,

$\qquad L_s = \dfrac{\lambda}{\mu - \lambda} = \dfrac{25}{24 - 25} = 25$ letters

$\therefore$ Total daily cost of waiting = Average no. of letters × Opportunity cost + Cost of equipment and salary

$\qquad = 25 \times 8 \times 0.8 + 50 = ₹\ 210$

Now, consider:

Model I:

$\qquad \lambda = 25, \quad \mu = 24 \times 1.5 = 36$

$\therefore \qquad L_s = \dfrac{\lambda}{\mu - \lambda} = \dfrac{25}{36 - 25} = 2.2727$

Average cost $= 2.2727 \times 8 \times 0.8 + 50 + 20 = ₹\ 84.$

Model II:

$\qquad \lambda = 25; \quad \mu = 24 \times 1.7 = 40.8$

$\qquad L_s = \dfrac{\lambda}{\mu - \lambda} = \dfrac{25}{40.8 - 25} = 1.5822$

$\therefore$ Average cost $= 1.5822 \times 8 \times 0.8 + 50 + 25$

$\qquad = ₹\ 85.12$

So in this case since the average daily cost of Model I is minimum, it is advisable to lease Model I.

Example 4.26: *Arrival of machinist of tool rib is considered to be Poisson's distributed at an average rate of 10 min. The length of service taken by tool grip operator to meet the needs of maintenance shop. The time is assumed to be exponentially distributed with mean 6 min. Find probability that machinist arriving at tool with grip will have to wait average length of queue that former and average time that operator spends in system.*

Manager of shop will install second booth when arrival till have to wait 10 min. or more for service. By how much must rate of arrival be increased in order to justify second booth. Estimate fraction of time tool grip operator will be idle. Probability that there will be 6 or more operators waiting in service.

Solution: $\lambda = 6/hour \qquad \mu = 10/hour$

$P_0 = $ No. of people in system

$\rho = \dfrac{\lambda}{\mu} = \dfrac{6}{10} = 0.6$

$\therefore \qquad P = (1 - P_0) = 0.4 \qquad \qquad \ldots (1)$

$L_q = \dfrac{\lambda^2}{\mu\,(\mu - \lambda)}$

$\quad = \dfrac{6^2}{10\,(10 - 6)} = 0.9$

Expected line length,

$L_s = L_q + \dfrac{\lambda}{\mu} = 0.9 + 0.6 = 1.5$

Expected waiting time $= \dfrac{L_q}{(\rho)^2} = 0.25$

$\omega_q = \dfrac{10}{60} = \dfrac{\lambda}{\mu}\left(\dfrac{1}{\mu - \lambda}\right)$

$\dfrac{10}{60} = \dfrac{\lambda}{10}\left(\dfrac{1}{10 - \lambda}\right)$

$\lambda = 1.42$

Increase in arrival time $= 1.42 - 6 = 4.583/hr.$

$\therefore \qquad \rho = \dfrac{1.42}{10} = 0.142$

$\qquad P = 1 - \rho = 1 - 0.142 = 0.858$

$\qquad P_n = \left(\dfrac{\lambda}{\mu}\right)^n$

Probability that there are more than 6 customer waiting in queue.

P (wait ≥ 5) $= [1 - (1 - \rho)\,(1 + \rho + \rho^2 + \rho^3 + \rho^4 + \rho^5)] = 0.0466.$

Example 4.27: *A xerox machine in an office is operated by a person who does other lot also. The average service time is 6 min./customer. On an average every 12 min. 1 customer arrives or Xeroxing. Find Xerox machine utilisation percentage of time that arrival has not to wait, average time spent by customer in queue average queue length. The arrival rate if management is willing to deploy person exclusively for Xeroxing when average time spent by customer exceeds 15 min.*

Solution: $\lambda = \dfrac{60}{12} = 5/hr.$

$\qquad \mu = \dfrac{60}{6} = 10/hr.$

$\qquad \rho = \dfrac{\lambda}{\mu} = \dfrac{5}{10} = 0.5$

$\qquad P_0 = 1 - \rho = 1 - 0.5$

$\qquad P_0 = 0.5$

Average time spent by customer in queue

i.e., $\omega_q = \dfrac{\lambda}{\mu}\left(\dfrac{1}{\mu - \lambda}\right) = \dfrac{5}{10}\left(\dfrac{1}{10 - 5}\right) = 0.1/hr.$

$\qquad = 0.6$ min.

$\qquad \omega_s = \dfrac{1}{\mu - \lambda} = \dfrac{1}{10 - 5} = 0.2/hr. = 0.2 \times 60$

$\qquad = 12$ min.

Average queue length,

$\qquad L_q = \dfrac{\lambda^2}{\mu\,(\mu - \lambda)} = \dfrac{5^2}{10\,(10 - 5)} = 0.5/hr$

Now, $\omega_q = 15$ min. $= 0.25$ hr.

$\qquad 0.25 = \dfrac{\lambda}{\mu}\left(\dfrac{1}{\mu - \lambda}\right)$

$\qquad 0.25 = \dfrac{\lambda}{10} = \left(\dfrac{1}{10 - \lambda}\right) = 7.14/hr.$

Example 4.28: *The trucks arrive randomly at a stock yard with 7 trucks/hr. A crew of 4 operative can load a truck in 5 min. Trucks waiting in queue to unload are paid a waiting charge at rate of ₹ 60/hr. operatives are paid a wage rate ₹ 20/hr. It is possible to augment crew strength to 2 or 3 of 4 operatives/crew. When unloading time will be 4 min. and 3 min./ truck respectively. Find optimal crew size.*

Solution:

Crew length,

$\qquad \lambda = 7$ trucks/hr.

$\qquad \mu = \dfrac{60}{6} = 10$ trucks/hr.

Waiting time of truck,

$$\omega_q = \frac{\lambda}{\mu}\left(\frac{1}{\mu - \lambda}\right) = \frac{7}{10}\left(\frac{1}{10 - 7}\right) = 0.233/hr.$$

$$= 14 \text{ min.}$$

Total waiting time for truck = No. of truck arriving/hr. × Waiting time

$$= 0.233 \times 7 = 1.633 \text{ hr.}$$

Waiting time cost/hr. ₹ 97.86 = 1.633 × 60

$\therefore$ Operative charges/hr. = 20 × 4 = 80

Total cost = 97.86 + 80 = ₹ 177.86

For 2 crew strength,

$$\lambda = 7 \text{ trucks/hr.}$$

$$\mu = \frac{60}{4} = 15 \text{ trucks/hr.}$$

$$\omega_q = \frac{7}{13}\left(\frac{1}{15 - 7}\right) = 0.0583$$

Total waiting time = 0.4081/hr.

Waiting time/hr. = 0.4081 × 60

$$= ₹ 24.486$$

Operative charges/hr. = 20 × 8 = ₹ 160

Total cost = ₹ 184.5

For 3 crew strength,

$$\lambda = 7 \text{ trucks/hr.}$$

$$\mu = \frac{60}{3} = 20 \text{ trucks/hr.}$$

$$\omega_q = \frac{7}{20}\left(\frac{1}{20 - 7}\right) = 0.025/hr.$$

Total waiting time = 1.175 hrs

Waiting time cost/hr. = ₹ 10.5

Operative charges/hr. = ₹ 240

Total cost = 250.5

Example 4.29: *Patient arrival at clinic at Poission rate of 40/hr. waiting row does not accommodate more than 14 patients. Examination time per patient is exponential with mean time is 20/hr.*

(i) Find mean effective arrival rate to clinic.

(ii) What is probability that arriving patient does not wait?

(iii) What is expected waiting time until patient discharge from clinic?

Solution: Data given:

$$\lambda = 40/hr.$$

$$\mu = 20/hr.$$

$$\therefore \quad \rho = \frac{40}{20} = 2$$

$$N = 15 \qquad \text{(1 patient inside clinic)}$$

(i) Effective arrival rate, $\lambda' = \lambda(1 - P_N)$

$$P_N = \delta^N \cdot P_0$$

$$P_0 = \left[\frac{1 - \delta}{1 - \delta^{N+1}}\right] = \left[\frac{1 - 2}{1 - 2^{16}}\right]$$

$$P_0 = 1.5259 \times 10^{-5}$$

$$\therefore \quad P_N = 2^{15} \times 1.529 \times 10^{-5} = 0.5$$

$$\lambda' = 40(1 - 0.5) = 20$$

(ii)
$$P_0 = \left[\frac{1}{1 + N}\right] \rho = 1 = \frac{1}{16} = 0.0625$$

(iii)
$$W_s = \frac{L_s}{\lambda'}$$

$$L_s = \frac{\rho\,[1 - (1 + N)'\,\rho^N + N\,\rho^{N+1}]}{(1 - \rho)(1 - \rho^{N+1})}$$

$$= \frac{2\,[1 - 16 \times 2^{15} + 15 \cdot 2^{16}]}{(1 - 2)(1 - 2^{16})}$$

$$\therefore \quad L_s = 14$$

$$\therefore \quad W_s = \frac{14}{20} = 0.7 \text{ hrs.}$$

Example 4.30: *A barber shop has space to accommodate only 10 customers and it can serve only 1 person at a time.*

If customer comes to his shop and finds it is full, he is going to next shop, customers randomly come.

Find:

(i) Probability that arriving customer will not wait (P_0).

(ii) Probability of N number of customer in system (P_N) and expected waiting time until customer is discharge from shop (ω_s).

Solution:

$$n = 10, \; N = 10^{+1} = 11,$$

$$\mu = 5 \text{ min./customer} = 60/5 = 12/hr.$$

$$\therefore \quad \rho = \frac{\lambda}{\mu} = \frac{10}{12} = 0.833$$

$$P_0 = \left[\frac{1 - \rho}{1 - e^{N+1}}\right] = \left[\frac{1 - 0.833}{1 - 0.833^{12}}\right]$$

$$\therefore \quad P_0 = 0.188$$

$$P_N = \delta^N \cdot P_0 = 0.025$$

$$\lambda' = \lambda(1 - P_N) = 0.975$$

$$\omega_s = \frac{L_s}{\lambda'}$$

$$L_s = \frac{\rho\,[1 - (1 + N)\,\rho^N + N\,\rho^{N+1}]}{(1 - \rho)(1 - \rho^{N+1})}$$

$$= 3.48$$

$$\therefore \quad \omega_s = \frac{3.48}{9.75}$$

$$= 0.357$$

Example 4.31: *If for a period of 3 hrs. in a day train arrives at the yard 25 min. (every 25 min.) but service time is 40 min. Calculate probability that yard is empty. Average no. of trains at yard. Capacity of yard is limited to 4 only.*

Solution:

$$\lambda = \frac{60}{25} = 24$$

$$\mu = \frac{60}{40} = 1.5, \ N = 4$$

$$\delta = \frac{\lambda}{\mu} = \frac{2.4}{1.5} = 1.6$$

Now the probability that yard is empty

$$\therefore \quad P_0 = \frac{(1-\rho)}{[1-(\rho)^{n+1}]} = \frac{1-1.6}{1}$$

$$P_0 = \frac{(1-\rho)}{1-(\rho)^{n+1}} = \frac{1-1.6}{1-(1.6)^{4+1}} = 0.063$$

Now average no. if trains at yard is

$$L_s = P_0 \cdot \sum_{N=0}^{4} N \cdot \rho^N$$

$$= P_0 [0 + 1.6 + 2(1.6)^2 + 3(1.6)^3 + 4(1.6)^4]$$

$$= 0.063 [0 + 1.5 + 5.12 + 12.288 + 26.2144]$$

$$= 0.063(45.22) = 2.85 \approx 3$$

$\therefore$ Average no. of trains at yard are 3.

Example 4.32: *If for a period of 2 hrs./day (9 am to 11 am) bus arrives at a station every 20 min. but service time is 40 min. Calculate probability that the place is empty. Average no. of bus at place line capacity of place is limited to 4 less only.*

Solution:

$$\lambda = 3/hr. \qquad \mu = \frac{60}{40} = 1.5/hr.$$

$$\delta = \frac{\lambda}{\mu} = \frac{3}{1.5} = 2$$

(i) Probability that the place is empty,

$$P_0 = \left[\frac{1-\rho}{1-\rho^{N+1}}\right] = \left[\frac{1-2}{1-2^{4+1}}\right]$$

$$= \frac{-1}{-31} = 0.0322$$

(ii) Average no. of buses at the place,

$$L_s = P_0 \sum_{n=0}^{N} n \times \rho^n$$

$$P_0 = \{0 + \rho + 2\rho^2 + 3\rho^3 + 4\rho^4\}$$

$$= 0.0322 \{0 + 2 + (2.2)^2 + (3.2)^3 + (4.2)^4\}$$

$$= 3.1566 \text{ buses}$$

Example 4.33: *At a car wash service facility car arrives for service in caution distinction with mean of 5/hr. Time for washing and cleaning each car has exponential distribution with mean of 10 min./car. The facility cannot handle more than 1 car at a time and has 5 parking spaces available.*

Find:

(i) Effective arrival rate.

(ii) Probability that car will get service immediately upon its arrival.

(iii) Expected no. of parking spaces occupied.

Solution:

Given data: $\lambda = 5, \quad \mu = 6$

(i) Effective arrival rate:

$$\rho = \frac{\lambda}{\mu} = \frac{5}{6} = 0.833$$

$$P_0 = \left[\frac{1-\rho}{1-\rho^{N+1}}\right] = \left[\frac{1-0.833}{1-(0.833)^6}\right] = 0.2507$$

$$P_n = \rho^N \cdot P_0 = (0.8339)^5 \times 0.2507 = 0.1005$$

$$\lambda' = \lambda(1 - P_n) = 5(1 - 0.1)$$

$$= 4.5 \text{ cars}$$

(ii) Probability that car will get immediate service,

$$P_0 = \left(\frac{\lambda}{\mu}\right)^0 \cdot P_0 = 0.2507.$$

(iii) Expected no. of parking spaces required,

$$L_s = P_0(0 + \rho + 2\rho^2 + 2\rho^3 + 4\rho^4 + 5\rho^5)$$

$$= 0.2507(0 + 0.833 + 2 \times (0.833)^2 + 3 \times (0.833)^3 + 4 \times (0.833)^4 + 5 \times (0.833)^5$$

$$= 1.9777$$

$$\therefore \quad L_q = L_s - \frac{\lambda'}{\mu} = 1.977 - \frac{4.5}{6} = 1.277 \approx 2.$$

Example 4.34: *In a machine shop there are 4 machines. Mean time between service requirement is 4 hrs. for each machine and forms exponential distribution. The main repair time is 2 hrs. and also follows same distribution pattern. Mechanic down-time cost 20/hr. and mechanic cost is 60/day. Find expected no. of operating machines. Determine expected downtime cost/day. What would be economical to engage 2 mechanics each repairing only 2 machines?*

Solution:

$$\text{Now,} \qquad \lambda = \frac{1}{2} = 0.5/hr.$$

$$\mu = \frac{1}{4} = 0.25/hr.$$

$$\therefore \qquad \delta = \frac{\lambda}{4} = 2$$

No. of machines = 4

$$P_0 = \cfrac{1}{\displaystyle\sum_{M=0}^{4} \frac{M!}{(M-N!)^n}\, e^n}$$

$$= \cfrac{1}{\dfrac{4!}{4!\,(e^0)}} + \frac{4!}{3!\,e^1} + \frac{4!}{2!\,e^2} + \frac{4!}{1/e^3} + \frac{4!}{e^4}$$

$$= \cfrac{1}{\dfrac{4!}{4!\,(e^0)} + \dfrac{4!}{3!\,(e^1)} + \dfrac{4!}{2!\,e^2} + \dfrac{4!}{11\,e^3} + \dfrac{4!}{e^4}}$$

$$= \frac{1}{1 + 1.4715 + 1.624 + 1.195 + 0.4396}$$

$$= 0.1745$$

Expected no. of operating machine:

$$L_s = M - \frac{\mu}{\lambda}(1 - P_0) = 4 - \frac{0.25}{0.5}(1 - 0.1745)$$

$$= 3.58 \approx 4$$

Now this is no. of operating machines.

Assume that 8 hrs. working in a day.

∴   No. of expected downtime cost/day

= Total working hrs. in day × No. of operating machine
× Expected No. of B.D. machine

$$= 8 \times 4 \times 20 = ₹\ 640/day$$

For M = 2:

$$P_0 = \cfrac{1}{\dfrac{2!}{2!\,e^0} + \dfrac{2!}{e^1} + \dfrac{2!}{e^2}} = \frac{1}{1 + 5.436 + 14.778}$$

$$P_0 = 0.04714$$

∴ $$L_s = 2 - 0.5\,(1 - 0.04714)$$

$$= 1.5 \approx 2$$

∴   Expected downtime cost/day

$$= 8 \times 2 \times 20 = ₹\ 320/day$$

Total cost of 2 mechanics = 60 × 2 + 320 = ₹ 440

Total cost for 1 mechanic is

60 + 640 = ₹ 700

It will be economical to engage 2 mechanic each repairing only 2 machines.

Example 4.35: *In a machine shop there are 2 identical machine products arrive for machine at average rate of 4/hr. Average machine time 24 min./product/month complains of loss of production time suggest the installation of third machine. What is the average waiting time/ product under present circumstances? What is it likely to be of third machine is installed.*

Solution:

$$\lambda = 4/hr.\ \mu = \frac{60}{24} = 2.5/hr.\ C = 2$$

Here, average no. of products waiting to be served (L_q)

$$L_q = L_s - \frac{\lambda}{\mu} = \frac{\lambda \cdot \mu \left(\dfrac{\lambda}{\mu}\right)^c}{(c-1)!\,(c\mu - \lambda)^2}\, P_0 + \frac{\lambda}{\mu}$$

$$\rho = \frac{\lambda}{\mu \times 2} = 0.8 = P_0$$

∴ $$L_q = \frac{4 \times 2.5 \left(\dfrac{4}{2.5}\right)^2}{1!\,(2 \times 2.5 - 4)^2} \times 0.8 + \frac{4}{2.5} = 0.4096$$

∴   Average waiting time/product:

$$\omega_q = \frac{L_q}{\lambda} = \frac{0.4096}{4} = 0.1024 = 6\ min.$$

Now, of third machine is installed, there will be

$$L_q = \frac{2 \times 2.5 \left(\dfrac{4}{2.3}\right)^3}{2!\,(3 \times 2.5 - 4)^2} \times 0.5$$

$$= 0.83$$

$$\omega_q = \frac{L_q}{\lambda}$$

$$= \frac{0.83}{4} = 0.20 = 3\ min.$$

Example 4.36: *A mechanic repairs 6 machines mean time between service requirements is 4 hrs. of each machine and form exponential distribution pattern. Repair time is 1.5 hrs. or (1.30) and also follow same distribution pattern machine down time cost ₹ 30/hr. Mechanic cost of ₹ 70/day find expected no. of operative machine. Expected domestic cost per day (working hrs.) would it be economical to engage two machines each repaired only by 3 mechanics.*

Solution:

$$\lambda = \frac{1}{4}\,/\,hr.$$

$$M = 6$$

$$\mu = \frac{1}{1.5} = 0.667/hrs.$$

$$R = 0.3748$$

$$P_0 = \cfrac{1}{\left[\dfrac{6!}{6!}(0.3748)^0 + \dfrac{6!}{5!}(0.3748) + \dfrac{6!}{4!}(0.3748)^2 + \dfrac{6!}{3!}(0.3748)^3 \right.}$$
$$\left. + \dfrac{6!}{2!}(0.3748)^4 + \dfrac{6!}{1!}(0.3748)^5 + \dfrac{6!}{0!}(0.3748)^6 \right]}$$

$$= 0.035$$

$$L_s = M - \frac{\mu}{\lambda}(1 - P_0) = 6 - \frac{0.667}{0.25}(1 - 0.035)$$

$$= 3.42$$

$$= 4 \text{ machines}$$

$\therefore$ Total machines under operating, $6 - 4 = 2$ machines.

Expected downtime/day = (Working hrs.) $\times$ (Expected breakdowns)

$$= 8 \times 4 = 32$$

Breakdown cost = $32 \times 30 = ₹\,960/\text{day}$

Total cost = $960 + 70 + 1 = ₹\,1030/\text{day}$

When there are 3 mechanics, each repairing 2 machines i.e.,

When M = 2:

$$P_0 = \cfrac{1}{\left[\frac{2!}{(2-0)}(0.3748)^0 + \frac{2!}{(2-1)}(0.3748)^1 + \frac{2!}{(2-2)}(0.3748)^2\right]}$$

$$= 0.492$$

Expected no. of machines in system,

$$L_s = M - \frac{\mu}{\lambda}(1 - P_0) = 2 - \frac{0.667}{0.25}(1 - 0.492)$$

$$= 0.644$$

$\therefore$ Expected downtime/day = (Working hrs./day) $\times$ (No. of machines)

$$= 8 \times (0.644 \times 2) = 10.3 \text{ hrs./day}$$

$\therefore$ Total cost with three mechanics is given as

Total cost = $3 \times 70 + 10.3 \times 30$

$$= 210 + 309 = ₹\,519$$

As total cost by using three mechanics is less than total cost by use of two mechanics, hence it is economical to engage three mechanics.

Example 4.37: *In a machine shop there are three identical machine products arrived for machine at average rate of 7 products/hr. Machining time is 21 min./product. Production Manager complains of production time and suggest installation of 4 machines. What is average waiting time product under present circumstances. What is it likely to be if fourth machine is installed. Assume position distribution for arrival pattern.*

Solution:

$$\lambda = \text{arrival rate} = 7/\text{hr.}$$

$$\mu = \text{service rate} = 2.85/\text{hr.}$$

$$M = 3/\text{machines}$$

$$\rho = 2.45$$

$$L_q = M - \left(\frac{\mu + \lambda}{\lambda}\right)(1 - P_0)$$

$$P_0 = \cfrac{1}{\displaystyle\sum_{n=0}^{M} \frac{M!}{(m-n)!}\left(\frac{\lambda}{\mu}\right)^n}$$

$$= \cfrac{1}{\left(\frac{3!}{3!} \times \rho^0 + \frac{3!}{1} \times (2.45) + \frac{3!}{1!} \times (2.45)^2 + \frac{3!}{0!} \times (2.45)^3\right)}$$

$$P_0 = 0.0075$$

$$L_q = 3 - \left(\frac{2.45 + 7}{7}\right) \times (1 - 0.0075) = 1.66$$

$$\omega_q = \frac{L_q}{\mu(1 - P_0)} = \frac{1.66}{2.85 \times (1 - 0.0075)} \times 60$$

$$= 33.94 \text{ min.}$$

Now, M = 4 machines

$$P_0 = \cfrac{1}{\left(\frac{4!}{4!} \times (2.45)^5 + \frac{4!}{3!} \times (2.45)^1 + \frac{4!}{2!} \times (2.45)^2 + \frac{4!}{2!} \times (2.45)^3 + \frac{4!}{0!} \times (2.45)^2\right)}$$

$$= 0.00077$$

$$L_q = 4 - \left(\frac{2.85 + 7}{7}\right) \times (1 - 0.00077) = 2.59$$

$$\omega_q = \frac{2.59}{2.85(1 - 0.00077)} \times 60 = 54.57 \text{ min.}$$

Example 4.38: *A telephone exchange has two long distance operators. Telephone company finds that during peak period long distance calls arrive at an average rate of 15/hr. The longest service of these calls is exponential distribution with mean length 5 min.*

(i) What is probability that a subscriber will have to wait for his long distance call during peak hours of a day?

(ii) If the subscriber will wait and serve in turn, what is expected waiting time.

Solution:

$$\lambda = 15/\text{hr}; \ \mu = \frac{60}{5} = 12/\text{hr.}$$

(i) Probability that subscriber will have to wait,

$$P_0 = \rho = \frac{\lambda}{\mu}$$

$$= \frac{15}{12} = 0.0625$$

$\therefore$ $P(\omega > 0) = 1 - P_0 = 1 - 0.0625 = 0.375$

(ii) Expected waiting time (ω_q),

$$\omega_q = \frac{\lambda}{\mu(\mu - \lambda)} = \frac{15}{(15 - 12)} = 0.41 \text{ hr.}$$

$$= 25 \text{ min.}$$

Example 4.39: *An Insurance Company has 3 claims adjusted in its main office. Customers-arrive @ 5/hr. for setting claims against the company. Service time is of 2.5 min. complaints are processed on FCFS basis. Calculate average no. of customers in the system average time customer spends in system average length of queue. Average waiting time for customer number of hrs./week spend on performing job.*

Solution:

Utilization rate,

$$\rho = \frac{\lambda}{k \cdot \mu} = \frac{5}{3 \times 2.4} = 0.674$$

∴ No. of hrs. each day Insurance company person spends doing his job 8 hrs.

$$x = \rho \times 8 = 0.094 \times 8 = 5.5556$$

On an average person is busy for 5 days

$$= 5 \times x = 27.78$$

$$P_0 = \cfrac{1}{\displaystyle\sum_{n=0}^{k-1} \frac{\left(\frac{\lambda}{\mu}\right)^n}{n!} + \frac{\left(\frac{\lambda}{\mu}\right)^c}{c!} \cdot \frac{c\mu}{c\mu - \lambda}}$$

$$= \cfrac{1}{\left[1 + \frac{\lambda}{\mu} + \frac{1}{2}\left(\frac{\lambda}{\mu}\right)^2\right] + \frac{\left(\frac{\lambda}{\mu}\right)^3}{6} \cdot \frac{3\mu}{3\mu - \lambda}}$$

$$= \cfrac{1}{\left[1 + \frac{5}{2.4} + \frac{1}{2}\left(\frac{5}{24}\right)^2\right] + \left[\frac{\left(\frac{5}{24}\right)^3}{6} \cdot \frac{3 \times 2.4}{3 \times 2.4 - 5}\right]} = 0.098$$

(i) Average no. of customer,

$$L_s = \frac{\lambda \cdot \mu \left(\frac{\lambda}{\mu}\right)^c}{[c - 1]! \, [c\mu - \lambda]^2} \cdot \delta_0 + \frac{\lambda}{\mu} = 2.8$$

(ii) Average time customer spends in the system,

$$\omega_s = \frac{L_s}{\lambda} = \frac{2.8}{5} = 0.56 \text{ hrs.} = 33.6 \text{ min.}$$

(iii) Average length of queue,

$$L_q = L_s - \frac{\lambda}{\mu} = 2.8 - \frac{5}{2.5}$$

∴ $$= 0.8$$

(iv) Average waiting time for a customer,

$$\omega_q = \frac{L_q}{\lambda} = \frac{0.8}{5} = 0.16 \text{ hrs.} = 9.6 \text{ min.}$$

Example 4.40: *Super market has two salesmen ringing up sales at the counter. If service time for each customer is exponentially with 4 min. and if people arrive in Poisson fashion at rate of 10/hr. What is expected percentage of idle time of each salesman? Find average queue length and average no. of unit in system.*

Solution:

$$P = (n \geq k)$$
$$K = \text{Parallel service channels} = 2$$
$$\lambda = 10/\text{hr.}$$
$$\mu = 13/\text{hr.}$$
$$1 - \rho = 0.67$$
$$\rho = \frac{10}{2 \times 15} = 0.33$$

$$P_0 = \cfrac{1}{\displaystyle\sum_{n=0}^{n=k-1} \frac{1}{n!}\left(\frac{\lambda}{\mu}\right)^n + \frac{1}{K!}\left(\frac{\lambda}{\mu}\right)^K \left(\frac{K \times \mu}{K \cdot \mu - \lambda}\right)}$$

$$= \cfrac{1}{\left[\frac{1}{0!}\left(\frac{10}{15}\right)^0 + \frac{1}{2!}\left(\frac{10}{15}\right)^2 \left(\frac{2 \times 15}{2 \times 15 - 10}\right) + \frac{1}{1!}\left(\frac{10}{15}\right)^1\right]} = 0.43$$

Probability that customer has to wait,

$$P(n \geq K) = \frac{\mu \left(\frac{\lambda}{\mu}\right)^K}{(K - 1)! \, (K\mu - \lambda)} \cdot P_0$$

$$= \frac{15\left(\frac{10}{15}\right)^2}{(2 - 1)! \, (2 \times 15 - 10)} \times 0.43 = 0.143$$

Probability of no customer in the system is P_0, which means that both salesmen are idle. Now, we have to set than one salesman is idle.

$$P_n = \frac{1}{n!}\left(\frac{\lambda}{\mu}\right)^n \cdot P_0 \quad (n < c)$$

$$P_1 = \frac{1}{11}\left(\frac{10}{15}\right)^1 0.43 = 0.287.$$

∴ Expected percentage of idle time of each salesman

$$= 2 \times P_0 + 1 \times P_1$$
$$= 2 \times 0.43 + 1 \times 0.287 = 1.147$$

Average no. of unit in system,

$$L_s = \frac{\lambda - \mu \left(\frac{\lambda}{\mu}\right)^K}{(K - 1)! \, (K\mu - \lambda)^2} P_0 + \frac{\lambda}{\mu}$$

$$= \frac{10 \times 15 \left(\frac{10}{15}\right)^2}{(2 - 1)! \, (2 \times 15 - 10)^2} \times 0.43 + \frac{10}{15}$$

$$= 0.738$$

Average queue length,

$$L_q = L_s = \text{(Average no. being served)}$$

$$= L_s - \frac{\lambda}{\mu}$$

$$= 0.738 - \left(\frac{10}{15}\right)$$

$$= 0.071.$$

QUESTIONS FOR PRACTICE

1. Explain with suitable examples about the queue. Why do you consider the study of waiting line as an important aspect? (Refer Section 4.1)

2. Explain with suitable examples about Poisson arrival pattern and exponential service pattern. (Refer Section 4.4.1)

3. Explain the various types of queues by means of a sketch and also give the situations for which each is suitable. (Refer Section 4.4)

4. Customers arrive at one window drive in a bank according to a Poisson distribution with a mean of 10 per hour. Service time per customer is exponential with a mean of 5 minutes. The space in front of the window, including that for the serviced car can accommodate a maximum, of three cars. Other cars can wait outside the space.

   (a) What is the probability that an arriving customer can drive directly to the space in front of the window?

   (b) What is the probability that an arriving customer will have to wait outside the indicated space?

   (c) How long an arriving customer is expected to wait before starting service?

   (d) How much space should be provided in front of the window so that all the arriving customers can wait in front of the window at least 90 percent of the time? (Refer Ex. 4.2)

5. A barber with a one-man shop takes exactly 25 minutes to complete one hair cut. If customers arrive in a Poisson fashion at an average rate of every 40 minutes. How long on the average must a customer wait for service? (Refer Ex. 4.5)

6. At a public telephone booth in a post office arrivals are considered to be Poisson with an average inter-arrival time of 12 minutes. The length of phone call may be assumed to be distributed exponentially with an average of 4 minutes. Calculate the following:

   (a) What is the probability that a fresh arrival will not have to wait for phone?

   (b) What is the probability that an arrival will have to wait more than 10 minutes before the phone is free?

   (c) What is the average length of queues that form from time to time?

   (d) What is the fraction of time is the phone busy?

   (e) What is the probability that an arrival that goes to the post office to make a phone call will take less than 15 minutes to complete his job?

   (f) The telephone company will install a second booth when convinced that an arrival would expect to have to wait at least 5 minutes for the phone? (Refer Ex. 4.2)

7. The mean rate of arrival of planes at an airport during the peak period is 20 per hour, but the actual number of arrivals in an hour follows the Poisson distribution. The airport can land 60 planes per hour on an average in good weather, or 30 per hour in bad weather, but the actual number landed in any hour follows a Poisson distribution with the respective averages. When there is congestion, the planes are forced to fly over the field in the stock awaiting the landing of other planes that arrived earlier.

   (a) How many planes would be flying over the field in the stack on an average in good weather and in bad weather?

   (b) How long a plane would be in the stack and the process of landing in good and bad weather?

   (c) How much stack and landing time to allow so that priority to land out of order would have to be requested only one time in twenty.

8. The belt snapping for conveyors in an open cast mine occur at the rate of 2 per shift. There is only one hot plate available for vulcanizing; and it can vulcanize on an average 5 belts snap per shift:

(a) What is the probability that when a belt snaps, the hot plate is readily available?

(b) What is the average number in the system?

(c) What is waiting time of an arrival?

(d) What is the average waiting time plus vulcanizing time?

9. Define: **[May 2017,Dec. 2011,8M]**

(i) Queue length

(ii) Traffic intensity

(iii) Service channels

(iv) Service in priority   (Refer Section 8.1.2.)

10. Explain : Single Channel Poisson Arrivals with exponential service, infinite population method.

(Refer Section 8.1.2.)   **[Dec. 2016, May 2012, 6M]**

11. A repair shop attended by a single mechanic has an average of 4 customers per hour who bring small appliances for repair. The mechanic inspects them for defects and quite often can fix them right away or otherwise render a diagnosis. This takes him 6 minutes on an average. Arrivals are Poisson and service time has the exponential distribution.

(i) Find the proportion of time during which the shop is empty.

(ii) Find the probability of finding at least one customer in the shop.

(iii) The average number of customers in the system.

(iv) The average time, including service, spent by a customer.   **[Dec 2015, May 2012, 10]**

(Refer Section 8.4.1 [Example 8.1].)

12. The type of policy to be established. Determine whether to go in for an additional tool store room attendant which will minimize the combined cost of attendant's idle time and the cost of workers waiting time. Charge of worker is ₹ 4 per hour and that of tool store room attendant is ₹ 0.75 per hour.

(Refer Section 8.4.1 [Example 8.1])

[May 2016, Dec 2012, 8M]

13. Workers come to tool store room to receive special tools for accomplishing a project. The average time between two arrivals is 60 seconds and the arrivals are in Poisson distribution. The average service time, is 40 seconds. Determine.   **[May 15, Dec 12,8M]**

(i) Average queue length.

(ii) Average length of non-empty queues.

(iii) Average number of workers in system including the worker being attended.

(iv) Mean waiting time of an arrival.

(v) Average waiting time of an arrival (worker) who waits and.

(vi) The type of policy to be established. Determine whether to go in for an additional tool store room attendant which will minimize the combined cost of attendant's idle time and the cost of workers waiting time. Charge of worker is ₹ 4 per hour and that of tool store room attendant is ₹ 0.75 per hour.

(Refer Section 8.4.1 [Example 8.1])

14. Explain the following with reference to queuing models: **[Dec 2015, May 2013, 6M]**

(1) M/M/1

(2) Traffic intensity

(3) Service discipline (Refer Section 8.1.2.)

15. In a railway marshalling yard, good's trains arrive at rate of 30 trains per day. Assuming that the inter arrival time follows an exponential distribution and the service time distribution is also exponential with an average 36 min. Calculate the following:

(1) The mean queue size.

(2) The probability that the queue size exceeds 10.

(3) If the input trains increase to an average 33 per day, what will be the change in (1) and (2)

[Dec. 2017,May 2013,10M]

(Refer Section 8.4.1 [Example 8.8].)

16. Define:　　　　　　　　　　**[May 16, Dec. 13, 8M]**

    (i) Queue length (ii) Traffic intensity (iii) Service channels (iv) Service in priority

    　　　　(Please refer to Sections 8.1.2 and 8.2.)

17. A super market has two sales girls at the sales counter. If the service time for each customer is exponential with a mean of 4 minutes and if people arrive in a Poisson fashion at the rate of 10 an hour, then calculate the　　　　**[May 17, Dec 13, 8M]**

(i) Probability that a customer has to wait for service.

(ii) Expected percentage of the idle time for each sales girl.

(iii) If a customer has to wait, what is the expected length of his waiting time?

　　　　(Please refer to Section 8.4.4 [Example 8.11])

5.1 INTRODUCTION

The problem of replacement arises when any one of the components of productive resources, such as machinery, building and men deteriorates due to time or usage. The examples are :

- A machine, which is purchased and installed in a production system, due to usage some of its components wear out and its efficiency is reduced.

- A building in which production activities are carried out, may leave cracks in walls, roof etc. and needs repair.

- A worker, when he is young, will work efficiently, as the time passes becomes old and his work efficiency falls down and after some time he will become unable to work.

5.1.1 Importance of Replacement Analysis

[Dec. 18, May 17, Dec. 09, May 12, 6M]

- In general, when any production facility is new, it works with full operating efficiency and due to usage or time, it may become old and some of its components wear out and the operating efficiency of the facility falls down. To regain the efficiency, a remedy by, namely **Maintenance** is to be attended.

- The act of maintenance consists of replacing the worn out part or oiling or overhauling or repair etc. In modern industrial scene, the presence of highly sophisticated machinery in manufacturing system will bother the manager, when any one of the facilities goes out of order or breakdown. Because of the breakdown of one of the facility, the entire production system may be affected.

- This is particularly true in batch manufacturing system and continuous manufacturing system. The loss of production hours is more expensive factor for the manufacturing unit. Hence, the management must take interest in maintaining the production facility properly, so that facility's available time will be more than the down time.

- All the production facilities are subjected to deterioration due to their use and exposure to the environmental conditions. The process of deterioration, if unchecked or neglected, it culminates and makes the facility useless. Hence, the management has to check the facilities periodically, and keep all the facilities in operating condition.

- Once the maintenance is attended, the efficiency may not be regained to previous level but a bit less than that of previous level. For example, if the operating efficiency is 95% and due to deterioration, the efficiency reduces to 90%, after maintenance, it may regain to the level of 93%. Once again due to usage the efficiency falls down and the maintenance is to be attended. This is an ongoing business of the management.

- After some time, the efficiency reduces to such a level, the maintenance cost will become very high and due to less efficiency the unit production cost will be very high and this is the time the management has to think of replacing the facility. This may be well explained by means of a Fig.. Referring to Fig. 5.1, the operating efficiency at the beginning is 95%. When first maintenance is attended, it is reduced to 93%. In the second maintenance it is reduced to 80%. Like this the facility deteriorates, and finally the operating efficiency reduces to 50%, where it is not economical to use the facility for further production, as the maintenance cost will be very high, and the unit production cost also increases, hence the replacement of the facility is due at this stage. In this chapter, we will discuss the mathematical models used for finding the optimal replacement time of facilities.

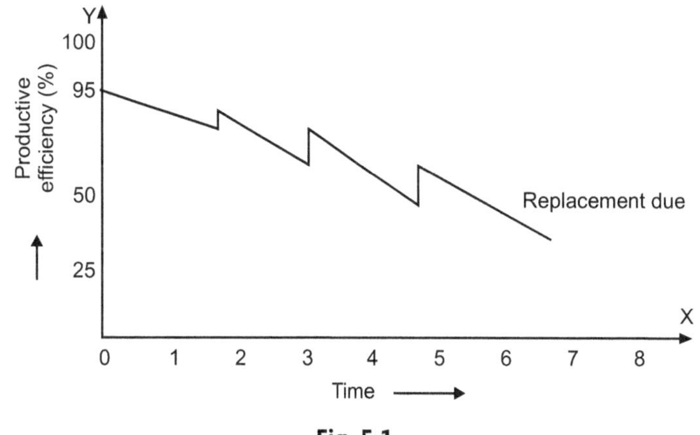

Fig. 5.1

- Thus the problem of replacement is experienced in systems where machines, individuals or capital assets are the main production or job performing units. The characteristics of these units is that their level of performance or efficiency decreases with time or usage and one has to formulate some suitable replacement policy regarding these units to keep the system upto some desired level of performance.

We May have to take Different Types of Decision such as :

- We may decide whether to wait for complete failure of the item (which may result in some losses due to deterioration or to replace earlier at the expense of higher cost of the item.

- The expensive item may be considered individually to decide whether we should replace now or, if not, when it should be reconsidered for replacement.

- Whether the item is to be replaced by similar type of item or by different type for example item with latest technology.

- The problem of replacement is encountered in the case of both men and machines. Using probability, it is possible to estimate the chance of death or failure at various ages. The main objective of replacement is to help the organization for maximizing its profit or to minimize the cost.

5.2 FAILURE MECHANISM OF ITEMS

[Dec. 17, May 12, 6M]

- The word failure has got a wider meaning in **Industrial Maintenance** than what it has in our daily life.

- We can categorize the failure in two classes. They are
  1. Gradual failure and
  2. Sudden failure.

1. Gradual Failure :

- In this class as the life of the machine increases or due continuous usage, due to wear and tear of components of the facility, its efficiency deteriorates due to which the management can experience :

  ➢ Progressive increase in maintenance expenditure or operating costs,

  ➢ Decreased productivity of the equipment and

  ➢ Decrease in the value of the equipment i.e. resale value of the equipment/facility decreases.

- Examples of this category are Automobiles, Machine tools etc.

2. Sudden Failure :

- In this case, the items ultimately fail after a period of time. The life of the equipment cannot be predicted and is some sort of random variable. The period between installation and failure is not constant for any particular type of equipment but will follow some frequency distribution, which may be :

Once again the sudden failure may be classified as:

(i)   Progressive failure,

(ii) Retrogressive failure and

(iii) Random failure.

(i) Progressive Failure : In this case, probability of failure increases with the increase in life of an item. The best example is electrical bulbs and computer components. It can be shown as in Fig. 5.2 (a).

(ii) Retrogressive Failure : Some items will have higher probability of failure in the beginning of their life, and as the time passes chances of failure becomes less. That is the ability of the item to survive in the initial period of life increases its expected life. The examples are newly installed machines in production systems, new vehicles, and infant baby (The probability of survival is very less in infant age, but once the baby get accustomed to nature, the probability of failure decreases). This can be shown as in Fig. 5.2 (b).

(iii) Random Failure : In this class, constant probability of failure is associated with items that fail from random causes such as physical shocks, not related to age. In such cases all items fail before aging has any effect. This can be shown as in Fig. 5.2. (c). Example is vacuum tubes.

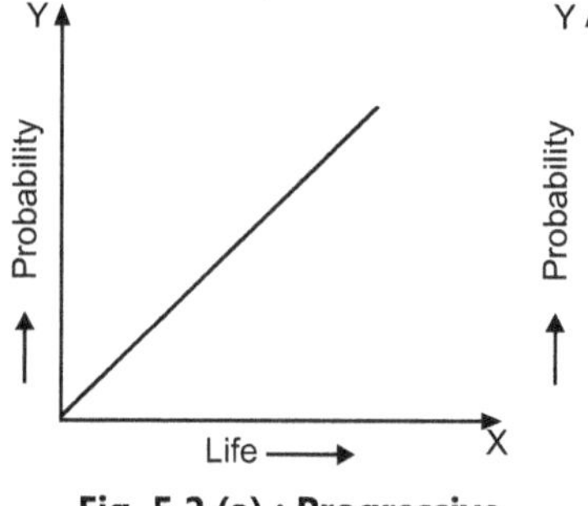

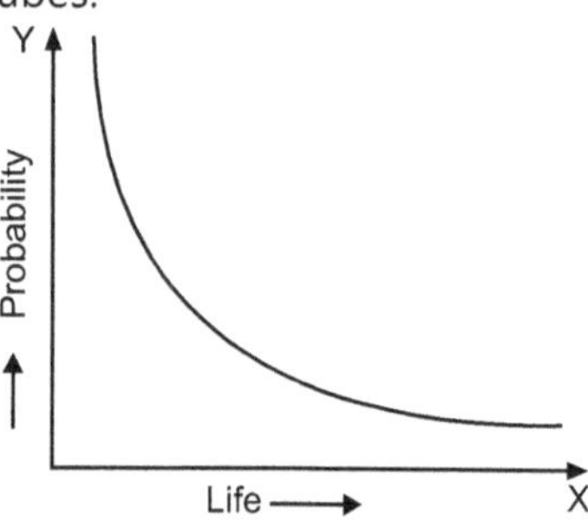

Fig. 5.2 (a) : Progressive failure-probability of failure increases with life of the item

Fig. 5.2 (b) : Retrogressive failure-probability of failure is more in early life of the item and then chance of failure decreases

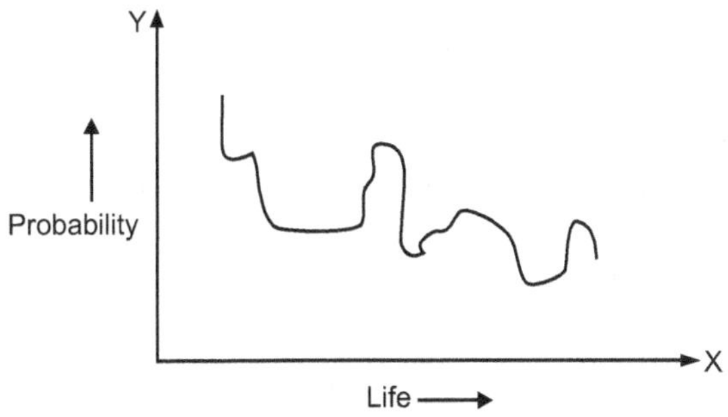

Fig. 5.2 (c) : Random failure

Item fail due to some random cause but not due to age.

The above may be shown as in Fig. 5.3.

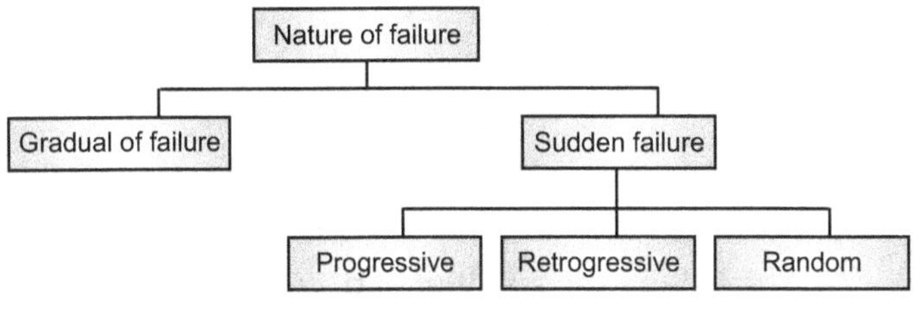

Fig. 5.3

All the above discussed points may be summarized as :

Table 5.1 : Summary of Three Stages of Maintenance

Phase	Type of Failure	Failure Rate	Causes of Failure	Cost of Failure	Suitable Maintenance Policy
Infant phase	Early failures	Decreasing Trend	Faulty design, erratic operation, installation errors, environmental problems.	Medium to high	Warrantee / guarantee by manufacturer
Youth phase	Random or chanced failures or rare event failures	Constant	Operational errors, heavy load, over run	Low to medium	Breakdown, predictive, preventive, repair maintenance etc.
Old age phase	Wear out or age failures due to wear and tear	Increasing	Wear, tear, creep, fatigue etc.	Low	Operate to fail and corrective maintenance
				High	Reconditioning or replacement

5.2.1 Costs Associated with Maintenance

- Our main aim in this chapter is to find optimal replacement period so as to minimize the maintenance cost. Hence, we are very much interested in the various cost associated with maintenance.

Various costs to be discussed are :

- **Purchase Cost or Capital Cost (C) :**
  This cost is independent of the age of the machine or usage of the machine. This is incurred at the beginning of the life of the machine, i.e. at the time of purchasing the machine or equipment. But the interest on the invested money is an important factor to be considered.

- **Salvage Value / Scrap Value / Resale Value / Depreciation (S) :**
  As the age of the machine increases, the resale value decreases as its operating efficiency decreases and the maintenance costs increases. It depends on the operating conditions of the machine and life of the machine.

- **Running Costs Including Maintenance, Repair and Operating Costs :**
  These costs are the functions of age of the machine and usage of the machine. As the usage increases or the age increases, due to wear and tear, many components fail to work and they are to be replaced. As the age increases, failures also increase and the maintenance costs goes on increasing. At some period the maintenance costs are so high, which will indicate that the replacement of the machine or equipment is essential.

  These costs can be shown by means of a curve as in Fig. 5.4.

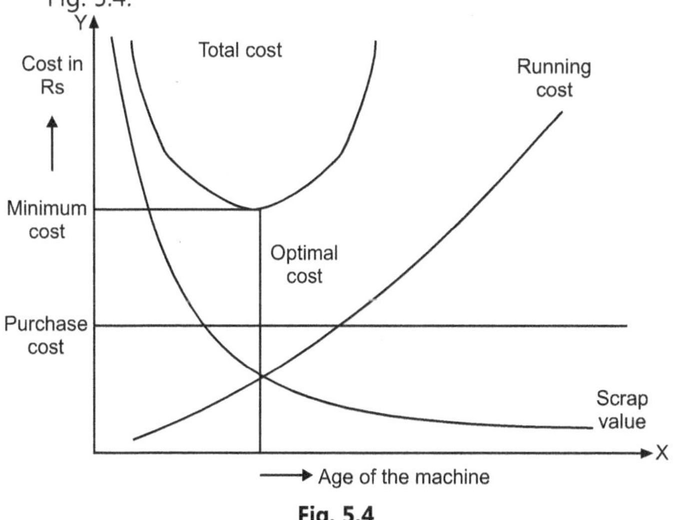

Fig. 5.4

5.3 TYPES OF REPLACEMENT EXAMPLES

[May 16, Dec. 07, May 12, 6M]

- One must remember that the study of Replacement of items is a field of application rather than a method of analysis. The study involves, the comparison of alternative replacement policies.

Various Types of Replacement Problems we Come Across in this Chapter are :

- Replacement of **Capital Equipment**, which looses its operating efficiency due to aging (passage of time), or due to continuous usage (due to wear and tear of components). Examples are Machine tools, Transport and other vehicles, etc. Here the system can maintain the level of performance by installing a new unit at the beginning of some unit of time (year, month or week) and decide to keep it up to some suitable period so as to minimize the operating and maintenance costs.

- In this case, the deterioration process is predictable and is represented by an increased maintenance cost and decreased in scrap cost and increased production cost per unit. In such cases, the optimum life of the item is determined on the assumption that increased age reduces efficiency.

- Deterministic models explain the problem and they are very much similar to that of inventory models where deterioration corresponds to demand against the desired level of efficiency (level of inventory). The cost of new item is similar to cost of replenishment of inventory and maintenance cost corresponds to cost of holding inventory.

- These Types of Problems are Solved by Two Methods. They are :
  - ➢ By calculating the cost per unit of time, without considering the money value. Here we calculate the total cost upto the period and divide by time unit (years, months, weeks etc.) to find the average cost to decide the period of replacement.
  - ➢ By taking the money value into consideration using present value concept to compare on a one number basis.

- Replacement of **Items that Fail Completely** all in a sudden in a random nature. We use **Group Replacement** or **Preventive Maintenance Technique** for these items and these are expensive to replace individually. Examples are Electric bulbs, Transistors, Electronic components etc. Here replacement of items are done in anticipation of failure, which is known as **Preventive Maintenance**. We assume that the items will have relatively constant efficiency until they fail or die. These models require the knowledge of statistics and stochastic process involving probability of failure. The replacement policy is formulated to balance the wasted life of items replaced before failure against the costs incurred when items fail in service.

- Replacement of **Human Beings in Organizations**, known as **Staffing Problem**, or known as **Human Resource Planning** or **Mortality and Staffing Problem**. This problem requires the knowledge of life distribution for service of staff in a system.

- Miscellaneous problems such as replacement of existing units due to availability of more effective and new and advanced technology. In these problems, replacement will become necessary due to research of new and advanced and more effective technology and old technology becomes out of date.

5.4 GENERAL APPROACH TO SOLUTION TO REPLACEMENT EXAMPLE

- Though it is not possible or it is difficult to predict the time of failure of an item exactly, likely failure pattern could be established by observation.

- Generating the probability distribution for the given situation and then using them in conjunction with relevant cost information we can formulate the optimum replacement policy. The information necessary to formulate optimum replacement policy is :

- Objective assessment of the probability of the item failing at a particular point of time.

- Assessments of the cost of replacement in terms of :
  - ➢ Actual cost of the item,
  - ➢ Direct costs of labour involved in replacement,
  - ➢ Costs of disruption in terms of lost production, lost orders etc.

5.5 REPLACEMENT OF ITEMS WHOSE EFFICIENCY REDUCES OR MAINTENANCE COST INCREASES WITH TIME OR DUE TO AGE AND MONEY VALUE IS NOT CONSIDERED

Costs to be Considered :

- Various cost items to be considered in replacement decisions are the costs that depend upon the choice or age of item or equipment. The costs those do not change with the age of the machine or item need not be taken into consideration. The replacement of items whose efficiency reduces with time is justified when the average cost per time period goes on reducing longer the replacement is postponed.

- However, there will come an age at which the rate of increase of running costs more than compensates the saving in average capital costs.

- At this age the replacement is justified. In the case of replacement of items whose efficiency deteriorates with time, the most important criteria to be considered is the measurement of efficiency. Consider a machine, in this case, the maintenance cost always increases with time and usage and a time comes when the maintenance cost becomes large enough, which indicates that it is better and economical to replace the machine with a new one.

- When we want to replace the machine, we may come across various alternative choices, where we have to compare the various cost elements such as running costs and maintenance costs to select optimal choice.

The Various Techniques we may Come Across to Analyze the Situation are :

- Replacement of items whose maintenance cost increases with time and *value of money remains same* during the period.

- Replacement of items whose maintenance cost increases with time and the *value of money also changes* with time.

- To compare alternative choices, use of *concept of present value.*

5.5.1 Replacement of Items whose Maintenance Cost Increases with Time and the Value of Money Remains Same During the Period

Points to Remember :

- If time is measured continuously, then the average annual costs will be minimized by replacing the machine or item, when the average cost to date becomes equal to the current maintenance cost.

- If time is measured in discrete units, then the average annual cost will be minimized by replacing the machine or item when the next period's maintenance cost becomes greater than the current average cost.

  Let,   C = Purchase cost or Capital cost of the item

           S = Scrap value or resale value of the item. It is assumed that this cost will remain constant over time.

SOLVED EXAMPLES

Example 5.1 : *A firm is thinking of replacing a particular machine whose cost price is ₹12,200. The scrap value of the machine is ₹200/-. The maintenance costs are found to be as follows :*

Table 5.2

Year	1	2	3	4	5	6	7	8
Maintenance Cost in ₹	220	500	800	1200	1800	2500	3200	4000

Determine when the firm should get the machine replaced.

[May 16]

Solution :

Table 5.3

Year (t) Y	u(t) Maintenance Cost (₹)	M(y) = y $\sum u(t)$ t = 1	C = Capital Cost in ₹	Scrap Cost (S) in ₹	T(y) = C − S + M(y)	Average Cost : G(y) = T(y)/y
1	2	34	–	5	6 = 4 − 5 + 3	7 = 6/1
1	220	220	12200	200	12220	12220
2	500	720	12200	200	12720	6360
3	800	1520	12200	200	13520	4506.67
4	1200	2720	12200	200	14720	3680
5	1800	4520	12200	200	16520	3304
6	2500	7020	12200	200	19020	3170
7	3200	10220	12200	200	22220	3174.29
8	4000	14220	12200	200	26220	3277.50

Replace the machine at the end of 6th year when the average annual maintenance cost is minimum.

Example 5.2 : *The initial cost of a machine is ₹ 6100 and its scrap value is ₹ 100. The maintenance costs found from experience are as follows :*   **[May 17]**

Table 5.4

Year	1	2	3	4	5	6	7	8
Annual Maintenance Cost in ₹	100	250	400	600	900	1200	1600	2000

When should the machine be replaced?

Solution :

The time period is discrete. We have to find the period when the average maintenance cost will be minimum.

Table 5.5

Years 't' = y	u(t) ₹	M(y) = y $\sum_{t=1}$ u(t)	T(y) = C − S + M(y) ₹	G(y) = T(y)/y ₹
1	100	100	6100	6100/1 = 6100
2	250	350	6350	6350/2 = 3175
3	400	750	6750	6750/3 = 2250
4	600	1350	7350	7350/4 = 1837.50
5	900	2250	8250	8250/5 = 1650
6	1200	3450	9450	**9450/6 = 1575**
7	1600	5050	11050	11050/7 = 1578.57
8	200	7050	13050	13050/8 = 1631.25

The annual average maintenance cost is minimum at the end of 6th year and it goes on increasing from 7th year. Hence, the machine is to be replaced at the end of 6th year.

Example 5.3 : *The maintenance cost and resale value per year of a machine whose purchase price is ₹ 7000/- is given below :*

Table 5.6

Year	1	2	3	4	5	6	7	8
Maintenance Cost in ₹	900	1200	1600	2100	2800	3700	4700	5900
Resale Value in ₹	4000	2000	1200	600	500	400	400	400

When should the machine be replaced?　　　**[May 16]**

Solution :

Table 5.7

Year (t) = y	Running Cost in u(y) in ₹	Cumulative Running Cost M(Y) in ₹	Resale Value S(y) in ₹	C − S (y) in ₹ C = 7000	T(y) = C − S (y) + M(y) in ₹	T(y)/y = G(y) Average Cost in ₹
1	900	900	4000	3000	3900	3900
2	1200	2100	2000	5000	7100	3550
3	1600	3700	1200	5800	9500	3166.67
4	2100	5800	600	6400	12200	3050
5	2800	8600	500	6500	15100	**3020**
6	3700	12300	400	6600	18900	3150
7	4700	17000	400	6600	23600	3371.43
8	5900	22900	400	6600	29500	3687.50

From the table we can see that the average cost is minimum at the end of the 5th year. Hence the machine may be replaced at the end of the 5th year.

Example 5.4 : *A fleet owner finds form his past records that the cost per year of running a truck and resale values whose purchase price is ₹6000/- are given as under. At what stage the replacement is due?*　　　**[May 17, Dec. 15 8M]**

Table 5.8

Year	1	2	3	4	5	6	7	8
Running Cost in ₹	1000	1200	1400	1800	2300	2800	3400	4000
Resale Value in ₹	3000	1500	750	375	200	200	200	200

Solution :

Let C = Capital cost = ₹ 6000/-, S (y) = Scrap value changes yearly, G(y) = Average yearly cost, u (t) = Annual maintenance cost.

Table 5.9

Year (t) = y	Running Cost in u(y) in ₹	Cumulative Running Cost M(Y) in ₹	Resale Value S(y) in ₹	C − S (y) in ₹	T(y) = C − S(y) + in ₹	T(y)/y = G(y) Average Cost in ₹
1	1000	1000	3000	3000	4000	4000
2	1200	2200	1500	4500	6700	3350
3	1400	3600	750	5250	8850	2950
4	1800	5400	375	5625	11025	2756
5	2300	7700	200	5800	13500	**2700**
6	2800	10500	200	5800	16300	2717
7	3400	13900	200	5800	19700	2814
8	4000	17900	200	5800	23700	2962

From the table above we can see that the G (y) is minimum at the end of 5th year. Hence, the truck is to be replaced at the end of 5th year.

Example 5.5 : *A Plant manager is considering the replacement policy for a new machine. He estimates the following costs in Rupees. Find an optimal replacement policy and corresponding minimum cost.*　　　**[Dec. 16]**

Table 5.10

Year	1	2	3	4	5	6
Replacement Cost at the Beginning of the Year (₹)	100	110	125	140	160	190
Salvage Value at the End of the Year : (₹)	60	50	40	25	10	0
Operating Costs (₹)	25	30	40	50	65	80

Solution :

In this problem, replacement cost at the beginning and the salvage value at the end of the year is given. If we subtract salvage value from the replacement cost we get the net value, i.e. C (Capital cost) − S (Resale value). Rest of the problem is worked as usual.

Table 5.11

Years (t) = y	Running Cost u(y) in ₹	Cumulative Running cost M(Y) in ₹	Resale Value S(y) in ₹	C − S (y) in ₹	T(y) = C − S(y) + M(y) in ₹	T(y)/y = G(y) Average Cost in ₹
1	25	25	60	100 − 60 = 40	65	65
2	30	55	50	110 − 50 = 60	115	**57.50**
3	40	95	40	125 − 40 = 85	180	60
4	50	145	25	140 − 25 = 115	260	65
5	65	210	10	160 − 10 = 150	360	72
6	80	290	0	190 − 0 = 190	480	80

From the table we see that the average annual maintenance cost is minimum at the end of 2nd year and is ₹ 57.50. Hence, the machine is to be replaced at the end of 2nd year.

Example 5.6 : *Machine A costs ₹ 45,000/- and the operating costs are estimated at ₹ 1000/- for the first year, increasing by ₹ 10,000/- per year in the second and subsequent years. Machine B costs ₹ 50000/and operating costs are ₹ 2000/- for the first year, increasing by ₹ 4000/- in the second and subsequent years. If we now have a machine of type A, should we replace it by B. If so when? Assume both machines have no resale value and future costs are not discounted.* **[Dec. 15]**

Solution :

Let us now calculate the average annual running cost for machine A and B in the table given below : (S = ₹ 0/-)

Table 5.12 : Machine A

Year (y)	Running Cost (₹) u(t)	Cumulative Running Cost in ₹ $\Sigma u(t) = M(y)$	Depreciation C − S	Total Cost TC = C − S + M(y)	Average Cost F(y) = TC/y
1	1000	1000	45000	46000	46000
2	11000	12000	45000	57000	28500
3	21000	33000	45000	78000	**26000**
4	31000	64000	45000	1,09,000	27200
5	41000	1,05,000	45000	1,50,000	30000
6	51000	1,56,000	45000	2,01,000	33500

As the annual maintenance cost is minimum at the end of 3rd year, the machine A is to be replaced at the end of 3rd year.

Table 5.13 : Machine B

Year (y)	Running Cost (₹) u(t)	Cumulative Running Cost in ₹ $\Sigma u(t) = M(y)$	Depreciation C − S	Total Cost TC = C − S + M(y)	Average Cost F(y) = TC/y
1	2000	2000	50000	52000	52000
2	6000	8000	50000	58000	29000
3	10000	18000	50000	68000	22667
4	14000	32000	50000	82000	20500
5	18000	50000	50000	1,00,000	20000
6	22000	71000	50000	1,22,000	20333

As the average annual maintenance cost is minimum at the end of 5th year, the machine is to be replaced at the end of 5th year.

When we compare machine A and machine B, the average annual maintenance cost of Machine B is less than that of A, the machine A should be replaced by machine B.

Now to find the time of replacement of machine A by machine B, the total cost of the machine A in the successive years is computed as given below :

Table 5.14

Year	1	2	3	4
Total Cost incurred in ₹	46000	57000 − 46000	78000 − 57000	1,09,000 − 78,000
Salvage Value at the End of the Year : (₹)	60	50	40	25
Operating Costs (₹)	25	30	40	50

The criterion is to replace machine A by machine B at the age when its running cost for the next year exceeds the lowest average running cost i.e. ₹ 20000 per year of machine B. From the calculations we can see that running cost of machine is in the third year, i.e. ₹ 21000/- is more than the lowest average running cost per year of machine B i.e. ₹ 20000/- at the end of fifth year. Hence, the machine A should be replaced by machine B after two years.

5.5.2 Replacement of Items whose Maintenance Costs Increases with Time and Value of Money also Changes with Time

1. Present Worth Factor

- Before dealing with this model, it is better to have the concept of **Present Value**. Consider replacement of items which involve huge expenditure, both initial value (Purchase price) and maintenance expenses. For a decision maker, there may be number of alternatives for replacement.

- But he always chooses the alternative, which minimizes the annual average cost. Here manager uses the concept of present value of money to select the alternatives. The present value of number of expenditures incurred over different periods of time represents their value at the current time.

- It is based on the fact that, one can invest money at an interest rate 'r' to produce the same amount of money at the end of certain time period or if an amount is to be spent in different years what is the worth of total expected expenses or its worth today? We can also think in another way.

- Suppose a businessman borrows money for his initial investment and working capital from various sources, he has to pay interest for the money he has borrowed. The amount of interest he has to pay depends on the rate of interest and the period for which he has borrowed (that is the period in which he has repaid the amount borrowed). The borrowed money is known as Principal (P), and the excess amount he has paid is known as Interest (i). The sum of both principal and the interest is known as Amount (A).

- If P is the principal, 'i' is the rate of interest, and A is the amount, then the amount at the end of 'n' years with compound interest is :

$$A = P(1 + i) n$$

OR　　　$P = (A)/(1 + i) n$ OR $P = A \times Pwf$

- **Where Pwf is Single Payment Present Worth Factor.** It is represented by 'v' and is also known as discount rate. Discount rate is always less than one. In other words we can say that 'v' or the present worth factor (Pwf) is present value of one rupee spent after 'n' years from now.

- Hence, P is known as the present worth of an amount A paid in 'n' years at interest rate 'i'. For calculation purpose present worth factor tables are available. Similarly, if R denotes the uniform amount spent at the end of each year and S is the total expenditure at the end of 'n' years, then

$$S = R \{ (1 + i) n - 1 \} / i \text{ OR}$$
$$R = (S \times i) / (1 + i) n$$
$$= \{ P(1 + i) n \times i \} / \{ (1 + i) n - 1 \} \text{ OR}$$
$$P = R \times \{P(1 + i) n - 1 \} / \{i(1 + i)n\}$$
$$= R \times (Pwf)$$

- **Pwf is Known as Uniform Annual Series Present Worth Factor.** In other words, suppose if we want ₹ 50,000/- for investment after 10 years, how much we save yearly, so that at the end of 10 years, we will have ₹ 50,000/- ready for investment. The discount rate to find this amount is known as Pwfs.

5.5.3 Replacement of Items whose Maintenance Cost Increases with Time and Money Value also Changes

This problem is complicated as the money value changes with time. This can be dealt under two different conditions:

- The maintenance cost goes on increasing with usage or age or time and then we have to find out optimum time of replacing the item. Here the value of money decreases with a constant rate which is known as its **Depreciation Ratio** or **Discounted Factor** and is represented by 'd'.

Here the money value changes can be understood as follows :

Suppose a person borrows ₹ 1000/- at an interest rate of 10% per year. After one year from now, he has to pay back ₹ 1100/-. That means to say today's ₹ 1000/- is equivalent to ₹ 1100/- after one year. OR ₹ 1100/- after one year is equivalent to ₹ 1000/- today. That is ₹ 1/- after one year from now is equal to 1000 / 1100 = (1.1) − 1 at present. This is known as **Present Value**.

To generalize, if the interest on ₹ 1/- is 'i' per year then the present value or present worth of ₹ 1/- after one year from now is 1 / (1 + i), this is the depreciation ratio, represented by 'd'. Similarly, the present worth of ₹ 1/- to be spent after 'n' years from now (the rate of interest is 'i') is $\dfrac{1}{(1 + i)^n}$.

- If a businessman takes a loan for a certain period at a given interest rate and agrees to pay it in a number of instalments, then we have to find the most suitable period during which the loan would be repaid.

Case I :

Let the equipment cost be ₹ A and the maintenance costs be ₹ $C_1, C_2, C_3,$ .... C_n (where, $C_n + 1 > C_n$) during the first, second and third years respectively upto 'n' years. If 'd' is the depreciation value per unit of money during a year then to find the optimum replacement policy, which minimizes the total of all future discounted costs.

It is assumed that the expenditure incurred at the beginning of the year and the resale value of the item is zero. Finding the total expenditure incurred on the equipment and its maintenance during the desired period and its present value solves the problem. The criterion is the period for which the total discounted cost is minimum will be the best period for replacement.

Example 5.7 : *The yearly cost of two machines A and B, when money value is neglected is shown in the table given below. Find their cost pattern if money value is 10% per year and hence find which machine is more economical.*

Table 5.15

Year	1	2	3
Machine A (₹)	1800	1200	1400
Machine B (₹)	2800	200	1400

Solution :

When the value of money is 10% per year, the discount rate, $d = (1/1 + i = 1/1 + 0.10 = 1/1.1 = 0.9091$

The discounted cost pattern of machines A and B are as below :

Table 5.16

Year	1	2	3	Total Cost (₹)
Machine A (₹)	1800	1200×0.9091	$1400 \times (0.0901)^2$	4047.94
Discounted cost		= 1090.90	= 1157.04	
Machine B (₹)	2800	200×0.9091	$1400 \times (0.0901)^2$	4138.86
Discounted cost		= 181.82	= 1157.04	

The table shows that the total cost of machine A is less than that of machine B. Hence, machine A is more economical when money value is changing.

Example 5.8 : *A machine costs ₹ 500/-. Operation and maintenance costs are zero for the first year and increase by ₹ 100/- every year. If money is worth 5% every year, determine the best age at which the machine should be replaced. The resale value of the machine is negligibly small. What is the weighed average cost of owning and operating the machine?*

Solution :

Discount rate = $d = 1/1 + i = 1/1 + 0.05 = 0.9524$.

Weighed Average Cost :

Table 5.17

Year of Service X	Maintenance Cost C_i ₹	Discount Factor d^{i-1}	Discounted Maintenance Cost $C_i \times d^{i-1}$ ₹	Total Cost $\sum_{i=1}^{X} A + C_i$ $\times d^{i-1}$ ₹	Cumulative Discount Factor $\sum_{i=1}^{X} d^{i-1}$	Weighed Average Annual Cost ₹ $G(x) = \dfrac{A + \sum_{i=1}^{X} C_i \times d^{i-1}}{\sum_{i=1}^{X} d^{i-1}}$
1	0	1.0000	0.0000	0.0000	1.0000	500
2	100	0.9524	95.2400	595.24	1.9524	304.88
3	200	0.9070	181.4000	776.64	2.8594	217.61 (Replace)
4	300	0.8638	259.14	1035.78	3.7232	278.20
5	400	0.8277	320.08	1364.86	4.5459	300.25

$200 < 217.61 < 300$ i.e. $G_3 = ₹\ 217.61 > C_3 = ₹\ 200.00$ and $G_3 = ₹\ 217.61 < G_4 = ₹\ 300$.

₹ 200 the running cost of 3rd year is less than weighed average of 3rd year and running cost of 4th year is greater than the weighed average of 3rd year, hence the machine is to be replaced at the end of 3rd year. The weighed average cost of running the machine is ₹ 217.61.

Example 5.9 : *Assume that present value of one rupee to be spent in a year's time is ₹ 0.90 and the purchase price A = ₹ 3000/-. The running cost of the equipment is given in the table below. When should the machine be replaced?*

[May 18]

Table 5.18

Year	1	2	3	4	5	6	7
Running Cost in ₹	500	600	800	1000	1300	1600	2000

Solution :

Given that A = ₹ 3000/-, Discount factor = d = 0.9

Table 5.19

Year of Service X	Maintenance Cost C_i ₹	Discount Factor d^{i-1}	Discounted Maintenance Cost $C_i \times d^{i-1}$ ₹	Total Cost $\sum_{i=1}^{X} A + C_i$ $\times d^{i-1}$ ₹	Cumulative Discount Factor $\sum_{i=1}^{X} d^{i-1}$	Weighed Average Annual Cost ₹ $G(x) = \dfrac{A + \sum_{i=1}^{X} C_i \times d^{i-1}}{\sum_{i=1}^{X} d^{i-1}}$
1	500	1.00	500	3500	1.0000	3500.00
2	600	0.90	540	4040	1.90	2126.32
3	800	0.81	648	4688	2.71	1729.89

...Conti.

4	1000	0.73	730	5418	3.44	1575.00
5	1300	0.66	858	6276	4.10	**1530.73 (Replace)**
6	1600	0.59	944	7220	4.69	1539.45
7	2000	0.53	1060	8280	5.22	1586.21

C_5 = ₹ 1300/- < G_5 = ₹ 1530.73 < C_6 = ₹ 1600/-. The machine is to be replaced at the end of 5th year.

Example 5.10 : *A manufacturer is offered two machines A and B. A has the cost price of ₹ 2,500/- its running cost is ₹ 400 for each of the first 5 years and increase by ₹ 100/- every subsequent year. Machine B having the same capacity as A and costs ₹ 1250/-, has running cost of ₹ 600/- for first 6 years, increasing thereby ₹ 100/- per year. Which machine should be purchased? Scrap value of both machines is negligible. Money value is 10% per year.* **[Dec. 16]**

Solution :

$$d = 1/1 + i$$
$$= 1/1 + 0.10$$
$$= 1/1.1 = 0.9091$$

Present worth of machine A is :

Table 5.20

Year of Service X	Maintenance Cost C_i ₹	Discount Factor d^{i-1}	Discounted Maintenance Cost $C_i \times d^{i-1}$ ₹	Total Cost $A + \sum_{i=1}^{X} C_i \times d^{i-1}$ ₹	Cumulative Discount Factor $\sum_{i=1}^{X} d^{i-1}$	Weighed Average Annual Cost ₹ $G(x) = \dfrac{A + \sum_{i=1}^{X} C_i \times d^{i-1}}{\sum_{i=1}^{X} d^{i-1}}$
1	400	1.0000	400.00	2900.00	1.0000	2900.00
2	400	0.9091	363.64	3263.64	1.9091	1709.45
3	400	0.8264	330.56	3594.20	2.7355	1313.84
4	400	0.7513	300.52	3894.72	3.4868	1116.90
5	400	0.6830	273.20	4167.92	4.1698	999.50
6	500	0.6209	310.45	4478.37	4.7907	934.80
7	600	0.5645	338.70	4817.07	5.3552	889.92
8	700	0.5132	359.24	5176.31	5.8684	881.92
9	800	0.4665	372.20	5549.51	6.3349	**875.86 (Replace)**
10	900	0.4241	381.69	5931.20	6.7590	877.35

As C_9 = ₹ 800 < G_9 = ₹ 875.86 < C_{10} = ₹ 900, it is replaced at the end of 9th year.

Replacement period for Machine B :

Table 5.21

Year of Service X	Maintenance Cost C_i ₹	Discount Factor d^{i-1}	Discounted Maintenance Cost $C_i \times d^{i-1}$ ₹	Total Cost $A + \sum_{i=1}^{X} C_i \times d^{i-1}$ ₹	Cumulative Discount Factor $\sum_{i=1}^{X} d^{i-1}$	Weighed Average Annual Cost ₹ $G(x) = \dfrac{A + \sum_{i=1}^{X} C_i \times d^{i-1}}{\sum_{i=1}^{X} d^{i-1}}$
1	600	1.0000	600.00	1850.00	1.0000	1850.00
2	600	0.9091	545.46	2395.46	1.9091	1254.75
3	600	0.8264	495.84	2891.30	2.7355	1056.95
4	600	0.7513	450.78	3342.08	3.4868	958.49
5	600	0.6830	409.80	3751.88	4.1698	899.77
6	600	0.6209	372.54	4124.42	4.7907	860.92
7	700	0.5645	395.15	4519.57	5.3552	843.96
8	800	0.5132	410.56	4930.13	5.8684	**840.11 (Replace)**
9	900	0.4665	419.85	5349.98	6.3349	844.52
10	1000	0.4241	424.10	5774.08	6.7590	854.28

C_8 = ₹ 800 < G_8 = ₹ 840.11 < C_9 = ₹ 900. The machine B is to be replaced at the end of 8th year.

Example 5.11 : *The cost of a new machine is ₹ 5000/-. The maintenance cost during the nth year is given by u_n = ₹ 500 (n − 1), where n = 1, 2, 3, … , n. If the discount rate per year is 0.05, after how many years will it be economical to replace the machine by a new one?* **[May 15]**

Solution :

Since, the discount rate is 0.05, d = 1/1 + 0.05 = 1/1.05 = 0.9523. The replacement period for the machine is :

Table 5.22

Year of Service X	Maintenance Cost C_i ₹	Discount Factor d^{i-1}	Discounted Maintenance Cost $C_i \times d^{i-1}$ ₹	Total Cost $A + \sum_{i=1}^{X} C_i \times d^{i-1}$ ₹	Cumulative Discount Factor $\sum_{i=1}^{X} d^{i-1}$	Weighed Average Annual Cost ₹ $G(x) = \dfrac{A + \sum_{i=1}^{X} C_i \times d^{i-1}}{\sum_{i=1}^{X} d^{i-1}}$
1	0.00	1.0000	0.00	5000	1.0000	5000.00
2	500	0.9523	476	5476	1.9523	2805.00
3	1000	0.9070	907	6383	2.8593	2232.00
4	1500	0.8638	1296	7679	3.7231	2063.00
5	2000	0.8227	1645	9324	4.5458	**2051.00 (Replace)**
6	2500	0.7835	1959	11283	4.3293	2117.00

C_5 = ₹ 2000/- < G_5 < ₹ 2051.00 < C_6 = ₹ 2500.00. Hence, the machine is to be replaced at the end of 5th year.

5.6 COMPARING OF REPLACEMENT ALTERNATIVES BY USING CRITERIA OF PRESENT VALUE

- Sometimes a businessman may come across a problem, when he want to purchase a machine or a vehicle for his business. He may have different models with comparatively equal price and with different maintenance and other expenses. In such cases, he can use present worth concept to select the right type of machine or a vehicle. Here the present value of all future expenditures and revenues is calculated for each alternative and the one for which the present value is minimum is preferred.

  Let, Q = Annual cost or purchase price in ₹

       i = Annual interest rate

       t = Period in years

       P = Principal amount in ₹

       S = Scrap value or salvage value in ₹

- The present value of total cost is given by

  $P + Q$ (Pwfs for i% interest rate for n years) $- S$ (Pwf for i% interest for n years)

- Similarly, if the annual operating costs vary for different years then the present value of these costs will be calculated on the basis of time period for which these expenditures are made. i.e. say for example, if Q_1, Q_2, Q_3, ... , Q_n are the operating costs for different years, then the present value of the operating cost during 'n' years will be given by

- Q_1 (Pwfs at i% interest for 1 year) $+ Q_2$ (Pwfs at i% for 2 years) $+ .... + Q_n$ (Pwfs at i% interest for n years).

- Now present value of total cost will be $P + Q_1$ (Pwfs at i% interest for 1 year) $+ Q_2$ (Pwfs at i% interest rate for 2 years) $+ Q_3$ (Pwfs at i% rate of interest for 3 years) $+ .... + Q_n$ (Pwfs at i% interest for n years) $- S$ (Pwf at i% interest for n years).

Example 5.12 : *An entrepreneur is considering purchasing a machine for his factory. The related data for alternative machines are as follows :*

Table 7.23

	Machine A	Machine B	Machine C
Present investment in ₹	10000	12000	15000
Total annual cost in Rs.	2000	1500	1200
Life of machine in years	10	10	10
Salvage value in Rs.	500	1000	1200

As an advisor of the company, you have been asked to select the best machine considering 12% normal rate of return per year.

Given that :

Present worth factor series @ 12% for 10 years is 5.650.

Present worth factor @ 12% for 10th year is 0.322.

Solution :

Table 5.24

	Machine A	Machine B	Machine C
1. Present investment in ₹	10000	12000	15000
2. Total annual cost in ₹	2000×5.650 = 11300	1500×5.650 = 8475	1200×5.650 = 6780
3. Present values of salvage value in ₹.	500×0.322 = 161	1000×0.322 = 322	1200×0.322 = 386.40
4. Total cost = 1 + 2 − 3	1000 + 1130 − 162 = ₹21139.00	12000 + 8475 − 322 = ₹20153.00	15000 + 6780 − 386.40 = ₹21393.60

Machine B is having less present value of total cost. Hence, to purchase the machine B.

5.7 REPLACEMENT OF ITEMS THAT FAIL COMPLETELY AND SUDDENLY AND ARE EXPENSIVE TO BE REPLACED

[Dec. 16, May 13, 6M]

- There are certain items or systems or products, whose probability of failure increases with time. They may work with designed efficiency throughout their life and if they fail to act they fail suddenly. The nature of these items is they are costly to replace at the same time and their failure affect the functioning of entire system.

- For example, resistors, components of air conditioning unit and certain electrical components. If we do not replace the item immediately, then loss of production, idle labour idle raw materials etc. are the results. It is evident failure of such items causes heavy losses to the organization.

- Such situations demand the formulation of a policy, which will help the organization to avoid losses. Sometimes we find it is better to replace the item before it fails so that the expected losses due to failure can be avoided. The following courses of action can be followed :

1. Individual Replacement Policy :

- This policy states that replace the item soon after its failure. Here the cost of replacement will be somewhat greater as the item is to be purchased individually from the seller as and when it fails. From the time of failure to the replacement, the system remains idle. More than that, as the item is purchased individually, the cost of the item may be more.

- In case, the component or the item is not available in the local market, we have to get it from other places, where the procurement cost may be higher for individual purchase. If the management wants to adopt this policy, it may have to waste its time and money also the losses due to failure.

2. **Group Replacement Policy :**

- If the organization has got the statistics of failure of the item, it can calculate the average life of the item and replace the item before it fails, so that the system can work without break. In this case, all the items, even they are in good working condition, are replaced at a stipulated period as calculated by the organization by using the group replacement policy. One thing we have to remember is that, in case any item fails, before the calculated group replacement period, it is replaced individually immediately after failure.

- Hence, this policy utilizes the strategy of both individual replacement and group replacement. The probability distribution of the failure of the item in a system can be determined by mortality tables for life testing techniques. Let us try to understand what a mortality table is.

5.7.1 Mortality Tables

- The *Mortality Theorem* states that a large population is subjected to a given mortality law for a very long period of time. All deaths are immediately replaced by births and there are no other entries or exits.

- Here age distribution ultimately becomes stable and that the number of deaths per unit of time becomes constant, which is equal to the size of the total population divided by the mean age at death. If we consider the problem of human population, no group of people ever existed under the conditions :

1. That all deaths are immediately replaced by births.

2. That there are no other entries or exists.

- These two assumptions help to analyze the situation more easily, by keeping virtual human population in mind. When we consider an industrial problem, deaths refer to item failure of items or components and birth refers to replacement by a new component.

- Mortality table for any item can be used to derive the probability distribution of life span. If M(t) represents the number of survivors at any time 't' and M(t − 1) is the number of survivors at the time (t − 1), then the probability that any item will fail in this time interval will be

$$\frac{\{M(t-1) - M(t)\}}{N} \qquad \text{... (5.1)}$$

where, N is the number of items in the system.

Conditional probability that any item survived up to age (t − 1) will die in next period, will be given by

$$\frac{\{M(t-1) - M(t)\}}{M(t-1)} \qquad \text{... (5.2)}$$

Example 5.13 : *Calculate the probability of failure of an item in good condition in each month from the following survival table :* **[May 17]**

Table 5.25

Month Number (t):	0	1	2	3	4	5	6	7	8	9	10
Original Number Of items Working at the End of Each Year	1000	940	820	580	400	280	190	130	70	30	0

Solution :

Here 't' is the number of month; M(t) is the number of items i.e. items in good condition at the end of tth month.

The probability of failure in each month is calculated as under :

Table 5.26

Year (t)	Items in Good Condition M(t)	Probability of Items that Fail in t^{th} Year = $\frac{\{M(t-1) - M(t)\}}{N}$
0	1000	–
1	940	(1000 − 940) / 1000 = 0.06
2	820	(940 − 820) / 1000 = 0.12
3	580	(820 − 580) / 1000 = 0.24
4	400	(580 − 400) / 1000 = 0.18
5	280	(400 − 280) / 1000 = 0.12
6	190	(280 − 190) / 1000 = 0.09
7	130	(190 − 130) / 1000 = 0.06
8	70	(130 − 70) / 1000 = 0.06
9	30	(70 − 30) / 1000 = 0.04
10	0	(30 − 00) / 1000 = 0.03

5.7.2 Group Replacement of Items

- A group replacement policy consists of two steps. Firstly, it consists of individual replacement at the time of failure of any item in the system and there is group replacement of existing live units at some suitable time.

- Here the individual replacement at the time of failure ensures running of the system, whereas group replacement after some time interval will reduce the probability of failure of the system. The application of such type of policy has to take into consideration the following points :

1. The rate of individual replacement during the period and

2. The total cost incurred due to individual and group replacement during the period chosen.

- This policy favors the group replacement, when the total cost is minimum and the period of replacement is known as *Optimal Period of Replacement.*

- The information required to formulate this policy are :
  (i) Probability of failure,

  (ii) Losses due to these failures,

  (iii) Cost of individual replacement, and

  (iv) Cost of group replacement.

  The procedure is explained in the worked examples.

- The group replacement policy states that Group replacement should be made at the end of ith period, if the cost of individual replacement for i^{th} period is greater than average cost per period by the end of the period 't' and one should not adopt a group replacement policy if the cost of individual replacement at the end of $(t-1)^{th}$ period is not less than the average cost per period through time $(t-1)$.

Example 5.14 : *A system consists of 10000 electric bulbs. When any bulb fails, it is replaced immediately and the cost of replacing a bulb individually is ₹ 1/- only. If all the bulbs are replaced at the same time, the cost per bulb will be ₹ 0.35. The percent surviving i.e. S(t) at the end of month 't' and P(t) the probability of failure during the month 't' are as given below. Find the optimum replacement policy.*

Table 5.27

t in Months	0	1	2	3	4	5	6
S(t)	100	97	90	70	30	15	0
P(t)	–	0.03	0.07	0.20	0.40	0.15	0.15

Solution :

The problem is to be solved in two stages :

(i) Policy of individual replacement and

(ii) Policy of group replacement.

As per the given data, we can see that no bulb will survive for more than 6 months. That is a bulb, which has survived for 5 months, is sure to fail during the sixth month. Though we replace the failed bulb immediately, it is assumed that the bulb fails during the month will be replaced just at the end of the month.

Let, N_i = Number of bulbs replaced at the end of ith month, then we can calculate different values of N_i

N_0 = Number of bulbs at the beginning
= 10000

N_1 = Number of bulbs replaced at end of first month

= Number of bulbs at the beginning × Probability that a bulb fails during 1st month of installation

= 10000×0.03 = **300**

N_2 = Number of bulbs to be replaced at the end of second month = (Number of bulbs at the beginning × Probability of failure during the second month) + (Number of bulbs replaced at the end of second month × Probability of failure during the second month)

$N_0 P_2 + N_1 P_1$ $(10000 \times 0.07) + (300 \times 0.03)$ = **709**

Similarly,

N_3 = $N_0 P_3 + N_1 P_2 + N_2 P_1$

= $10000 \times 0.20 + 300 \times 0.07 + 709 \times 0.03$

= **2042**

N_4 = $N_0 P_4 + N_1 P_3 + N_2 P_2 + N_3 P_1$

= $10000 \times 0.40 + 300 \times 0.20$

= $709 \times 0.07 + 2042 \times 0.03$

= **4171**

N_5 = $N_0 P_5 + N_1 P_4 + N_2 P_3 + N_3 P_2 + N_4 P_1$

= $10000 \times 0.15 + 300 \times 0.40 + 709 \times 0.20 + 2042 \times 0.07 + 4171 \times 0.03$

= **2030**

N_6 = $N_1 P_5 + N_2 P_4 + N_3 P_3 + N_4 P_2 + N_5 P_1$

= $10000 \times 0.15 + 300 \times 0.15 + 709 \times 0.40 + 2042 \times 0.20 + 4171 \times 0.07 + 2030 \times 0.03$

= **2590**

From the above we can see that the failures increases from 300 to 4171 in 4th month and then decreases. It is very much common in any system that failure rate increases and then decreases and finally after certain period, it stabilizes, when the system attains steady state.

Now let us work out the expected life of each bulb which is = $\odot$ $x_i P_i$, where x_i is the month and P_i is corresponding probability of failure.

$\text{©} x_i P_i = 1 \times 0.03 + 2 \times 0.07 + 3 \times 0.20 \times 4$
$$\times 0.40 + 5 \times 0.15 + 6 \times 0.15$$

= 4.02 months

If the average life of a bulb is 4.02 months, the average number of replacements every month = Number of bulbs in the system/average life of the bulb. = 10000/4.02 = **2488 bulbs**. As the cost of individual replacement cost is ₹ 1/- per bulb, on an average, the organization has to spend 2488 × ₹ 1/- = ₹ 2488/- per month.

Now let us work the cost of group replacement :

Table 5.28

End of the Period	Total Cost of Group Replacement in ₹ Individual Replacement Cost + Group Replacement Cost	Cost per Month Total Cost/Period ₹ / Month
1	$300 \times 1 + 10000 \times 0.35 = 3800$	3800/1 = 3800
2	$(300 + 709) \times 1 + 10000 \times 0.35 = 4509$	4509/2 = 2254.50
3	$(300 + 709 + 2042) \times 1 + 10000 \times 0.35 = 6551$	6551/3 = 2183.66
4	$(300 + 709 + 2042 + 4771) \times 1 + 10000 \times 0.35$ $= 10722$	10722/4 = 2680.50
5	$(300 + 709 + 2042 + 4771 + 2030) \times 1 + 10000 \times 0.35 = 12752$	12752/5 = 2550.40
6	$(300 + 709 + 2042 + 4771 + 2030 + 2590)$ $\times 1 + 10000 \times 0.35 = 15342$	15342/6 = 2557.00

In the above table in column No. 2 it is shown the cost of individual replacement at the end of month plus group replacement of all the bulbs at the end of month. The minimum cost of group replacement i.e. ₹ 2183.66 is at the end of third month. This is compared with individual replacement cost per month, which is ₹ 2488/- per month. Hence, replacement of all the bulbs at the end of third month is more beneficial to the organization. Hence, optimal replacement policy is replace all the bulbs at the end of third month.

Example 5.15 : *Truck tyres, which fail in service, can cause expensive accidents. It is estimated that a failure in serviced results in an average cost of ₹1000/- exclusive of the cost of replacing the burst tyre. New tyres cost ₹400/- each and are subjected to mortality as in table on next page. If the tyres are to be replaced after a certain fixed mileage or on failure (whichever occurs first), determine the replacement policy that minimizes the average cost per mile. Mention the assumptions you made to arrive at the solution.*

[May 16]

Table 5.29 : Table Showing the Truck Tyre Mortality

Age of Tyre at Failure (Miles)	Proportion of Tyre
≤ 10000	0.000
10001 – 12000	0.020
12001 – 14000	0.035
14001 – 16000	0.063
16001 – 18000	0.100
18001 – 20000	0.220
20001 – 22000	0.345
22001 – 24000	0.205
24001 – 26000	0.012
Total	**1.000**

Solution :

Assumptions :

(1) The failure occurs at the midpoint of the range given in the table i.e. exactly at 11000, 13000, 15000, etc.

(2) Initially let there be 1000 tyres.

(3) Upto the age of 10000 miles proportion of tyres fails = 0. The proportion of failure from 11000 to 13000 miles is 0.030. Assume that in this period the average cost of maintaining is ₹ 1000 /-. If in this period a tyre bursts, the cost will be ₹ 1400/-. Thus, the cost of individual replacement will be ₹ 1400/-. The cost of group replacement is given as ₹ 400/- per tyre. Hence, the number of tyres fail and are to be replaced is :

$$N_0 = 1000$$
$$N_1 = N_0 P_1 = 1000 \times 0.020 = \textbf{20}$$
$$N_2 = N_0 \times P_2 + N_1 P_1$$
$$= 1000 \times 0.035 + 20 \times 0.020$$
$$= 35 + 0.04 = 35.04$$
$$= \text{Approximately } \textbf{35}$$
$$N_3 = N_0 \times P_3 + N_1 P_2 + N_2 P_1$$
$$= 1000 \times 0.063 + 20 \times 0.035 + 35 \times 0.02$$
$$= 63 + 0.7 + 0.7 = 64.4 = \textbf{64}$$
$$N_4 = N_0 \times P_4 + N_1 \times P_3 + N_2 \times P_2 + N_3 \times P_1$$
$$= 1000 \times 0.1 + 20 \times 0.063 + 35 \times 0.035 + 64 \times 0.02$$
$$= 100 + 1.26 + 1.23 + 1.28 = 103.73$$
$$= \textbf{104}$$
$$N_5 = N_0 \times P_5 + N_1 \times P_4 + N_2 \times P_3 + N_3 P_2 + N_4 \times P_1$$
$$= 1000 \times 0.220 + 20 \times 0.100 + 35 \times 0.063 + 64 \times 0.035 + 104 \times 0.020$$
$$= 220 + 2 + 2.205 + 2.24 + 2.08$$

$$= 224.205$$
$$= \textbf{App. 224}$$

$$N_6 = N_0 \times P_6 + N_1 \times P_5 + N_5 \times P_4 + N_3 \times P_3 + N_2 + P_2 + N_1 \times P_1$$
$$= 1000 \times 0.345 + 20 \times 0.220 + 35 \times 0.1 + 64 \times 0.063 + 104 \times 0.02$$
$$= 345 + 4.4 + 3.5 + 4.032 + 3.64 + 4.48$$
$$= 714.45 = \textbf{App. 714}$$

$$N_7 = N_0 \times P_7 + N_1 \times P_6 + N_2 \times P_5 + N_3 \times P_4 + N_4 \times P_3 + N_5 \times P_2 + N_6 \times P_1$$
$$= 1000 \times 0.205 + 20 \times 0.345 + 35 \times 0.220 + 64 \times 0.1 + 104 \times 0.063 + 224 \times 0.035 + 714 \times 0.02$$
$$= 205 + 7.3 + 7.7 + 6.4 + 6.24 + 7.84 + 14.28 = 254.76 = \textbf{App. 255}$$

$$N_8 = N_0 \times P_8 + N_1 \times P_7 + N_2 \times P_6 + N_3 \times P_5 + N_4 \times P_4 + N_5 \times P_3 + N_6 \times P_2 + N_7 \times P_1$$
$$= 1000 \times 0.012 + 20 \times 0.205 + 35 \times 0.345 + 64 \times 0.220 + 104 \times 0.10 + 224 \times 0.063 + 714 \times 0.035 + 255 \times 0.02$$
$$= 12 + 4.1 + 12.075 + 14.08 + 10.4 + 14.112 + 24.99 + 5.1$$
$$= 96.857 = \textbf{App. 97}$$

Expected average life of a tyre is $\sum\limits_{i=1}^{i=inf} i p_i$

p_i = Time period, p is the probability of failure.

$$\Sigma_i\, p_i = 1 \times 0.020 + 2 \times 0.035 + 3 \times 0.063 + 4 \times 0.100 + 5 \times 0.220 + 6 \times 0.345 + 7 \times 0.205 + 8 \times 0.012$$
$$= 0.02 + 0.07 + 0.189 + 0.40 + 1.1 + 2.07 + 1.435 + 0.096 = \textbf{5.38}$$

Average number of failures

$$= (1000 / 5.38) = 185.87$$
$$= \text{App. 186 tyres.}$$

Cost of individual replacement per period

$$= 186 \times 1400 = 260400/\text{-}$$

Cost of individual replacement per mile

$$= (234 \times 1400) / 2000$$
$$= 260400 / 2000$$
$$= ₹\,\textbf{130.20}$$

Replacement by Group Replacement Policy :

Table 5.30

End of the Period in Miles	Total Cost of Group Replacement = Individual Replacement + Group Replacement (₹)	Average Cost per Period in Miles (₹)
11000 – 13000 (1)	$1000 \times 400 + 20 \times 1400 = 428000$	428000
13000 – 15000 (2)	$(1000 + 20) \times 400 + 35 \times 1400$ $= 408000 + 49000$ $= 457000$	457000/2 = 228500
15000 – 17000 (3)	$(1000 + 20 + 35) \times 400 + 64 \times 1400$ $= 422000 + 89600 = 511600$	511600/3 = 170533
17000 – 19000 (4)	$(1000 + 20 + 35 + 64) \times 400 + 104 \times 1400$ $= 447600 + 145600$ $= 593200$	593200/4 = 148300
19000 – 21000 (5)	$(1000 + 20 + 35 + 64 + 104) \times 400 + 224 \times 1400$ $= 489200 + 313600$ $= 802800$	802800/5 = 160560
21000 – 23000 (6)	$(1000 + 20 + 35 + 64 + 104 + 224) \times 400 + 714 \times 1400$ $= 578800 + 996600$ $= 1598400$	1598400/6 = 263066
23000 – 25000 (7)	$(1000 + 20 + 35 + 64 + 104 + 224 + 714) \times 400 + 255 \times 1400$ $= 864400 + 357000$ $= 1221400$	1221400/7 = 174485.70
25000 – 27000 (8)	$(1000 + 20 + 35 + 64 + 104 + 224 + 714 + 225) \times 400 + 97 \times 1400$ $= 966400 + 135800$ $= 1102200$	1102200/8 = 137775

Now we know that the individual replacement cost is ₹ 130.20. Where as even at the end of 8th period, the average cost per period in miles is 137775 / 1000 = ₹ 137.775 which is higher than ₹ 130.02. Hence, it is better to stick to individual replacement policy.

Example 5.16 : *Find the cost per period of individual replacement policy of an installation of 300 bulbs, given the following :*

(1) Cost of individual replacement of bulb is ₹ 2/- per bulb.

(2) Conditional probability of failure of bulbs is as follows :

[May 15]

Table 5.31

Weekend	0	1	2	3	4
Probability of Failure	0	0.1	0.3	0.7	1.0

Solution :

If p_i is the probability of failure of bulbs then :

$p_0 = 0$, $p_1 = 0.1$, $p_2 = 0.3 - 0.1 = 0.20$, $p_3 = 0.7 - 0.3 = 0.4$, and $p_4 = 1 - 0.7 = 0.3$

Since sum of probabilities is unity, all probability higher than pa must be zero, i.e. a bulb that has been survived up to 4th week, is sure to fail at the end of fourth week. Let us find the average life of a bulb, which is given by

$$\sum_{i=1}^{i=\inf} ip_i = 1 \times 0.1 + 2 \times 0.2 + 3 \times 0.4 + 4 \times 0.3$$

$$= 2.9 \text{ weeks.}$$

Average number of failures per week is $300/2.9 = 103.448$

$$= \text{App. } 103$$

Cost of individual replacement is ₹ $2 \times 103 = ₹ 206$.

Number of bulbs to be replaced at the end of every week is

$$N_0 = 300$$
$$N_1 = N_0 \times P_1 = 300 \times 0.1 = 30$$
$$N_2 = N_0 \times P_2 + N_1 \times P_1$$
$$= 300 \times 0.2 + 30 \times 0.1$$
$$= 60 + 3 = 63$$
$$N_3 = N_0 \times P_3 + N_1 \times P_2 + N_2 \times P_1$$
$$= 300 \times 0.4 + 30 \times 0.2 + 63 \times 0.1$$
$$= 120 + 6 + 6.3 = 132.3$$
$$= \textbf{App. 132}$$
$$N_4 = N_0 \times p_4 + N_1 \times p_3 + N_2 \times p_2 + N_3 \times p_1$$
$$= 300 \times 0.3 + 30 \times 0.4 + 63 \times 0.2 + 132 \times 0.1$$
$$= 90 + 12 + 12.6 + 13.2 = 127.8$$
$$= \textbf{App. 128}$$

The number failures increase upto 3rd week and then it reduces. This is very much common in all the systems that after sometime the system reach steady state.

Example 5.17 : *A typing pool of a large organization employs 100 copy typists. The distribution of length of service is given below :*   ***[Dec. 17]***

Table 5.32

Duration of Employment in Years	1	2	3	4	5 or more
Proportion of Employees that Leave in that Year of Employment	30%	40%	20%	10%	0%

Assuming that an employee leaving is replaced by another at the end of the year, determine :

(1) *The number of staff who leaves in each of the first 8 years of the department's existence, assuming it stared with 100 employees and this total number does not change.*

(2) *The number leaving each year when the steady state situation is reached, and*

(3) *The total annual cost of recruiting staff in the steady state if replacement of each new copy typist costs ₹200.*

Solution :

Note that there are 100 copy typists in the beginning and no person stays more than five years. Hence, let us calculate the number of persons leaving the organization and new employees employed every year.

Table 5.33

Year	Number of Employees Leaving at the End of The Year	No. of New Employees Employed
0	Nil	$N_0 = 100$
1	$N_0 \times P_1 = 100 \times 0.30 = 30$	$N_1 = 30$
2	$N_0 \times P_2 + N_1 \times P_1 = 100 \times 0.4 + 30 \times 0.30 = 49$	$N_2 = 40 + 9$ $= 49$
3	$N_0 \times P_3 + N_1 \times P_2 + N_2 \times P_1$ $= 100 \times 0.2 + 30 \times 0.4 + 49 \times 0.3 = 46.7$	$N_3 = 20 + 12 + 14.7$ $= 46.7$
4	$N_0 \times P_4 + N_1 \times P_3 + N_2 \times P_2 \times N_3 \times P_1$ $= 100 \times 0.1 + 30 \times 0.2 + 49 \times 0.4 + 46.7 \times 0.3$ $= 10 + 6 + 19.6 + 14.1 = 49.7$	$N_4 = 49.7$
5	$N_0 \times P_5 + N_1 \times P_4 + N_2 \times P_3 + N_3 \times P_2 + N_4 \times P_1$ $= 100 \times 0 + 30 \times 0.1 + 49 \times 0.2 + 46.7 \times 0.4$ $+ 49.6 \times 0.3$ $= 0 + 3 + 9.8 + 18.68 + 14.88$ $= 46.36$	$N_5 = 46.36$ $= 46.4$
6	$N_0 \times P_6 + N_1 \times P_5 + N_2 \times P_4 + N_3 \times P_3 + N_4 \times P_2 \times N_5 \times P_1$ $= 100 \times 0 + 30 \times 0 + 49 \times 0.1 + 46.7 \times 0.2$ $+ 49.6 \times 0.4 + 46.4 \times 0.3$ $= 0 + 0 + 4.9 + 9.34 + 19.84 + 13.92 = 48$	$N_6 = 48$
7	$N_0 \times P_7 + N_1 \times P_6 + N_2 \times P_5 + N_3 \times P_4 + N_4 \times P_3 \times N_5 \times P_2 + N_6 \times P_1$ $= 100 \times 0 + 30 \times 0 + 49 \times 0 + 46.7 \times 0.1$ $+ 49.6 \times 0.2 + 46.4 \times 0.4 + 48 \times 0.3$ $= 0 + 0 + 0 + 4.67 + 9.92 + 18.56 + 14.4 = 47.55$	$N_7 = 47.6$

...Conti.

8	$N_0 \times P_8 + N_1 \times P_7 + N_2 \times P_6 + N_3 \times P_5 + N_4 \times P_4 \times N_5$ $\times P_3 + N_6 \times P_2 + N_7 \times P_1$	N_7 = 47.6
	$= 100 \times 0 + 30 \times 0 + 49 \times 0 + 46.7 \times 0 + 49.6$ $\times 0.1 + 46.4 \times 0.2 + 48 \times 0.4 + 47.06 \times 03$	
	$= 0 + 0 + 0 + 0 + 4.96 + 9.28 + 19.2 + 14.12$	
	$= 47.56$	

Expected length of service of a copy typist in the organization = $©' \, x_i \cdot p_i$ = $1 \times 0.3 + 2 \times 0.4 + 3 \times 0.2 + 4 \times 0.1$ = 0.3 + 0.8 + 0.6 + 0.4 = 2.1 years. Hence, average number of employees leaving the organization at the end of the year = 100 / 2.1 = 47.62 employees. Annual cost of replacing a copy typist in steady state = Cost of replacement × Average number replaced = ₹ 200 × 47.62 = ₹ 9524/-.

5.8 STAFFING EXAMPLE

- The replacement model may be well applied to *Manpower Planning*, where one can plan well in advance the requirement of different types of staff personnel or skilled/unskilled personnel. Here personnel are also considered as elements replaced for some reason or the other.

- Any organization requires at various period of time different types of personnel due to retirement, persons quitting the job in search of better jobs, vacancies arising due to death of personnel, termination, resignation etc.

- Therefore to maintain suitable strength of staff members in a system there is a need to formulate some useful recruitment policy. In this case, we assume that the life distribution for the service of staff in a system is known.

Example 5.18 : *A research team is planed to raise its strength to 50 chemists and then to remain at that level. The wastage of recruits depends on their length of service, which is as follows :*   ***[May 15]***

Table 5.34

Year	Total Percentage Who have Left up to the End of the Year	Year	Total Percentage Who have Left up to the End of the Year
1	5	6	73
2	36	7	79
3	56	8	87
4	63	9	97
5	68	10	100

What is the recruitment per year necessary to maintain the strength? There are 8 senior posts for which the length of service is the main criterion for promotion. What is the average length of service after which new entrant can expect his promotion to one of these posts.

Solution :

Table 5.35

Year	Number of Persons who Leave at the End of the Year	Number of Persons in Service at the End of the Year 100 – (2)	Probability of Leaving at the End of the Year (2)/100	Probability at the in Service at the End of the Year (3)/100
(1)	(2)	(3)	(4)	(5)
0	0	100	0	1.00
1	5	95	0.05	0.95
3	36	64	0.36	0.64
4	56	44	0.56	0.44
5	63	37	0.63	0.37
6	73	27	0.73	0.27
7	79	21	0.70	0.21
8	87	13	0.87	0.13
9	97	3	0.97	0.03
10	0	0	1.00	0.00
	Total	436		

From the given data, we can find the probability of a chemist leaving during a certain year. The person who is joining the organization will not continue after 10 years. And we know that mortality table for any item can be used to derive the probability distribution of life span by $\{M(t - 1) - M(t)\} \, N$.

The required probabilities are calculated in the table above. The column (3) shows that a recruitment policy 9 of 100 every year, the total number of chemists serving in the organization would have been 436. Hence, to maintain strength of 50 chemists, then the recruitment should be : (100 × 50) / 436 = 11.5 or approximately 12 chemists per year. As per life distribution of service 12 chemists are to be recruited every year, to maintain strength of 50 chemists. Now referring to column (5) of the table above, we find that number of survivals after each year. This is given by multiplying the various values in column (5) by 12 as shown in the table given below :

Table 5.36

Number of Years	0	1	2	3	4	5	6	7	8	9	10
Number of Chemists in Service	12	11	7	5	4	4	3	2	2	0	0

As there are 8 senior posts, from the table we find that there are 3 persons in service during the 6th year, 2 in 7th year, and 2 in 8th year. Hence, the promotion for new recruits will start from the end of fifth year and will continue upto sixth year.

(OR it can be done in this way : if we recruit 12 persons every year, then we want 8 seniors. Suppose we recruit 100 every year, then we shall require $(8 \times 100)/12 = 66.4$ or approximately 64 seniors. It is seen from the first table above that required number of persons would be available, if we promote them at the end of fifth year.)

Example 5.19 : *A firm is considering replacement of machine, the maintenance cost and resale value per year of the machine whose purchase price is ₹ 7000 is given below :*　　**[May 17]**

Table 5.37

Years	1	2	3	4	5	6	7	8
Maintenance Cost in (₹)	900	1200	1600	2100	2800	3700	4700	5900
Resale Value in (₹)	4000	2000	1200	600	500	400	400	400

When should the machine be replaced?　　**[May 17, Dec. 08]**

Solution : C = ₹ 7,000.

Table 5.38

Year	Resale Cost S	C − S	Maintenance Cost	Σ Maintenance Cost	Total Cost 1 + 2	Average Annual Cost B/A
(A)	(1)			(2)	(B)	
1	4000	3000	900	900	3900	3900
2	2000	5000	1200	2100	7100	3550
3	1200	5800	1600	3700	9500	3166.67
4	600	6400	2100	5800	12200	3050.00
5	500	6500	2800	8600	15100	3020.00
6	400	6600	3700	12300	18900	3150.00
7	400	6600	4700	17000	23600	3371.43
8	400	6600	5900	22900	29500	3687.5

As the average annual cost is minimum for 5th year. The machine should be replaced after 5 years.

Example 5.20 : *A firm has a machine whose purchase price is ₹1,00,000. Its running cost and resale price (₹) at the end of different years are as follows :*

Table 5.39

Year	1	2	3	4	5	6
Running Cost	7,500	8,500	10,000	12,500	17,500	27,500
Resale Cost	85,000	76,500	70,000	60,000	40,000	15,000

(1) Obtain the economic life of the machine and the minimum average cost.

(2) The firm has obtained a contract to supply the goods produced by machine, for the period of five years from now. After this time period, the firm does not intended to use the machine. If the firm has a machine of this type which is one year old, what replacement policy should it adopt if it intends to replace the machine not more than once?　　**[May 17, Dec. 09]**

Solution : C = 1,00,000, S = Resale cost,

Table 5.40

Year (n)	Depreciation (C − S)	Annual Maintenance	Summation of Maintenance	Total Cost (2 + 4)	Average Annual Cost [5/1]
1	2	3	4	5	6
1	15,000	7,500	7,500	22,500	22,500
2	23,500	8,500	16,000	39,500	19,750
3	30,000	10,000	26,000	56,000	18666.67
4	40,000	12,500	38,500	78,500	19,625
5	60,000	17,500	56,000	1,16,000	23,200
6	85,000	27,500	83,500	1,68,500	28,083.34

(a) The minimum average cost is 18,666.67 corresponds to 3rd year. Therefore, economic life is 3 years.

(b) Just calculate the yearly cost (and communication cost) of keeping one year old machine in year 1-5.

Table 5.41

Year	Maintenance	Depreciation	Total Cost	Cumulative Cost
1	8,500	8,500	17,000	17,000
2	10,000	6,500	16,500	33,500
3	12,500	10,000	22,500	56,000
4	17,500	20,000	37,500	93,500
5	27,500	25,000	52,500	1,46,000

If keeping old machine for zero year and new for 5 years then

$$\text{Cost} = (0 + 1,16,000)$$
$$= ₹\ 1,16,000$$

Keeping old machine for 1 year and new for 4 years

$$\text{Cost} = (17,000 + 78,500)$$
$$= ₹\ 95,500$$

Keeping old machine for 2 years and new for 3 years

$$\text{Cost} = (33,500 + 56,000)$$
$$= ₹\ 89,500$$

Keeping old machine for 3 years and new for 2 years

$$\text{Cost} = (56,000 + 39,500)$$
$$= ₹\ 95,500$$

Keeping old machine for 4 years and new for 1 year

$$\text{Cost} = (93,500 + 22,500)$$
$$= ₹\ 1,16,000$$

Keeping old machine for 5 years and do not buy new one

$$\text{Cost} = (1,46,000 + 0)$$
$$= ₹\ 1,46,000$$

Thus, keep the old machine running for 2 years and then buy a new machine in next year.

Example 5.21 : *Find the cost per period of individual replacement policy of an installation of 300 bulbs given the data :*

(1) *Cost of replacing an individual bulb is ₹ 2.00.*

(2) *Conditional probability of failure is given below :*

Table 5.42

Week No.	0	1	2	3	4
Conditional Probability of Failure	0	0.1	0.3	0.7	1.0

(3) *Also calculate number of bulbs that would fail during each of four weeks.*   ***[May 17, Dec. 10]***

Solution :

(1) Calculating no. of bulbs that would fail in each of the week :

$$P_0 = 0$$
$$P_1 = 0.1$$
$$P_2 = 0.2$$
$$P_3 = 0.4$$
$$P_4 = 0.3$$

(2)

Table 5.43

Week	Expected no. of Failure	No. of Failures
0	$N_0 = 300$	300
1	$N_1 = N_0 P_1$	30
2	$N_2 = N_0 P_2 + N_1 P_1$	63
3	$N_3 = N_0 P_3 + N_1 P_2 + N_3 P_1$	132.3
4	$N_4 = N_0 P_4 + N_1 P_3 + N_2 P_2 + N_4 P_4$	128

(3) Average life :

$$\sum_{i=0}^{n} i P_i = 2.9$$

(4) Average number of failures per week

$$= \frac{N}{\text{Average life}} = \frac{300}{2.9}$$
$$= 103.45$$

(5)
$$\text{Cost} = 2 \times 103.45$$
$$= ₹\ 206.89$$

Example 5.22 : *The following mortality rates have been observed for a certain type of light bulbs.*

Table 5.44

Week	1	2	3	4	5
% failing by end of week	10	25	50	80	100

There are 1000 bulbs in use and it cost ₹ 2 to replace an individual bulb which are burnt out. If all bulbs were replaced it would cost 50 paise per bulb. It is proposed to replace all bulbs at fixed intervals whether or not they have burnt out. At what interval should all the bulbs be replaced?

[Dec. 15, May 10]

Solution : Let P_i be the probability that a light bulb which was new when placed in position for use fails during i^{th} week of its life.

$$P_1 = \text{Probability of failure in } 1^{st} \text{ week} = 0.1$$
$$P_2 = \text{Probability of failure in } 2^{nd} \text{ week} = 0.15$$
$$P_3 = \text{Probability of failure in } 3^{rd} \text{ week} = 0.25$$
$$P_4 = \text{Probability of failure in } 4^{th} \text{ week} = 0.3$$
$$P_5 = \text{Probability of failure in } 5^{th} \text{ week} = 0.2$$

$$\text{Sum of all probabilities} = \underline{1}$$

Hence, all further probabilities i.e. P_6, P_7 ... will be zero.

$$N_1 = \text{No. of replacement at end or } i^{th} \text{ week}$$
$$= \mathbf{1000}$$
$$N_0 = N_0$$

$$N_1 = N_0 \cdot P_1 = 1000 \times 0.1 = \mathbf{100}$$

$$N_2 = N_0 \cdot P_2 + N_1 \cdot P_1$$
$$= 1000 \times 0.15 + 100 \times 0.1$$
$$= \mathbf{160}$$

$$N_3 = N_0 \cdot P_3 + N_1 \cdot P_2 + N_2 \cdot P_1 = 100 \times 0.25$$
$$+ 100 \times 0.15 + 160 \times 0.1 = \mathbf{281}$$

$$N_4 = N_0 \cdot P_4 + N_1 \cdot P_3 + N_2 \cdot P_2 + N_3 \cdot P_1$$
$$= \mathbf{377}$$

$$N_5 = N_0 \cdot P_5 + N_1 \cdot P_4 + N_2 \cdot P_3 + N_3 \cdot P_2$$
$$+ N_4 \cdot P_1 = \mathbf{350}$$

$$N_6 = N_1 \cdot P_5 + N_2 \cdot P_4 + N_3 \cdot P_3 + N_4 \cdot P_2$$
$$+ N_5 \cdot P_1 = \mathbf{230}$$

$$N_7 = N_2 \cdot P_5 + N_3 \cdot P_4 + N_4 \cdot P_3 + N_5 \cdot P_2$$
$$+ N_6 \cdot P_1 = \mathbf{286}$$

Mean Age of bulbs $= 1 \times P_1 + 2 \times P_2 + 3 \times P_3 + 4 \times P_4 + 5 \times P_5 = \mathbf{3.35 \ weeks}$

Number of failures in each week $= \dfrac{1000}{3.35} = \mathbf{299 \ per \ week}$

Cost of replacing 299 bulbs each week $= 299 \times 10 = 2990$

Table 5.45

End of Week	Cost of Individual Replacement	Total Cost of Group Replacement	Average Cost
1	$100 \times 10 = 1000$	$1000 \times 4 + 100 \times 10$ $= 5000$	5000
2	$160 \times 10 = 1600$	$5000 + 1600 = 6600$	3300
3	$281 \times 10 = 2810$	$6600 + 2810 = 9410$	3136.67
4	$377 \times 10 = 3770$	$3770 + 9410 = 13180$	3295

Cost of individual replacement in 4th week exceeds the average cost of 3 weeks. Hence, it will be optimal to replace all bulbs in 3rd week otherwise cost will increase.

QUESTIONS FOR PRACTICE

1. What is replacement? Explain by means real world examples. **[May 17]**

2. Explain different types of replacement problems by giving examples. **[Dec. 16]**

3. (a) Write a brief note on replacement.

 (b) The cost of maintenance of equipment is given by a function of increasing with time and its scrap value is constant. Show that replacing the equipment when the average cost to date becomes equal to the current maintenance cost will minimize the average annual cost.

4. A firm is considering when to replace its machine whose price is ₹ 12,200/-. The scrap value of the machine is ₹ 200/- only. From past experience maintenance cost of machine is as under. **[May 15]**

Year	1	2	3	4	5	6	7	8
Maintenance cost in ₹	200	500	800	1200	1800	2500	3200	4000

5. The following is the cost of running a particular car to date and the forecast into the future. Assume that a similar car will replace the car, when is the best time to replace it and what will be the average yearly running cost?

Year	Resale Value at the End of the Year ₹	Petrol and Tax During the Year ₹	All other Running Cost during the Year ₹
0	700	–	–
1	625	90	10
2	575	90	30
3	550	90	50
4	500	90	70
5	450	90	90
6	450	90	110
7	350	90	130
8	300	90	150

6. Machine B costs ₹ 10,000/-. Annual operating costs are ₹ 400/- for the first year and they increase by ₹ 800/- each year. Machine, A which is one year old, costs ₹ 9000/- and the annual operating costs are ₹ 200/- for the first year and they increase by ₹ 2000/- every year. Determine at what time is it profitable to replace machine A with machine B. (Assume that machines have no resale value and the future costs are not discounted).

7. A firm pays ₹ 10,000/- for its automobiles. Their operating and maintenance costs are about ₹ 2,500/- per year for the first two years and then go up by ₹ 1500/- approximately per year. When should such vehicles be replaced? The discount rate is 0.9.

[May 16]

8. The cost of new machine is ₹ 4000/-. The maintenance cost of 'n' th year is given by $R_n = 500 (n - 1)$ where $n = 1, 2, 3, ..., n$. Suppose that the discount rate per year is 0.05. After how many years will it be economical to replace the machine by a new one?

9. The following mortality rates have been observed for a certain type of light bulbs :

Week	1	2	3	4	5
Per Cent Failing by the Weekend	10	25	50	80	100

There are 1000 bulbs in use and it costs ₹ 2/- to replace an individual bulb, which has burnt out. If all bulbs were replaced simultaneously, it would cost 50 paise per bulb. It is proposed to replace all the bulbs at fixed intervals, whether or not they have burnt out, and to continue replacing burnt out bulbs as they fail. At what intervals should all the bulbs be replaced?

10. The following mortality rates have been found for a certain type of coal cutter motor :

Week	10	20	30	40	50
Total % Failure upto end of 10 Weeks Period	5	15	35	65	100

If the motors are replaced over the week and the total cost is ₹ 200/-. If they fail during the week the total cost is ₹ 100/- per failure. Is it better to replace the motors before failure and if so when?

11. Explain how the theory of replacement is used in following problems :
    (i) Replacement of items that fail completely.
    (ii) Replacement of items whose maintenance cost varies with time.   **[May 16, Dec 11, 8M]**

12. A JCB excavator operator purchases the machine of ₹ 15,00,000. The operating cost and the resale value of machine is given below :
    [May 16, Dec. 15 8M, Dec. 11, 8M]

Year	1	2	3	4	5	6	7	8
Operating Cost (₹)	30,000	32,000	36,000	40,000	45,000	52,000	60,000	70,000
Resale value in (Lacks of ₹)	12	10	8	5	4.5	4	3	2

When should the machine be replaced ?

13. Discuss the replacement policy for the items that fail suddenly.   **[Dec. 17, May 12, 6M]**

14. A firm is thinking of replacing a particular machine whose cost price is ₹ 12,200. The scrap price of this machine is only ₹ 200. The maintenance costs are found to be as follows :   **[10]**

Year	1	2	3	4	5	6	7	8
Maintenance Cost (₹)	220	500	800	1200	1800	2500	3200	4000

Determine when the firm should get the machine replaced.   **[May 12, 10]**

15. A manufacturer is offered 2 machines A & B. A has cost of ₹ 2500, its running cost is ₹ 400 for each of the first 5 years and increases by ₹ 100 every subsequent year. Machine B having the same capacity as A, costs ₹ 1250, has running cost of ₹ 600 for 6 years increasing by ₹ 100 per year thereafter. If moneys worth 10% per year. Which machine should be purchased? Scrap value of both machines is negligibly small.   **[May 17, Dec. 12, 16]**

16. A manufacture is offered two machines A and B. A is priced at ₹ 5000 and running cost are estimated at ₹ 800 for each of the first five years,' increasing by ₹ 200 per year in the sixth and subsequent years. Machine B, which has the same capacity as A, cost ₹ 2500 but will have running cost of ₹ 1200 per year for six years, increasing by ₹ 200 per year thereafter. If money is worth 12% per year, which machine should be purchased? (Assume that the machines will eventually be sold for scrap at negligible price).   **[Dec. 15, May 13,10]**

17. Explain how the theory of replacement is used in following problems :   **[Dec. 13, 6M]**
    (i) Replacement of items that fail completely.
    (ii) Replacement of items whose maintenance cost varies with time.

18. A machine costs ₹ 500. Operation and maintenance costs are zero for the first year and increased by ₹ 100 every year. If money is worth 5% every year, determine the best age at which the machine should be replaced. The resale value of the machine is negligibly small. What is the weighted average cost of owing and operating the machine?

19. Following figures are related to toy manufacturing company.   **[May 18, 6M]**

    Variable cost per unit = ₹ 8/-

    Selling price per unit = ₹ 14/-

    Total units sold = ₹ 50,000/-

    Fixed cost = ₹ 12,000/-

    Calculate :

    (a) P/v ration,

    (b) BEP in units

    (c) Margin of safety.

20. Explain how the theory of replacement is used in following problems :

    (i)   Replacement of items that fail completely.

    (ii) Replacement of items whose maintenance cost varies with time.   **[May 2016, Dec 2011, 8]**

    (Refer Section 5.3, Section 5.5.1.)

21. A JCB excavator operator purchases the machine of ₹ 15,00,000. The operating cost and the resale value of machine is given below :   **[May 16, Dec 11, 15,8]**

Year	1	2	3	4	5	6	7	8
Operating Cost (₹)	30,000	32,000	36,000	40,000	45,000	52,000	60,000	70,000
Resale value in (Lacks of ₹)	12	10	8	5	4.5	4	3	2

    When should the machine be replaced ?

    (Refer Section 5.5.1 [Example 5.3])

    [Ans. Replaced after 3 years]

22. Discuss the replacement policy for the items that fail suddenly.   **[Dec. 17, May 12, 6M]**

23. A firm is thinking of replacing a particular machine whose cost price is ₹ Rs 12,200. The scrap price of this machine is only ₹ Rs 200. The maintenance costs are found to be as follows :

Year	1	2	3	4	5	6	7	8
Maintenance cost (₹)	220	500	800	1200	1800	2500	3200	4000

Determine when the firm should get the machine replaced.   **[May 2012, 10]**

(Refer Section 5.5.1 [Example 5.1].)

24. A manufacturer is offered 2 machines A & B. A has cost of ₹ 2500, its running cost is ₹ 400 for each of the first 5 years and increases by ₹ 100 every subsequent year. Machine B having the same capacity as A, costs ₹ 1250, has running cost of ₹ 600 for 6 years increasing by ₹ 100 per year thereafter. If moneys worth 10% per year. Which machine should be purchased? Scrap value of both machines is negligibly small.   **[May 17, Dec 12, 16]**

(Refer Section 5.5.3 [Example 5.7].)

26. A manufacture is offered two machines A and B. A is priced at ₹ 5000 and running cost are estimated at ₹ 800 for each of the first five years,' increasing by ₹ 200 per year in the sixth and subsequent years. Machine B, which has the same capacity as A, cost ₹ Rs 2500 but will have running cost of ₹ 1200 per year for six years, increasing by ₹ 200 per year thereafter. If money is worth 12% per year, which machine should be purchased? (Assume that the machines will eventually be sold for scrap at negligible price).

          [Dec. 2015, May 2013, 10]

(Refer Section 5.5.2 [Example 5.10])

27. Explain how the theory of replacement is used in following problems: **[Dec 2013, 6M]**

    (i)   Replacement of items that fail completely.

    (ii) Replacement of items whose maintenance cost varies with time.

    (Please refer to Sections 5.7 and 5.5.2.)

28. A machine costs Rs. 500. Operation and maintenance costs are zero for the first year and increased by Rs. 100 every year. If money is worth 5% every year, determine the best age at which the machine should be replaced. The resale value of the machine is negligibly small. What is the weighted average cost of owing and operating the machine?

          [May 2016, Dec 2013, 10]

(Please refer to Section 5.5.3 [Example 5.8])

◈ ◈ ◈

CHAPTER 6
INVENTORY MODELS

6.1 INTRODUCTION

- Dictionary meaning of inventory is a detailed list of movable goods. It also means `stock on hand' at a given time. Inventory in a wider sense can be defined as any idle resource of an enterprise. But it is commonly used to indicate materials i.e. raw materials, finished, semi-finished, packing, spares and others stocked in order to meet an expected demand or distribution.

- Inventory is almost necessary to maintain some inventories for smooth functioning of an organisation.

- Requirements to keep inventories are as follows:

  ➢ It takes time to complete one operation and more product from one stage to another.

  ➢ To run manufacturing operations economically.

  ➢ To take care of uncertainties-uncertainty in demand from customers and uncertainty in procuring material in time.

  ➢ To reduce clerical costs and to take advantages of discounts, transportations etc.

6.2 TYPES OF INVENTORIES

1. **Raw Materials:** Basic unfabricated materials which have not undergone for any operation as they are supplied from vendor directly e.g., round bars, angles channels, pipes etc.

2. **Brought-Out Parts:** These parts refer to those finished parts, sub-assemblies which are purchased from outside as per specification.

3. **Work-In-Process (WIP):** It is the items or materials in partially completed condition of manufacturing e.g., subassemblies.

4. **Finished Goods Inventories:** Completed products ready for dispatch or waiting for sales.

5. **Maintenance, Repairs and Operating Stores:** These are not the items which actually do not from the part of final product but are consumed in production process e.g., oil, grease etc.

6. **Tools Inventory:** It includes market procured tools as well as special tools manufactured for the purpose of manufacturing single product also.

7. **Miscellaneous Inventories:** Office stationeries, consumable stores.

- Another way of classification of inventories is:

  1. **Lot Size Inventory:** Inventories procured for particular lot size.

  2. **Transportation Inventory:** If material is in transportation for longer time, that also indicates inventory.

  3. **Anticipation Inventory:** Those are built-up for a particular season when demand is high and remains demandable only for shorter duration.

  4. **Fluctuation Inventories:** These are the inventories as a result of non-accurate sales forecast and so as to satisfy variations in demand.

6.3 CONCEPT OF INVENTORY CONTROL

- Inventory means all movable items in store either ready for sale or for consumption in course of production with a view to convert them into finished stocks for sale.

- Inventories form an integral part of working capital and requires a considerable investment. Hence, it is necessary to have a control over the inventories.

- Inventory control is a system that ensures provision of required quantity of inventories of required quality at the required time with minimum amount of capital investment. Thus, function of inventory control is to obtain maximum inventory turnover with sufficient stock to meet all requirements.

  Inventory control basically deals with two problems:

  1. When the order should be placed ?

  2. How much should be ordered ?

- These questions can be formulated with the help of model. Inventory control ultimately balances the loss due to non-availability of an item and cost of carrying the stock of an items. It aims at maintaining optimum level of stock of goods required by company at minimum cost to company.

6.4 OBJECTIVES OF INVENTORY CONTROL

- To make less financial investment in inventory.
- To ensure adequate supply of products to customer and avoid shortages.
- Efficient purchasing, storing, consumption, accounting for materials etc.
- Stock recording and timely action for replenishment.
- To provide reserve stock for accommodating variations in lead times of delivery of materials.
- Scientific short term and long-term planning of materials.
- To ensure timely delivery of goods and services for improvement in customers relationship.
- To minimize the risk of stock-out.
- Efficient utilization of working capital.
- To minimize loss due to deterioration, obsolescence damage and preliferage.
- Economy in purchasing and to minimize possibility of duplicating the orders.

6.5 INVENTORY COSTS   [DEC. 2007, 2008]

The inventory costs represent cost associated with functions of inventory system. Optimal inventory policy will be one which minimizes total cost:

The Following are Major Components of Inventory Costs:

- **Ordering Costs (C_o):** This represents expenses involved in placing an order with outside supplier. This is predominant where inventory is replenished. This includes costs involved in processing and ordering for purchase, expediting over-due orders, receiving consignment and inspection. This is costs in rupees per order.
- **Set-Up Cost (C_o):** When items are produced within organisation, it is set-up cost. It is the internal product cost in changing over existing production run to produce the ordered items. This includes costs involved in requisition, preparing shop order, scheduling work, pre-production set-up, tools and die preparations, expediting and lot inspection. It is cost in ₹ 1 order.
- **Carrying Cost (C_c):** It is cost of holding and storage of inventory. It is proportional to amount of inventory and time over which it is held.

  It consists of:

  1. Storage and handling
  2. Interests of funds raised for inventory
  3. Insurance
  4. Obsolescence and deterioration.

5. Stock and record keeping. It considers effect of increased clerical and other costs in maintaining stock, record keeping and stock checking. Carrying cost is expressed as cost per unit time. This is also represented as percentage of average annual investment in inventory, (I). Then $C_c = C_u \cdot I$ (Where C_u is unit cost and I is in decimal fraction).

- **Shortage Cost (C_s):** It is less to firm due to non-availability of item, when it is required.
- **Unit Cost (C_u):** It refers to normal cost of inventory item per unit. It is the purchase price of item if it is bought from outside.
- The inventory decisions are based on total inventory costs. The carrying cost is directly proportional to order quantity (Q). If the order quantity is increased, number of orders per unit time will be less and hence ordering cost will decrease. It is inversely proportional to order quantity (Q). The total inventory cost = carrying cost + ordering cost.
- The following Fig. 6.1 shows inventory cost, known as Inventory cost curve.

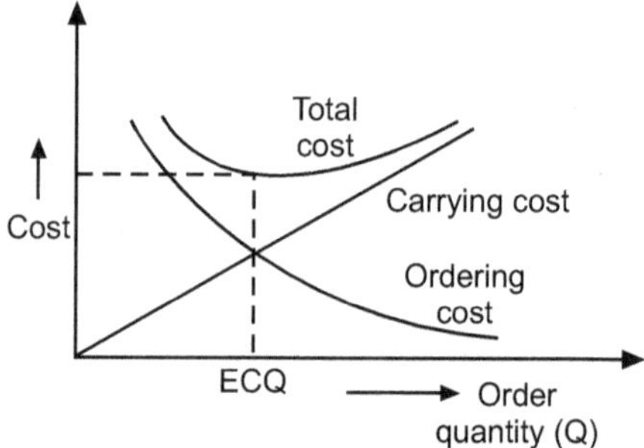

Fig. 6.1: Inventory cost curve

- **Economic Order Quantity (EOQ):** From inventory cost curve, total cost decreases with increase in an ordering quantity and starts increasing thereafter. The ordering quantity (Q) for which total cost is minimum is called as EOQ. In case of purchasing, it is called as EOQ. In case of production of an item within an organisation, it is called as Economic batch size.
- **Lead-Time (L):** The time period from point of time when an order for purchasing item is triggered, until the item has been arrived and taken into stock, is called as lead-time.

  It includes time required for:

  1. In house order preparation,
  2. Order transmission to supplies,
  3. Manufacturing and assembly,
  4. Goods transit time from supplier
  5. In house preparation of inspection, storage and safety, stock entry. Lead-time is independent of order quantity.

- **ABC Analysis:** In industries, a huge variety of items are required for production. Hence, it is required to establish effective control on all inventory items. One has to establish priorities for different items with regard to extent and nature of control required in every respect to each categories of items. Hence, inventory items are classified on basis of cost of inventory of different items in stores. This technique is known as 'Always Better Control' or ABC analysis.

[Dec. 2007, May 2011]

- **A-Items:** In industrial scenario, it is noticed that a top high value items (say 10 to 20%) would account a heavy portion of total cost consumption value (say 70 to 75%). These are called as A-items. These items requires detailed and rigid control and need to be stocked in less quantities. These items should be procured frequently.

- **B-Items:** These items are of 10-15% of total items and represent 10 to 15% of total expenditure on material. The control is not too rigid as A-items.

- **C-Items:** These are numerous (70-80%) and inexpensive 5-10% of the total expenditure on material) and represents non-important items. "C" items should be procured infrequently and in sufficient quantities. The following points represent comparison between A, B and C items.

Sr. No.	A-Items	B-Items	C-Items
1.	Strict control	Moderate control	Less control
2.	No safety stock	Low safety stock	High safety stock
3.	Ordering is frequent	Orders are less in number	Ordering is bulk and non-frequent
4.	Stock record to be kept upto date	Not strict control	No such computation
5.	Value analysis is required	Value analysis is favourable	May be done.

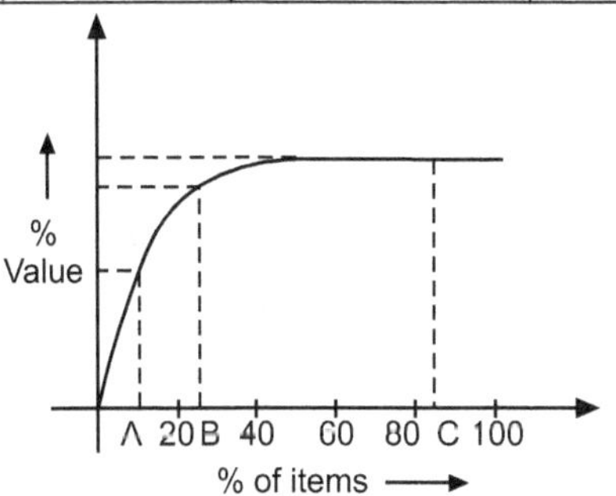

Fig. 6.2

Procedure for ABC Analysis:

1. Calculate total inventory investment by multiplying unit cost with number of units consumed.

2. Arrange all items in descending order of total inventory investment.

3. Compute cumulative total number of items and their usage value obtained in above step.

4. Find percentage of cumulative value to total consumption.

5. Percentage of items v/s value graph is plotted.

6. Include the items in group A in range of 0 to 75% cumulative %.

7. Include the items in group B in range of 75 to 90% cumulative %.

8. Remaining items are included in group C.

Advantages:

- This will establish selective control and focuses on few items.

- This will reduce clerical costs and help in better planning.

- Equal attention to all items is not economical and feasible. This will help to concentration such items which are having high usage value.

Limitations:

- The analysis depends only on annual consumption value and does not rely on unit cost.

- Some items irrespective of their consumption are important and may create bottlenecks. This may result suddenly and no provisions are made in this regard in above category.

- The result of ABC analysis should be reviewed periodically and updated regularly.

Re-Order Point (ROP):

- It acts as an indicator, to know, when purchase order is to be placed. The re-order point (ROP) indicates stock level when it is reached, order will be triggered.

6.6 INVENTORY CONTROL MODELS

A broad classification of EOQ models is as shown in Fig. 6.3.

Lists of Symbols Used:

C = Purchase cost per unit (₹ /unit)

C_o = Ordering (or set-up) cost per order (₹ /order)

C_c = Cost of carrying one unit in inventory (₹ /unit time)

r = Cost of carrying one rupee's worth of inventory per time period (Percent/time)

C_s = Shortage cost per unit per time (₹ /unit-time)

D = Demand rate, units per time

Q = Order quantity i.e., number of units ordered per order (units)

R_{OL} = Reorder level

Q' = Optimum order quantity

L_T = Replenishment lead time (time)

n = Number of orders per time period (orders/time)

t = Reorder cycle time

t_p = Production (period/time)

P = Production rate i.e., the rate at which quantity (q) is added to inventory R (quantity/time)

TC = Total inventory cost (₹)

TVC = Total variable inventory cost (₹)

TVC' = Optimum total variable inventory cost CRS.

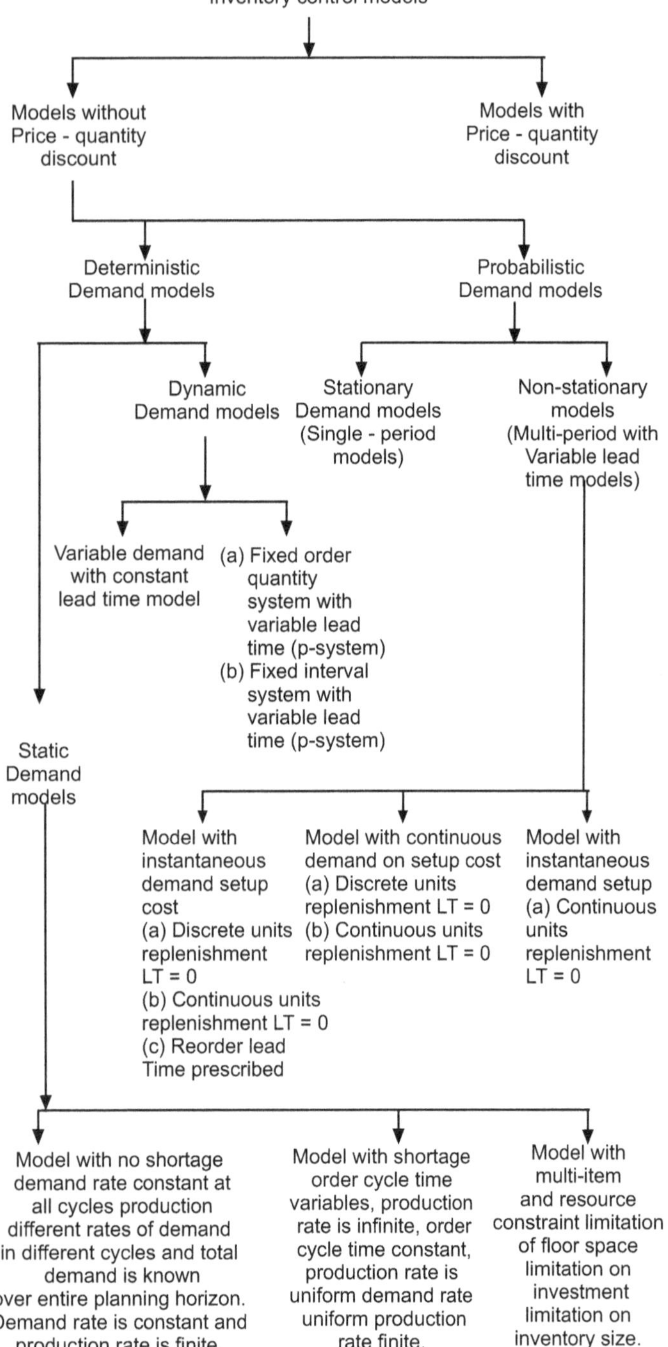

Fig. 6.3

- **Deterministic Inventory Control Models:** In deterministic inventory control models, demand is deterministic, constant and it is known.

6.7 DETERMINISTIC INVENTORY MODELS WITH NO SHORTAGE

Model I (a): The Economic Lot Size Model with Constant Demand:

Assumptions: **[Dec. 2009, May 2010, 2011]**

- Demand rate is constant and known throughout reorder cycle time.

- Production rate (r_p) is infinite i.e., replenishment is instantaneous.

- Lead time (L_T) is constant and known with certainty i.e., no shortage of inventory ($C_S = 0$).

- Purchase price (C) is constant as quantity discount is not available.

- Unit costs of carrying inventory (C_c) and ordering costs (C_o) are known and constant.

From above assumption, $Q = D \times t$.

The following Fig. 6.4 shows behavioural of above inventory control system.

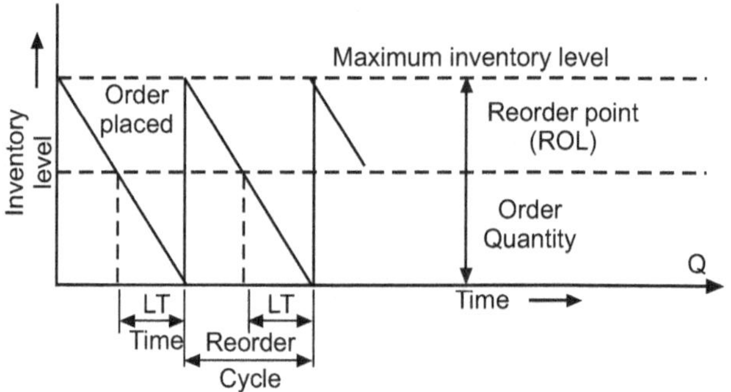

Fig. 6.4: Inventory model with constant demand and instantaneous supply

Total variable inventory cost = Carrying cost + Ordering cost

= (Average inventory level) × (Carrying cost/unit/year) + {(Number of order placed per year) × (Ordering cost/order)}

$$= \frac{I_{max} + I_{min}}{2} \cdot C_c + \frac{D}{Q} \cdot C_o$$

$$\therefore \quad TVC = \frac{Q}{2} \cdot C_c + \frac{D}{Q} \cdot C_o$$

Graphically, cost components are as shown in the following Fig. 6.5.

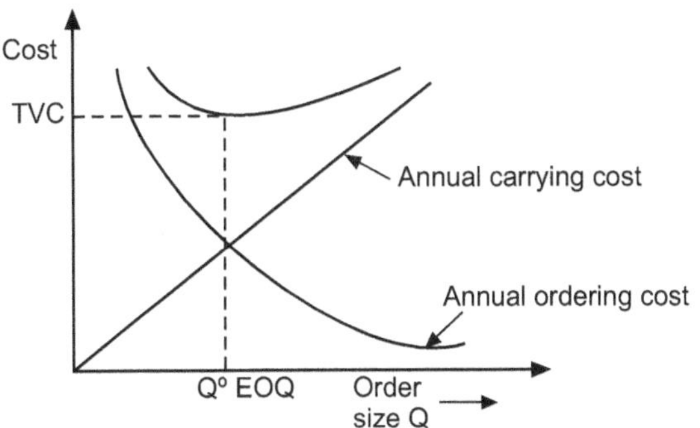

Fig. 6.5: Relationship between EOQ and TVC

The total variable inventory cost is minimum at a point where total inventory carrying cost becomes equal to total ordering cost.

$$\frac{D}{Q} \cdot C_o = \frac{Q}{2} \cdot C_c$$

$$\therefore \quad Q^2 = \frac{2DC_o}{C_2}$$

$$\therefore \quad Q' \, (EOQ) = \sqrt{\frac{2DC_o}{C_c}}$$

$$= \sqrt{\frac{2 \times \text{Annual demand} \times \text{Ordering cost}}{\text{Carrying cost}}}$$

This is also known as Wilson or Harris lot size formula.

Other Important Formulae: **[Dec. 2008, May 2010]**

1. Optimal length of inventory replenishment cycle time (t'). Since,

$$Q' = D \cdot t'$$

$$t' = \frac{Q'}{D}$$

$$= \frac{1}{D} \times \sqrt{\frac{2DC_o}{C_c}}$$

$$t' = \sqrt{\frac{2C_o}{DC_c}}$$

$$= \sqrt{\frac{2 \times \text{Ordering cost}}{\text{Annual demand} \times \text{Carrying cost}}}$$

2. Optimal number of orders to be placed in given time period,

$$N' = \frac{\text{Annual demand}}{\text{Optimal order quantity}}$$

$$= \frac{D}{Q'}$$

$$\therefore \quad = D \times \frac{1}{\sqrt{2DC_o/C_c}}$$

$$= \sqrt{\frac{DC_c}{2 \times C_o}}$$

$$N' = \sqrt{\frac{\text{Annual demand} \times \text{Carrying cost}}{2 \times \text{Ordering cost}}}$$

3. Optimal (minimum) total variable inventory cost (TVC'):

$$TVC' = \frac{D}{Q'} C_o + \frac{Q'}{2} C_c$$

$$= \frac{D}{\sqrt{\frac{2DC_o}{C_c}}} \cdot C_o + \sqrt{\frac{2DC_o/C_c}{2}} \times C_c$$

$$= \frac{D}{\sqrt{\frac{2DC_o}{C_c \cdot C_o}}} + \frac{\sqrt{2DC_o/C_c^2}}{C_c \times 4}$$

$$= \frac{1}{\sqrt{\frac{2D}{C_c\,C_o\,D^2}}} + \sqrt{\frac{DC_o C_c}{2}}$$

$$= \sqrt{\frac{DC_o C_c}{2}} + \sqrt{\frac{DC_o C_c}{2}}$$

$$= 2\sqrt{\frac{DC_o C_c}{2}}$$

$$= \sqrt{2DC_o C_c}$$

$$TVC' = \sqrt{2 \times \text{Annual demand} \times \text{Ordering cost} \times \text{Carrying cost}}$$

SOLVED EXAMPLES

Example 6.1: *The production department for a company requires 3600 kg of raw material for manufacturing a particular item/year. It has been estimated that cost of placing an order is ₹ 36 and cost of carrying an inventory is 25% of investment in inventories. The price is ₹ 10/kg. The purchase manager wishes to determine an ordering policy for raw material.*

Solution: $D = 3600$ kg per year

$C_o = 36$ ₹ /order

$C_c = 25\% = 10 \times 0.25 = ₹\ 2.5$ kg/year

(i) The optimal lot size is given by

$$Q' = \sqrt{\frac{2DC_o}{C_c}} = \sqrt{\frac{2 \times 3600 \times 36}{2.5}}$$

$$= 321.99 \text{ kg/order}$$

(ii) The optimal order cycle time,

$$t' = \frac{Q'}{D} = \frac{321.99}{3600} = 0.894 \text{ year}$$

(iii) The minimum yearly variable inventory cost

$$TVC' = \sqrt{2DC_o\,C_c} = \sqrt{2 \times 3600 \times 36 \times 2.5}$$

$$= ₹\ 804.98/\text{year}$$

(iv) The minimum yearly total cost,

$$TC' = TVC' + DC$$

$$= 804.98 + (3600 \times 10)$$

$$= 36804.98/\text{year}$$

Model I (b): Economic Lot Size with Different Rates of Demand in Different Cycles:

1. Assumptions are same as that of last model except that demand is constant with different rates in different cycles. The total demand is specified over total time period, T.

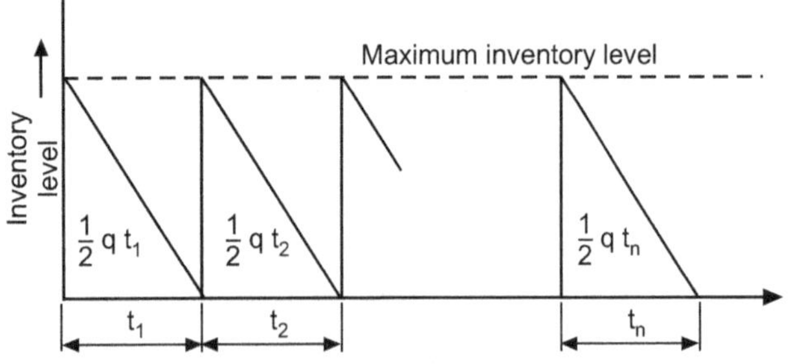

Fig. 6.6

If t_1 ... t_2 ... t_n are the time of successive cycle and D_1, ... D_2, ... D_3, ... D_n are demand rates than total period T is given by

$$T = t_1 + t_2 + t_3 + ... + t_n$$

2. Assume that each time a fixed quantity say q is produced. Number of production cycles in total time period T will be n = D/q, where D is total demand in time period T. Thus, the inventory carrying cost and set-up cost for time period T will be.

$$\text{Carrying cost} = \frac{1}{2}q \cdot t_1 \cdot C_c + \frac{1}{2}q \cdot t_2 \cdot C_c + ...$$

$$+ \frac{1}{2}q \cdot t_n \cdot C_c$$

$$= \frac{1}{2}q \cdot C_c (t_1 + t_2 + t_3 .... + t_n) = \frac{1}{2}q \cdot C_c \cdot T$$

$$\text{Set-up cost} = \frac{D}{q}C_o$$

Total year variable inventory cost is given by

$$TVC = \frac{1}{2}q \cdot C_c \cdot T + \frac{D}{q}C_o$$

For optimum value of q which minimizes TVC, equating set-up cost and carrying cost, we have

$$\frac{1}{2}q\, C_c\, T = \frac{D}{q} \cdot C_o$$

$$q^2 = \frac{2DC_o}{TC_c}$$

$$q' = \sqrt{\frac{2DC_o}{TC_c}}$$

$$\therefore \quad TVC = \frac{1}{2}\sqrt{\frac{2DC_o}{TC_c}} \cdot C_c \cdot T + \frac{D}{\sqrt{\dfrac{2DC_o}{TC_c}}}C_o$$

$$TVC = \sqrt{\frac{2C_oC_cD}{T}} \quad \text{(Optimal cost)}$$

Example 6.2: *A manufacturing company purchases 9000 parts of a machine for its annual requirements, ordering one month's requirement at a time. Each part costs ₹ 20. The ordering cost per order is ₹ 15 and carrying charges are 15% of average inventory/year. Suggest a economical purchasing policy for company. What advice would you offer and how much would it save the company per year?*

Solution:

Given:

$$D = 9000 \text{ parts/year}$$
$$C_o = 15/\text{order}$$
$$C = 20/\text{part}$$
$$C_c = 15\% \text{ of investment in inventories}$$
$$= 20 \times 0.15 = 3.0/\text{part/year}$$

Optimal lot size,

$$Q' = \sqrt{\frac{2DC_o}{C_c}} = \sqrt{\frac{2 \times 9000 \times 15}{3}}$$
$$= 300 \text{ units}$$

Optimal order cycle time

$$t' = \frac{Q'}{D} = \frac{300}{9000} = \frac{1}{30} \text{ years for 12 days}$$

The minimum yearly total variable inventory cost,

$$TVC = \sqrt{2DC_oC_c} = \sqrt{2 \times 9000 \times 15 \times 3} = 900$$

But if company follows policy of ordering each month, Annual ordering cost (C_o) will = 12 × 15 = 180 ₹ . Thus, lot size (Q') of inventory each month = 9000/12 = 750 parts.

Average inventory at any point of time $= \frac{Q}{2} = 375$ parts.

Carrying cost at any time = 3 × C_c = 3 × 375 = 1125.

Minimum yearly total variable inventory cost,

$$TVC' = 180 + 1125 = 1305$$

Hence, if company orders 300 parts after every 12 days instead of 750 parts every month. There will be a net saving of ₹ (1305 – 900) = ₹ 405 per year.

Model I (c): Economic Lot Size Model with Finite Replenishment Rate:

Assumptions made are same as that Model I(a) except that of instantaneous replenishment (Supply). Additional assumptions are as follows:

1. Supply is continuous and constant until Q units are supplied to stock and then it stops.

2. The production rate (P) per unit of time is greater than demand rate d i.e., p > d,

3. Production begins immediately after production set-up.

The following Fig. 6.7 shows model of this category:

t_p is the time required to produce amount Q at a rate P, then we have

$$Q = P \cdot t_p \ \text{ or } \ t_p = \frac{Q}{P}$$

In production, t_p is inventory increasing at rate of P and simultaneously decreased at rate of d. Thus, inventory accumulates at rate of p – d units. Hence, maximum inventory level reached at the end of t_p will be

$$E_{max} = \text{Inventory accumulation rate} \times \text{Production time}$$
$$= (p - d)\, t_p = (p - d)\, Q/P$$
$$= (1 - d/p)Q$$

Since, minimum inventory level, $I_{max} = 0$

$$\text{Average inventory} = \frac{Q}{2}\left(1 - \frac{d}{p}\right)$$

The total annual carrying cost is average inventory level multiplied by carrying cost per unit of average inventory

$$\text{carrying cost} = \frac{Q}{2}\left(1 - \frac{d}{p}\right)C_c$$

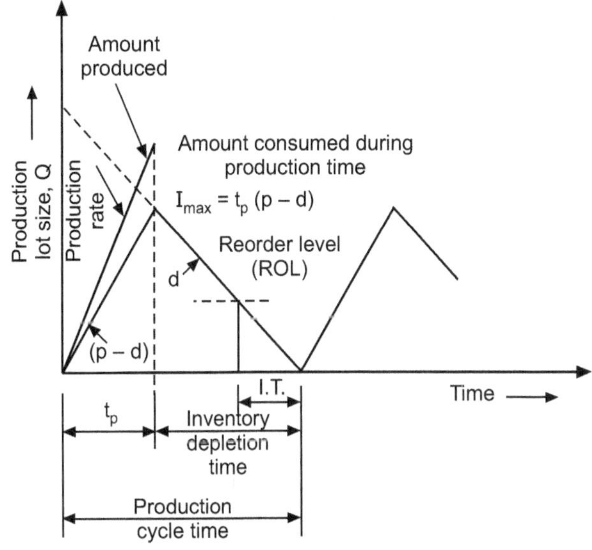

Fig. 6.7 : Finite replenishment model

$$\text{Set-up cost} = \frac{D}{Q} \cdot C_o$$

Hence, total variable inventory cost,

$$TVC = \frac{Q}{2}\left(1 - \frac{d}{p}\right)C_c + \frac{D}{Q} \cdot C_c$$

Since, set-up costs decreases and carrying costs increases when production quantity (Q) increases, a minimum total inventory variable costs occur when two costs are equal.

$$\text{Set-up cost} = \text{Carrying cost}$$
$$\frac{D}{Q} \cdot C_o = \frac{Q}{2}\left(1 - \frac{d}{p}\right)C_c$$

$\therefore$ EOQ,

$$Q' = \sqrt{\frac{2DC_o}{C_c}\left(\frac{P}{p - d}\right)}$$

On substituting, value of Q' in equation of TVC, we get,

$$TVC' = \sqrt{2DC_o\, C_c \left(1 - \frac{d}{p}\right)}$$

This represents optimal cost.

Optimal length of each lot size production run,

$$t'_p = \frac{Q'}{P} = \frac{1}{P} \times \sqrt{\frac{2DC_o}{C_c}\left(\frac{P}{p - d}\right)}$$

Optimal production cycle time,

$$t' = \frac{Q'}{D} = \frac{1}{D} \times \sqrt{\frac{2DC_o}{C_c}\left(\frac{P}{p - d}\right)}$$
$$= \sqrt{\frac{2C_o}{DC_c}\left(\frac{P}{p - d}\right)}$$

Optimal number of production runs year,

$$N' = \frac{D}{Q'} = D \times \frac{1}{\sqrt{\dfrac{2DC_o}{C_c}\left(\dfrac{P}{p - d}\right)}}$$
$$N' = \sqrt{\frac{DC_o}{2C_c}\left(\frac{p - d}{p}\right)}$$

Example 6.3: *A contractor has to supply 10000 bearings per day to an automobile manufacturer. When he started production runs, he can produce 25000 bearings per day. The cost of holding a bearings in stock for a year is ₹ 2 and the set-up cost of a production run is ₹ 180. How frequently should production run be made ?*

Solution: Given data:

$$C_p = 1800/\text{Production run},$$
$$C_c = ₹\ 2/\text{ year},$$
$$P = 25000 \text{ bearing/day},$$
$$d = 10000 \text{ bearing/day},$$
$$D = 10000 \times 300 = 3000000 \text{ units/year}$$

(Assuming 300 working days in a year).

(i)   Economic lot size,

$$Q' = \sqrt{\frac{2DC_o}{C_c}\left(\frac{P}{p - d}\right)}$$
$$= \sqrt{\frac{2 \times 30000000 \times 1800}{2} \times \left(\frac{25000}{25000 - 10000}\right)}$$
$$= 104446 \text{ bearings}$$

(ii) Frequency of production run,

$$t' = \frac{Q}{d} = \frac{104446}{10000} = 10.44 \text{ days}$$

6.8 DETERMINISTIC INVENTORY MODELS WITH SHORTAGES

Model II: (a) Economic Lot Size Model with Constant Demand and Variable Order Cycle Time.

This model is based upon all assumptions of model I (a), except that inventory system run out of stock for certain period i.e. shortages are allowed.

The notations used in this model are:

t_i = Time between receipt of an order and when inventory level drops to zero

t_b = Time during which back order or stockouts will occur

t = Total cycle time = $t_i + t_b$

R = Number of units that are back ordered.

The following Fig. 6.8, describes changes in inventory level with time.

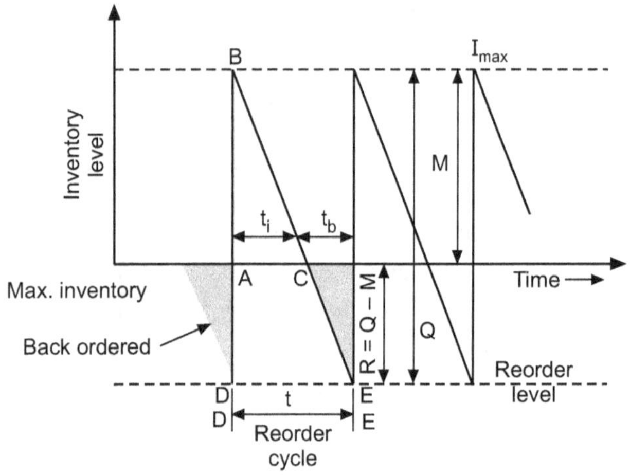

Fig. 6.8: Inventory control model with shortages

$$\text{TVC} = \text{Ordering cost} + \text{Carrying cost} + \text{Shortage cost}$$

$$\text{Time period in days} = \frac{\text{Total units}}{\text{Demand in units/day}}$$

$$t_i = \frac{M}{D}$$

The total cycle time,

$$t = \frac{Q}{D}$$

The time for back orders

$$= t_b = \frac{Q - M}{D}$$

Average inventory level

$$= \frac{\text{Average level over } t_i + \text{Average level over } t_b}{t}$$

$$\therefore \quad \text{Average inventory level} = \frac{(M/2)\, t_i + Q}{t} = \frac{M}{2}\frac{t_i}{t}$$

Put $\quad t_i = \dfrac{M}{D}$ and $\dfrac{Q}{D} = t$

$$\text{Average inventory level} = \frac{M^2}{2Q}$$

$$\text{Carrying cost} = \frac{M^2}{2Q} \cdot C_c$$

Similarly, average back order level can be determined by dividing area under triangle CFE by cycle time t, i.e.,

Average number of units on back order,

$$= \frac{(\text{Average level over } t_i) + (\text{Average level over } t_b)}{t}$$

$$= \frac{Q \cdot t_i + [(Q - M)/2]\, t_b}{t}$$

$$= \left(\frac{Q - M}{2Q}\right)^2$$

$$\left(\text{Putting } t_b = \frac{Q - M}{D} \text{ and } t = \frac{Q}{D}\right)$$

$$\therefore \quad \text{Shortage cost} = \frac{(Q - M)^2}{2Q} \cdot C_s$$

Hence, total variable inventory cost,

$$\text{TVC} = \frac{D}{Q} C_o + \frac{M^2}{2Q} C_c + \frac{(Q - M^2)}{2Q} C_s$$

TVC is function of Q, M

$$\frac{\partial(\text{TVC})}{\partial Q} = 0$$

or $\quad \dfrac{\partial(\text{TVC})}{\partial M} = 0$

$\therefore$ Solving, we get,

$$Q' = \sqrt{\frac{2DC_o}{C_c}\left(\frac{C_c + C_s}{C_s}\right)}$$

and $\quad M' = \sqrt{\dfrac{2DC_p}{C_c}\left(\dfrac{C_s}{C_c + C_s}\right)}$

$$\therefore \quad \text{TVC} = \sqrt{2DC_c\, C_o \left(\frac{C_s}{C_c + C_s}\right)} \quad \text{(Optimal cost)}$$

Other Important Formulae:

1. Optimal amount back ordered in units:

$$R' = Q' - M' = Q\left(\frac{C_s}{C_c + C_s}\right)$$

2. Total cycle time:

$$t = \frac{Q'}{D}\sqrt{\frac{2C_o}{DC_c}\left(\frac{C_c + C_s}{C_s}\right)}$$

Example 6.4: *A commodity is to be supplied at a constant rate of 200 units per day. Supplies of any amount can be had at any required time, but each ordering costs ₹ 50; cost of holding commodity in inventory is ₹ 2.00 per unit per day while delay in supply of item induces a penalty of ₹ 10 per unit per day. Find optimal policy where t is reorder cycle period and Q is inventory level after reorder. What would be best policy, if penalty cost becomes in ∞ ?*

Solution: Given data:

$$D = 200 \text{ units/day,} \qquad C_c = 2/\text{unit/day,}$$
$$C_o = 50/\text{order,} \qquad C_s = 10/\text{unit/day}$$

(a) Optimal order quantity,

$$Q' = \sqrt{\frac{3DC_o}{C_c}\left(\frac{C_c + C_c}{C_s}\right)}$$

$$= \sqrt{\frac{2 \times 200 \times 50}{2}\left(\frac{2 + 10}{10}\right)}$$

$$= 108.5 \text{ units}$$

(b) Recorder cycle time $= \dfrac{109.5}{200}$

$$= 0.547 \text{ day} \qquad \left(\because t' = \frac{Q'}{D}\right)$$

Thus, optimal order quantity of 10.5 units to be supplied after every 0.547 days. If penalty cost, $C_s = \infty$, the expression Q' will become,

$$Q' = \sqrt{\frac{3DC_o}{C_c}\left(\frac{C_c + C_c}{C_s}\right)}$$

$$= \sqrt{\frac{2DC_o}{C_c}}$$

$$= \sqrt{\frac{2 \times 200 \times 50}{2}}$$

$$= 100 \text{ units}$$

$$t' = \frac{Q'}{D} = \frac{100}{200} = 0.5 \text{ day}$$

Model II (b): Economic Lot Size Model with Constant Demand and Fixed Reorder Cycle Time.

Let the reorder cycle time, t is fixed. Q is fixed lot size to meet demand for period t. The following Fig. 6.9 represents this model.

As shown in Fig. 6.8 for last model, the amount M(< Q) is planned to meet demand during time, t_1 = MID. Since, reordering (or set-up) cost and time, t are constant,

$$TVC (M) = \text{Carrying cost} + \text{Shortage cost}$$

$$TVC (M) = \frac{M^2}{2Q} C_c + \frac{1}{2Q} (Q - M^2) \cdot C_s$$

To calculate M,

$$\frac{d}{dM} (TVC) = 0$$

$$\therefore \frac{d}{dM} (TVC) = \frac{M}{Q} \cdot C_c + \frac{1}{Q} (Q - M) \cdot C_s (-1) = 0$$

$$M = \left(\frac{C_s}{C_c + C_s}\right) Q = \left(\frac{C_s}{C_c + C_s}\right) \cdot D \cdot t$$

Put value of M in TVC equation, we get,

$$TVC' = \left(\frac{C_c \cdot C_s}{C_c + C_s}\right) D \cdot t \text{ (Optimal costs)}$$

Example 6.5: *A part is to be supplied at a constant rate of 25 units per day. A penalty cost is being charged at rate of ₹ 10 per unit per day late for missing the scheduled delivery date. The cost of carrying the commodity in inventory is ₹16 per unit per month. The production process is such that each month (30 days) a batch of items is started and are available for delivery any time after end of month. Find optimal level of inventory at beginning of each month.*

Solution: Given data:

$$D = 25 \text{ units/day,}$$
$$C_s = \frac{16}{30} = 0.53 \text{ unit/day}$$
$$C_s = 10 \text{ unit/day}$$
$$t = 30 \text{ days}$$

Optimum level of inventory

$$M = \left(\frac{C_s}{C_c + C_s}\right) D \cdot t$$

$$= \left(\frac{10}{0.53 + 10}\right) (25) (30) = 712 \text{ units}$$

Model II (c): Economic Lot Size Model with Finite Replenishment Rate and Shortage Allowed.

This model is based on same assumption as that of Model I (c) except that shortages are allowed. The inventory system is shown in Fig. 6.9.

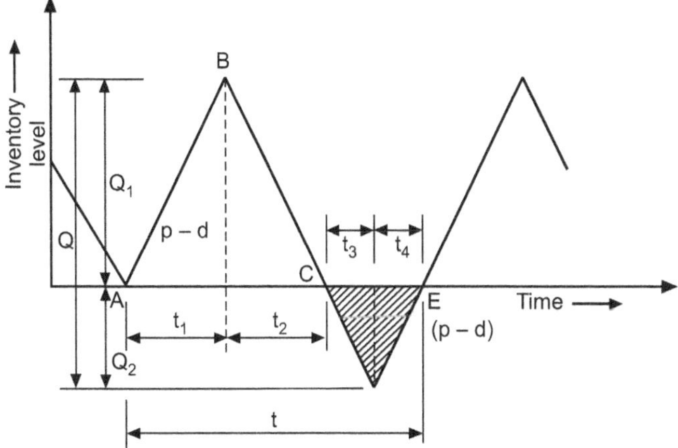

Fig. 6.9: Inventory model with finite replenishment and shortages allowed

In this case,

$$TVC = \text{Set-up cost} + \text{Carrying cost} + \text{Shortage cost}$$

As per Model I (c), Maximum inventory level Q_1, reached at end of time t_1 is given by

$$Q_1 = (p - d)/t_1$$

After time t_1, the stock Q_1 is used up during t_2,

$$\therefore \quad Q_1 = D \cdot t_2$$

During time t_3 shortage accumulate at rate of D. Thus, maximum shortage occurred is given by

$$Q_2 = D \cdot t_3$$

After time t_3, production starts and therefore, shortage starts reducing at rate of

$$= (p - d)\, t_4$$

i.e. $\quad Q = p - d$

$$\text{Average inventory} = \frac{1}{2}\frac{Q_1\,(t_1 + t_2)}{t}$$

$$\text{Average inventory} = \frac{1}{2}\frac{Q_2\,(t_3 + t_4)}{t}$$

Production cycle,

$$t = t_1 + t_2 + t_3 + t_4$$

$$= \frac{Q_1}{p - d} + \frac{Q_1}{d} + \frac{Q_2}{d} + \frac{Q_2}{p - d}$$

$$= Q_1\left(\frac{1}{p - d} + \frac{1}{d}\right) + Q_2\left(\frac{1}{d} + \frac{1}{p - d}\right)$$

$$= \frac{P}{d(d - d)}\,(Q_1 + Q_2)$$

If Q is lot size, then,

$$Q_1 = Q - Q_2 - d \cdot t_1 - d \cdot t_4$$

$$= Q - Q_2 - d\left(\frac{Q_1}{p - d} + \frac{1}{p - d}\right)$$

$$= (Q - Q_2)\,(p - d/p)$$

$$\text{or} \qquad = Q_1 + Q_2$$

$$= \left(\frac{p - d}{P}\right)Q$$

Put $Q_1 + Q_2$ in production cycle equation,

$$t = \frac{P}{d(p - d)} \times \frac{(p - d)}{P} \cdot Q$$

$$= \frac{Q}{d}$$

Hence $TVC = \dfrac{d}{Q} \cdot C_o + \dfrac{1}{2}\dfrac{Q_1\,(t_1 + t_2)}{t}$

$$\cdot C_c + \frac{1}{2}\frac{Q_2\,(t_3 + t_4)}{t} \cdot C_s$$

$$= \frac{d}{Q} \cdot C_o + \frac{1}{2Q} \cdot \frac{P}{2Q}\left(Q_1^2 \cdot C_s \cdot Q_2^2 \cdot C_s\right)$$

$$\therefore \quad TVC = \frac{d}{Q} \cdot C_o + \frac{1}{2Q}\left(\frac{P}{p - d}\right)$$

$$\left[C_s\left(\frac{p - d}{P} \cdot [Q_1 - Q_2]^2 + Q_2^2 \cdot C_s\right)\right]$$

The necessary conditions of optimum TVC are:

$$\frac{\partial(TVC)}{dQ_2} = \frac{1}{2Q}\left(\frac{P}{p - d}\right)2C_c$$

$$\left\{\frac{p - d}{P} \cdot (Q - Q_2) \times (-1) + 2Q_2\,C_s\right\} = 0$$

$$\therefore \quad C_c\left\{\frac{Q_2}{Q\left(1 - \dfrac{d}{p}\right)} - 1\right\} + \frac{Q_2 C_s}{Q\left(1 - \dfrac{d}{p}\right)} = 0$$

$$\therefore \quad Q_2' = \frac{C_c Q}{(C_c + C_s)}\left(1 - \frac{d}{p}\right)$$

and also, $\dfrac{\partial(TVC)}{\partial Q} = \dfrac{d}{Q^2}\,C_o - C_c\left[\dfrac{1}{2}\dfrac{p - d}{2} - \dfrac{Q_2^2}{2Q^2\left(\dfrac{p - d}{P}\right)}\right]$

Solving we get, $\dfrac{d}{Q^2} \cdot C_o \cdot \dfrac{C_s}{2}\left(1 - \dfrac{d}{P}\right)\left(1 - \dfrac{C_s}{C_c + C_s}\right)$

$$Q' = \sqrt{\frac{2C_o\,(C_c + C_s)}{C_c \cdot C_s\left(\dfrac{P_d}{p - d}\right)}}$$

$$\therefore \quad Q' = \sqrt{\frac{2DC_o}{C_s}\left(\frac{P}{p - d}\right)\left(\frac{C_c + C_s}{C_s}\right)}$$

This is equation for optimal production lot size,

$$Q_2' = Q'\left(1 - \frac{d}{p}\right)\left(\frac{C_c}{C_c + C_s}\right)$$

This is expression for optimal level of shortage.

Other formulae:

1.   Production cycle time,

$$r = \frac{Q}{D} = \sqrt{\frac{2C_o}{D_c}\left(\frac{P}{p - d}\right)\left(\frac{C_c + C_s}{C_c}\right)}$$

2.   Optimal inventory level,

$$Q_1' = \left(\frac{p - d}{P}\right)Q' - Q_2'$$

$$= \sqrt{\frac{2DC_o}{C_c}\left(1 - \frac{d}{P}\right)\left(\frac{C_s}{C_c + C_s}\right)}$$

3.   Total minimum inventory cost,

$$TVC = \sqrt{2DC_oC_c\left(1 - \frac{d}{P}\right)\left(\frac{C_s}{C_c + C_s}\right)}$$

Example 6.6: *The demand for an item in a company is 18000 units per year, and the company can produce the item at a rate of 3000 per month. The cost of one set-up is ₹ 500 and the holding cost of 1 unit per month is paise. The shortage cost of one unit is ₹ 240/year. Determine optimum manufacturing quantity and number of shortages. Also determine the manufacturing time and time between setups.*

Solution:

$$D = 18000 \text{ units/year, } = 1500 \text{ units/month,}$$

$$P = 3000 \text{ units/month, } C_c = ₹\ 0.1/\text{month/unit,}$$

$$C_o = 500/\text{set-up, } \quad C_s = 240/\text{year} = 20/\text{month.}$$

(1) Optimal lot size,

$$Q' = \sqrt{\frac{2DC_o}{C_c}\left(\frac{P}{P-D}\right)\left(\frac{C_c + C_s}{C_s}\right)}$$

$$= \sqrt{\frac{2 \times 1500 \times 300}{0.15} \times \left(\frac{3000}{3000-1500}\right)\left(\frac{0.15 + 20}{20}\right)}$$

$$= 4.489 \text{ units}$$

(2) Optimal number of shortages,

$$Q'_2 = \frac{C_c}{C_c + C_s}\left(1 - \frac{d}{P}\right) \cdot Q'$$

$$Q' = \frac{0.15}{0.15 + 20}\left(1 - \frac{1500}{3000}\right) \times 4489 = 17 \text{ units}$$

(3) Production time,

$$t_1 = \frac{Q}{P} = \frac{4489}{3000} = 1.50 \text{ months}$$

(4) Production cycle time,

$$t = \frac{Q'}{P} = \frac{4489}{1500} = 3 \text{ months}$$

6.9 MULTI-ITEM INVENTORY MODELS WITH CONSTRAINTS

- In this, we have to calculate EOQ for each item separately which will minimize total inventory cost under specified constraints.

- The following assumptions and notations are used to develop inventory models.

Assumptions:

1. Production is instantaneous with no lead time,

2. Demand is uniform and deterministic,

3. Shortages are not allowed.

Notations:

$$n = \text{Total no. of items being controlled simultaneously}$$

$$f_i = \text{Storage space required per unit of item } i \ (i = 1, 2,.... \ n)$$

$$W = \text{Total warehouse capacity}$$

$$\lambda = \text{Non-negative Lagrange multiplier}$$

$$D_i = \text{Annual demand for } i_{th} \text{ item}$$

$$Q_i = \text{Quantity of } i_{th} \text{ item stocked at beginning of cycle}$$

$$M = \text{Upper limit of average no. of all units}$$

$$C_i = \text{Cost per unit of item } i \ (i = 1, 2, ... \ n)$$

$$F = \text{Total funds available for inventory.}$$

Model III (a): EOQ Model with Warehouse Capacity Constraint:

$$\text{Total floor area required} = \sum_{i=1}^{n} f_i Q_i \le Q$$

This problem will become, minimization of total variable inventory cost for each item together under warehouse capacity constraint.

$$\text{Minimum TVC} = \sum_{i=1}^{n} \frac{D_i}{Q_i} C_o^{(i)} + \sum_{i=1}^{n} \frac{Q_i}{2} C_c^{(i)}$$

Subject to constraints of $\sum_{i=1}^{n} f_i Q_i \le W$ and $Q_i \ge 0$, $i = 1, 2, ... , n$

Let, λ be non-negative Lagrange multiplier. The Lagrangian function is given by

$$L(Q_i, \lambda) = \text{TVC}$$

$$= \sum_{i=1}^{n} C_o^{(i)} + \sum_{i=1}^{n} \frac{Q_i}{2} - (C_c)^{(i)}$$

$$+ \lambda \left\{ \sum_{i=1}^{n} f_i Q_i - W \right\}$$

The necessary conditions for L to be minimum are

$$\frac{\partial L}{\partial Q_i} = \frac{\partial L}{\partial \lambda}$$

$$\sum_{i=1}^{n} f_i Q_i - W \le 0 \quad \text{or} \quad \lambda \left\{ \sum_{i=1}^{n} f_i Q_i - W \right\} = 0 \ \lambda \ge 0$$

The minimum value of TVC can be obtained by $\lambda = 0$ and $\lambda > 0$

For $\lambda = 0$, we have,

$$\frac{\partial L}{\partial Q_i} = -\frac{D_i}{Q_i^2} C_o^{(i)} + \frac{1}{2}(C_c)^{(i)} = 0$$

or Q_i $\sqrt{\dfrac{2D_i C_o^{(i)}}{C_c^{(i)}}}$

For $\lambda \neq 0$,

$$\frac{\partial L}{\partial Q} = -\frac{D_i}{Q_i^2} C_o^{(i)} + \frac{1}{2} C_c^{(i)} + \lambda f_i = 0$$

$$\therefore \quad \frac{\partial L}{\partial \lambda} = \sum_{i=1}^{n} f_i \cdot Q_i - W = 0$$

Solving above two equations,

$$Q_i = \sqrt{\frac{2D_i C_o^{(i)}}{C_c^{(i)} + 2\lambda f_i}} \quad \text{where, } i = 1, 2, \ldots, n$$

and $\displaystyle\sum_{i=1}^{n} f_i Q_i = W$

Example 6.7: *A small shop produces three machine parts I, II and III in lots. The shop has only 650 sq. feet of storage space. The appropriate data for three items are represented in the following table:*

Table 6.1

Item	I	II	III
Demand rate (units/year)	5000	2000	10000
Procurement cost (₹/order)	100	200	75
Cost per unit (₹)	10	15	5
Floor space required (Sq. feet/unit)	0.70	0.80	0.40

The shop uses an inventory carrying charge of 20% of average inventory valuation per year. If stockout is not allowed, determine optimal lot size for each item under given storage constraint.

Solution: For, $\lambda = 1$, computing Q_1' where $i = 1, 2, 3$ for each item,

$$Q_1' = \sqrt{\frac{2 \times 5000 \times 100}{0.20 \times 10 + 2 \times 0.70}} = 542 \text{ units}$$

$$Q_2' = \sqrt{\frac{2 \times 2000 \times 200}{0.20 \times 15 + 2 \times 0.80}} = 417 \text{ units}$$

$$Q_3' = \sqrt{\frac{2 \times 10000 \times 75}{0.20 \times 5 + 2 \times 0.40}} = 913 \text{ units}$$

Since, $\displaystyle\sum_{i=1}^{3} f_i Q_2' = 0.7 \times 542 + 0.8 \times 417 + 0.4 \times 9.13$

$= 1078 > 650$ sq. feet

(Available storage space)

For, $\lambda = 5.4$ we get, $Q_1' = 324$, $Q_2' = 263$, $Q_3 = 531$ units and floor space required = 649 Sq. feet, which is available.

Model III (b): EOQ Model with Investment Non-Inventory Constraint:

If cost per unit of an item i is C_i and F be the total funds available to be invested on inventory,

$$\sum_{i=1}^{n} C_i Q_i \leq F$$

$$\therefore \quad \text{Minimum TVC} = \sum_{i=1}^{n} \frac{D_i}{Q_i} C_o^{(i)} + \sum_{i=1}^{n} \frac{Q_i}{2} C_c^{(i)}$$

Subject to constraint, $\displaystyle\sum_{i=1}^{n} C_i Q_i \leq F$ and $Q_i \geq 0$

Let λ be non-negative Lagrange multiplier, then

$$L(Q_i, \lambda) = \sum_{i=1}^{n} \frac{D_i}{Q_i} C_o^{(i)} + \sum_{i=1}^{n} \frac{Q_i}{2} (C_c)^{(ii)}$$

$$+ \lambda \left\{ \sum_{i=1}^{n} C_i Q_i - F \right\} \text{ for } \lambda \geq 0$$

The necessary conditions for L to be optimum are:

$$\frac{\partial L}{\partial Q_i} = -\frac{D_i}{Q_i^2} C_o^{(i)} + \frac{1}{2} C_c^{(i)} + \lambda C_i = 0$$

$$\frac{\partial L}{\partial \lambda} = \sum_{i=1}^{n} C_i Q_i - F = 0$$

Solving above two equations, we get,

$$Q_i' = \sqrt{\frac{2D_i C_o^{(i)}}{C_c^{(i)} + 2\lambda C_i}} \text{ and } \sum_{i=1}^{n} C_i Q_i = F$$

Example 6.8: *A shop produces three items in lots. The demand rate for each item is constant and can be assumed to be deterministic. No back orders are allowed. The pertinent data for items is been given in the following table:*

Table 6.2

Item	I	II	III
Carrying cost (₹)	20	20	20
Set-up cost (₹)	50	40	60
Cost per unit (₹)	6	7	5
Year demand rate (units)	10000	12000	7500

Determine approximately economic order quantity for three items subjected to condition that total value of average inventory level does not exceed ₹1000.

Solution: For, k = 1

$$Q_1' = \sqrt{\frac{2 \times 10000 \times 50}{20 + 2 \times 6}} = 177 \text{ units}$$

$$Q_2' = \sqrt{\frac{2 \times 12000 \times 40}{20 + 2 \times 7}} = 168 \text{ units}$$

$$Q_3' = \sqrt{\frac{2 \times 7500 \times 60}{20 + 2 \times 5}} = 173 \text{ units}$$

Investment over average inventory at any time is given by

$$\sum_{i=1}^{3} C_i \left(\frac{Q_i}{2}\right) = 6 \times \frac{177}{2} + 7 + \frac{168}{2} + 5 \times \left(\frac{173}{2}\right)$$

$$= 1551 > 1000$$

For $\lambda = 4.7$, $Q_1 = 114$, $Q_2 = 105$, $Q_3 = 116$

Average inventory = 999.50 which is closer to ₹ 1000.

Model III (c): EOQ Model with Average Inventory Constraint:

1. Average number of units in stock of an item i is $\frac{Q_i}{2}$, and it is required that the total average number of units of all items held in inventory should not exceed to number M.

2. We can explain that

$$\frac{1}{2} \sum_{i=1}^{n} Q_i \leq M$$

3. Problem of minimizing total variable inventory cost under total average inventory constraint can be formulated as

$$\text{Minimum TVC} = \sum_{i=1}^{n} \frac{D_i}{Q_i} C_o^{(i)} + \sum_{i=1}^{n} \frac{Q_i}{2} C_c^{(i)}$$

This equation is subjected to constraints of $\frac{1}{2} \sum_{i=1}^{n} Q_i \leq M$

and $Q_i \geq 0$

We can construct Lagrangian function as

$$L(Q_i, \lambda) = \sum_{i=1}^{n} \frac{D_i}{Q_i} C_o^{(i)} + \sum_{i=1}^{n} \frac{Q_i}{2} C_c^{(i)} + \lambda \left\{ \frac{1}{2} \sum_{i=1}^{n} Q_i - M \right\}; \lambda \geq 0$$

4. The necessary conditions for L to be minimum are

$$\frac{\partial L}{\partial Q_i} = -\frac{D_i}{Q_i^2} C_o^{(i)} + \frac{1}{2} C_c^{(i)} + \frac{1}{2}; 1, 2, \ldots, n$$

$$\frac{\partial L}{\partial \lambda} = \frac{1}{2} \sum_{i=1}^{n} Q_i - M = 0$$

Solving above two equations,

$$Q_i' = \sqrt{\frac{2 D_i C_o^{(i)}}{C_o^{(i)} + \lambda}}; i = 1, 2, \ldots, n$$

and $\sum_{i=1}^{n} Q_i = 2M$

Model III (d): EOQ Model 3 with Total Number of Orders Constraints:

This approach is used where ordering cost per order and carrying cost per unit per time period are not known.

Assumptions :

* Demand is constant,

* Stockouts are not permitted,

* Orders are given in lots

* Ordering cost and carrying costs are same.

The total no. of orders per year $\left(N = \frac{D}{Q}\right)$ and also number of orders per year

$$= N \times \frac{\sqrt{DC}}{\Sigma\sqrt{DC}}$$

where,   DC = Demand in rupees

   N = Specified orders

Example 6.9: *A company has to purchase four items A, B, C and D for three years. The projected demand and unit price (in ₹) are follows:*

Table 6.3

Item	Demand (units)	Unit price (₹)
A	60000	3
B	40000	2
C	1200	24
D	5000	4

If a company wants to restrict total number of orders to 40 for all the four items, how many orders should be placed for each item ?

Solution: Allocation of 40 total orders per year to four items is given below:

Table 6.4

Item	Demand units D	Units Price (₹)	$\sqrt{DC}$	$\dfrac{\sqrt{DC}}{\Sigma\sqrt{DC}}$	Number of orders per year
A	60000	3	424.6	0.416	0.416 (40) = 17
B	40000	2	282.84	0.277	0.227 (40) = 11
C	1200	24	169.70	0.166	0.166 (40) = 6.64 = 7
D	5000	4	141.42	0.138	0.138 (40) = 5
			1010.00		39.88

6.10 THE EOQ MODELS WITH QUANTITY DISCOUNTS　　　[Dec. 2009, May 2011]

- Quantity discounts often are offered for externally purchased items to encourage buyers to purchase more units of an item.

- Such economics sale of items may result in different unit costs for different production lot sizes.

Quantity discounts are of following types:

1. All units discounts.

2. Incremental quantity discounts.

Assumptions:

- Demand is uniform and deterministic,

- Shortages are not allowed,

- Replenishment is instantaneous.

Model IV (a): EOQ Model with All Unit Discounts Available:

- If there are several price breaks, 0, b_1, b_2 .... b_i and the ordered quantity Q lies in discount interval, say $b_{i-1} \le Q \le b_i$, then price per unit for Q units C_i , where $C_i < C_{i-1}$.

- Hence, total annual cost for a discount schedule that has price-breaks at specified quantities includes item. Costs as specified below,

$$T_{Ci} = D_{Ci} + \frac{D}{Q} C_o + \frac{D}{2}$$

$$= D_{Ci} + \frac{D}{Q} C_o + \frac{D}{2} (C_i) \times r$$

Fig. 6.10 shows the total annual cost function, $T_C(Q)$. For any purchase quantity Q in range $0 \le Q < \infty$.

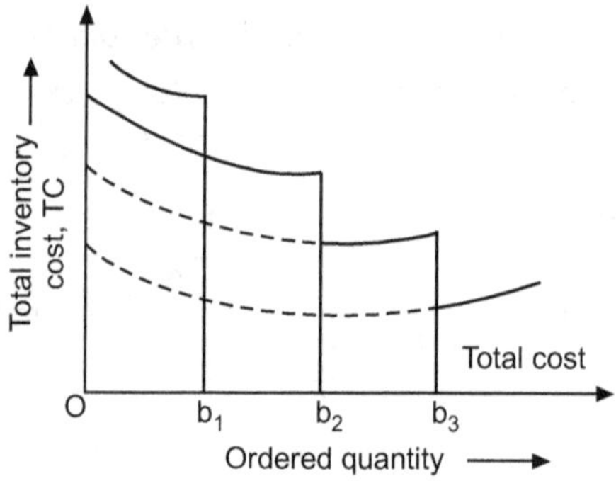

Fig. 6.10: EOQ model with quantity discount

$$T'_{ci}(Q) = \text{Model purchase cost of item} +$$

Total minimum variable inventory cost

$$= D_{Ci} + TVC' = \sqrt{QD \cdot C_r^{(i)}} + DC_i$$

Model with One Price Break: Suppose the following price discount schedule is quoted by a supplier in which a price break occurs at quantity b_1.

Example 6.10: *The annual demand of a product is 10000 units. Each unit costs ₹ 100, if orders placed in quantities below 200 units, but for orders of 200 or above the price is ₹ 95. The annual inventory holding costs is 10% of value of item and ordering cost is ₹ 5 per order. Find economic lot size.*

Solution: Given:

$$D = 10000 \text{ units/year,}$$

$$C_o = ₹ 5/\text{order,}$$

$$r = 10\% \text{ of price of an item} = 0.10 ₹$$

The unit cost for range of quantities:

Table 6.5

Quantity	Price per unit (₹)
$0 < Q_1 < 200$	100.00
$200 \le Q_2$	95.00

The optimum order quantity Q'_2 based on price $C_2 = ₹ 95$.

$$Q'_2 = \sqrt{\frac{2DC_o}{C_2 \times r}} = \sqrt{\frac{2 \times 10000 \times 5}{95 \times 0.10}}$$

$$= 103 \text{ units (Approximately)}$$

Since, $Q' < b (= 200)$ hence calculating Q'_1, we get,

$$Q_1 = \sqrt{\frac{2DC_o}{C_1 \times r}}$$

$$= \sqrt{\frac{2 \times 10000 \times 5}{100 \times 0.10}}$$

$$= 100 \text{ units}$$

As $Q_1' < b$ (= 200), comparing $T_c(Q_i)$ and $T_c(b = 200)$

$$T_C\left(Q_1'\right) = D_{C_1} + \frac{D}{Q'}C_o + \frac{Q'}{2}(C_1 \times r)$$

$$= 10000 \times 100 + \frac{10000}{100} \times 5$$

$$+ \frac{100}{2}(100 \times 0.10)$$

$$= ₹\,1001000$$

$$T_C(b) = D_{C2} + \frac{D}{b}C_o + \frac{b}{2}(C_2 \times r)$$

$$= 10000 \times 95 + \frac{10000}{200} \times 5$$

$$+ \frac{200}{2}(95 \times 0.10)$$

$$= ₹\,1001250$$

Since, $T_C Q_i < T_C(b)$, therefore, the optimal order quantity is $Q'' = Q_1' = 100$ units.

6.11 DYNAMIC DEMAND INVENTORY CONTROL MODELS

Terminologies Used:

1. **Reorder Level:** Reorder level (ROL) = $D \times LT$

   where, D = Demand rate during lead time (in units/time period)

            LT = Average lead time.

2. **Service Level:** The probability with which an organisation will be able to supply stock to its customer from the current stock, Is referred as service level.

   Service level (SL)

   $$= \left\{\frac{\text{Expected number of units out of stock annually}}{\text{Total annual demand}}\right\}$$

3. **Buffer Stock:** If demand rate and/or lead time are not known with certainty, additional stocks in form of safety stock, reserve stock and also known as buffer stock is stored.

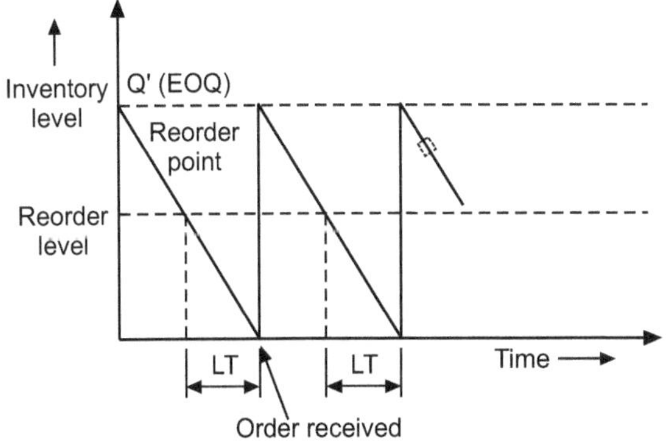

Fig. 6.11: Recorder level

Buffer stock = (BS)

       = Average demand × Average lead time

e.g., If demand rate is 200 units/month, the normal and maximum lead time are 10 days and 30 days respectively, then buffer stock can be computed as

$$\text{Buffer stock} = \frac{\text{Demand rate}}{\text{Maximum lead time}} \times (\text{Average lead time})$$

$$\text{Buffer stock} = \frac{200}{30} \times \left(\frac{10 + 30}{2}\right)$$

$$= 133.33 \text{ units}$$

When no stockouts are desired, buffer stock is given by

   Buffer stock = (Maximum demand during LT)

            − (Average demand during LT)

$$= D_{\text{maximum}} \times LT - (D_{\text{average}} \times LT)$$

$$= (D_{\text{maximum}} - D_{\text{average}}) \times LT$$

When demand rate varies about average demand ($D_{\text{avg.}}$) during a constant lead time (LT),

Reorder level (ROL) = $D_{\text{avg}} \times LT$

To avoid shortage buffer stock is kept.

Reorder level (ROL) = $D_{\text{avg}} \times LT + BS$

Example 6.11: *The information provided for un item is as follows:*

   Annual demand = 12000 units

     Ordering cost = ₹ 60/order

     Carrying cost = 10%

     Units cost of item = ₹ 10 and lead time = 10 days.

These are 300 working days of a year. Determine EOQ and number of orders/year. In past two years, use rate has gone as high as 50 units/day. For a reordering system based on inventory level.

Calculate buffer stock. What should be reorder level at this buffer stock ? What would be carrying costs for a year ?

Solution: Given data:

     $D = 12000$ units/year ; $C_o = ₹\,60$/order,

     $C = ₹\,101$ items, $C_c = 10\%$ of ₹ 10 = ₹ 1 per unit year

and $LT = 10$ days

Hence,

$$\text{EOQ (Q')} = \sqrt{\frac{2DC_o}{C_c}}$$

$$= \sqrt{\frac{2 \times 12000 \times 60}{1}}$$

$$= 1200 \text{ units}$$

$$\text{Number of orders} = N = \frac{D}{Q'}$$

$$= \frac{12000}{1200} = 10 \text{ per year}$$

Average consumption

$$= \frac{12000}{30} = 40 \text{ units/day}$$

and, Maximum consumption

$$= 50 \text{ units/day}$$

$$\text{Buffer stock} = (50 - 40) \times LT$$

$$= 10 \times 10 = 100 \text{ units}$$

Reorder level = Average demand during LT + Buffer stock

$$= 40 \times 10 + 100 = 500 \text{ units}$$

Average inventory level

$$= \text{Buffer stock} + \frac{Q'}{2}$$

$$= 100 + \frac{1200}{2}$$

$$= 700 \text{ units}$$

Inventory carrying cost/year = ₹ (700 × 10% of C)

$$= 700 \times 1$$

$$= ₹ 700$$

6.11.1 Dynamic Inventory Control Model Approaches

- It has two approaches which are:

  1. Fixed order quantity (Q-system).

  2. Periodic review (P-system).

1. Fixed Order Approach (Q-System):

- In this, as soon as inventory level drops to a point equal to or less than reorder level, a replenishment order for fixed quantity equivalent to Q' is placed.

- Formulae used are:

$$Q' = \sqrt{\frac{2DC_o}{C_c}}$$

ROL = Average demand during lead time + Reserve stock + Safety stock.

Reserve stock (₹) = Standard deviation of demand during lead time × Service level factor.

Safety stock = Average demand during maximum delay in lead time × Probability of such delay.

- Following Fig. 6.12 shows operation of Q-system for a single item where demand during lead time (DDLT) is known only to extent of its probability distribution.

Following Example Illustrates Q-System:

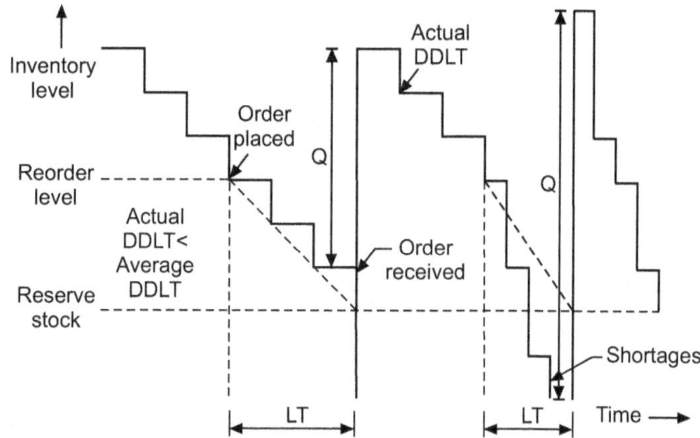

Fig. 6.12: System with probabilistic demand

2. Periodic Review Approach:

- In P-system, orders for replenishment are placed after reviewing inventory level at fixed intervals of time. This is time (t) between orders are determined and held constant for that period.

- The following Fig. 6.13 shows P-system, For P-system,

$$Q' = \sqrt{\frac{2DC_o}{C_c}}$$

and optimal time between orders is

$$t' = \frac{Q'}{D}$$

$$t' = \frac{1}{D} \times \sqrt{\frac{2DC_o}{C_c}}$$

$$= \sqrt{\frac{2C_o}{DC_c}}$$

where, D = Annual demand

C_c = Carrying cost per unit/year

C_o = Ordering cost per orders

t = Interval roman between roman successive orders

Example 6.12: *A firm has experienced probability distribution for inventory demand during lead time of 9 days is given below:*

Table 6.6

Number of Units	Probability
51 – 60	0.01
61 – 70	0.04
71 – 80	0.11
81 – 90	0.20
91 – 100	0.29
101 – 110	0.20
111 – 120	0.10
121 – 130	0.04
131 – 140	0.01

(i) *Determine reserve stock when no stockout is desired.*

(ii) *Determine reserve stock and reorder level corresponding to given service level of 95%.*

Solution: The calculations of maximum and average demand during given lead time.

Table 6.7

Number of Units	Mid-Point (m)	Probability (f)	Probability of Corresponding Demand	$f \times m$
51 – 60	55.5	0.01	0.01	0.555
61 – 70	65.5	0.04	0.05 (0.04 + 0.01)	2.620
71 – 80	75.5	0.11	0.16	8.305
81 – 90	85.5	0.20	0.36	17.100
91 – 100	95.5	0.29	0.65	27.695
101 – 110	105.5	0.20	0.85	21.100
111 – 120	115.5	0.10	0.95	11.550
121 – 130	125.5	0.04	0.99	5.020
131 – 140	135.5	0.01	1.00	1.355
		1.00		95.300

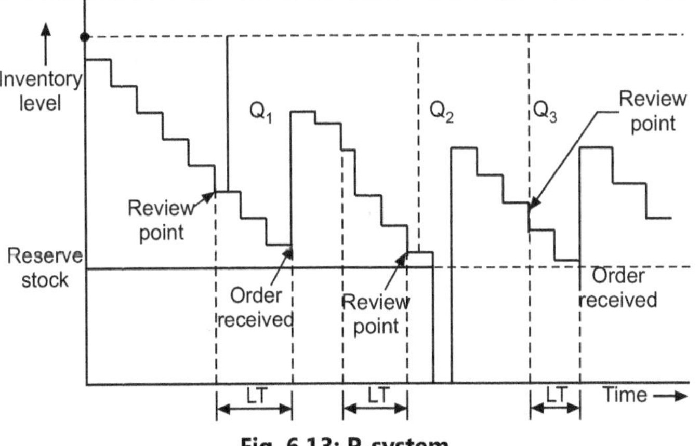

Fig. 6.13: P-system

Average demand during lead time, DDLT = 95.30 units and maximum DDLT = 140 units.

(1) Reserve stock when no stockout is desired,

Reserve stock = 140 – 95.30 = 44.70 units

(2) By interpolation, we have,

P_{95} (95th percentile) =

$$l + \left(\frac{i \times \dfrac{N}{100} - \text{Cumulative frequency}}{f} \right) \times h$$

where, t = 95 and N = 1

$$= 110.5 + \frac{0.95 - 0.85}{0.10} \times 10$$

$$= 120.50 \text{ units}$$

Thus, reserve stock when a 95% service level is desired.

Reserve stock = (Maximum DDLT for 95% service level) – (Average DDLT)

$$= 120.50 - 95.30 = 25.20 \text{ units}$$

Example 6.13: *A company produces an item for which,*

Annual demand	=	1000 units
Standard deviation of demand/week	=	10 units
Cost per unit of item	=	₹ 5
During cost/order	=	150
Inventory carrying cost	=	30%
Maximum delay	=	3 weeks
Service level	=	95%
Average lead time	=	4 weeks
Probability of delay	=	0.30

Determine buffer stock, reserve stock, safety stock and desirable maximum inventory level. For this item in both Q and P-systems.

Solution: (i) Q-system

$$EOQ(Q') = \sqrt{\frac{2DC_o}{C_c}}$$

$$= \sqrt{\frac{2 \times 100 \times 150}{5 \times 0.30}}$$

$$= 447 \text{ units}$$

Buffer stock = Average demand during lead time

$$= \left(\frac{1000}{52} \right) \times 4 = 77 \text{ units}$$

Safety stock = Average demand during maximum delay × Probability of delay

$$= \left(\frac{1000}{52}\right) \times 3 \times 0.30$$

$$= 17 \text{ units}$$

Reverse stock = Z × Standard deviation of demand during average lead time

$$= Z \times \sqrt{LT}$$

$$= 1.65 \times 10\sqrt{4} = 33 \text{ units} \quad [\text{For SL} = 95\%]$$

Reorder point = (Average demand during lead time) + (Safety + Reverse) stock

$$= 77 + (17 + 33) = 127 \text{ units}$$

For P-system:

$$\text{Review period} = \frac{EOQ}{b} = 447 \times 52 = 23.444 \text{ weeks}$$

The review period of 23.444 weeks is rounded for 23 weeks as inventory carrying cost of 23 weeks will be less than that of 24 weeks.

Buffer stock = Average demand during (t' + Lead time)

$$= \frac{1000}{52} \times (23 + 4) = 519 \text{ units}$$

$$\text{Safety stock} = \frac{1000}{52} \times 3 \times 0.30 = 17 \text{ units}$$

$$\text{Reverse stock} = 1.65 \times 10 \times \sqrt{27} = 87 \text{ units}$$

Target inventory = D × (t' + LT) + (Safety stock) + (Reverse stock)

$$= \frac{1000}{52} \times (23 + 4)(17 + 86)$$

$$= 622 \text{ units}$$

Example 6.14: *The annual demand for a product is 3600 units, with an average 12 units/day. The lead time is 10 days. The order processing cost is ₹ 200.00 per order and annual inventory carrying cost is 20% of value of money in form of material. The unit price of product is ₹ 250.*

(1) What will be economic-order quantity?

(2) Find purchase cycle time.

(3) Find annual total cost including material cost.

Solution: Given:

$$D = 3600 \text{ units/year}$$

$$LT = 10 \text{ days}$$

$$C_c = 20\% \text{ of investment in inventories}$$

$$= 0.20 \times 250 = 50 \text{ per unit per year}$$

$$C_o = ₹ \ 200 \text{ per order}$$

(1) The economic order quantity

$$= Q' = \sqrt{\frac{2DC_o}{C_o}} = \sqrt{\frac{2 \times 3600 \times 200}{50}}$$

$$= 169.70 \text{ units per order}$$

(2) Purchase cycle time,

$$t' = \frac{Q'}{D} = \frac{169.70}{3600} = 0.047140 \text{ year}$$

Also 12 units/day

$$\therefore \quad t' = \frac{Q'}{D} = \frac{169.70}{3600} = 14.14 \text{ day}$$

The minimum variable inventory cost (yearly)

$$TVC' = \sqrt{DC_oC_c} = \sqrt{2 \times 3600 \times 200 \times 50}$$

$$= 8485.28 \ ₹/\text{year}$$

The total annual cost per year,

$$TC' = TVC' + DC = 8485.28 + 3600 \times 250$$

$$= 908485.28 \ ₹/\text{year}$$

Example 6.15: *A demand of an item is 160000 units/annum. The item cost is ₹ 20/unit. The cost of setting order is ₹ 50/order. The inventory carrying cost is 5% of inventory value. The stock outcost may be ₹ 2/unit.*

Calculate:

(1) Optimal order quantity, if stockout is permitted,

(2) The units back ordered at shortage cost indicated,

(3) The savings/loss occurred if no stockouts are permitted.

Solution: Given:

$$D = 160000 \text{ units/annum,}$$

$$C = 20/\text{unit,}$$

$$C_o = 50/\text{order,}$$

$$C_c = 5\% \text{ of inventory value} = 0.05 \times 20,$$

$$= 1 \ ₹/\text{unit 1 annum,}$$

and $\quad C_s = ₹ \ 2/\text{unit.}$

(1) Optimal order quantity,

$$Q' = \sqrt{\frac{2DC_o}{C_c}\left(\frac{C_c + C_s}{C_s}\right)}$$

$$= \sqrt{\frac{2 \times 160000 \times 50}{1}\left(\frac{1 + 2}{1}\right)}$$

$$= 4898.97 \text{ units}$$

(2) Optimal stock level,

$$M' = \sqrt{\frac{2DC_o}{C_c}\left(\frac{C_s}{C_o + C_s}\right)}$$

$$= \sqrt{\frac{2 \times 160000 \times 50}{1}\left(\frac{2 + 1}{1}\right)}$$

$$= 3265.98 \text{ units}$$

Average number of units on Backorder

$$= \frac{(Q - M)^2}{2Q} = \frac{(4898.97 - 3265.98)^2}{2 \times 4898.97}$$

$$= 272.165 \text{ units}$$

(3) Inventory cost if stockouts are permitted,

$$TVC' = \sqrt{2DC_oC_c\left(\frac{C_s}{C_c + C_s}\right)}$$

$$= \sqrt{2 \times 160000 \times 50 \times 1\left(\frac{2}{1 + 2}\right)}$$

$$= ₹\,3265.98$$

Inventory costs of stockouts are not permitted,

$$TVC' = \sqrt{2DC_oC_c} = \sqrt{2 \times 160000 \times 50 \times 1}$$

$$= ₹\,4000$$

Savings on inventory if stockouts are permitted.

$$= (4000 - 3265.98)$$

$$= 734.04 \text{ units}$$

Example 6.16: *Annual demand for an item is 10000 units. Costs per unit is ₹100. The supplier has agreed to provide a discount of 5% if order exceeds 250 units. Costs of holding inventory is 20% of cost of material per unit/year Cost of ordering is ₹50 order, find economic order quantity.*

Solution: Given:

$$D = 10000 \text{ units,}$$

$$C = 100,$$

$$r = 0.20,$$

$$C_o = 50/\text{order}$$

$$C_c = C \times r = 0.20 \times 100 = 20$$

Optimal order quantity without discounts,

$$Q' = \sqrt{\frac{2DC_o}{C_c}} = \sqrt{\frac{2 \times 10000 \times 50}{20}}$$

$$= 223.63$$

With discount,

$$C = 100 - 100 \times 0.005 = 95$$

$$\therefore \quad C_c = C \times r = 0.20 \times 95 = 19$$

$$Q' = \frac{2DC_o}{C_c} = \sqrt{\frac{2 \times 10000 \times 50}{19}}$$

$$= 229.41 \text{ units}$$

Example 6.17: *The following is information with regard to an item dealt by purchase manager.*

$$\text{Annual requirement} = 12000$$

$$\text{Order processing cost} = ₹\,12/\text{order}$$

$$\text{Inventory carrying cost} = 20\% \text{ of average inventory}$$
$$\text{value per year}$$

$$\text{Price paid} = ₹\,250$$

The purchase manager is considering possibility of allowing stockout to occur. He has estimated that annual stockout will be 25% of value of inventory.

(1) *What should be optimum no. of units of product he should buy in one lot of stock, if stock-out are not permitted ?*

(2) *What quantity of item should be ordered if stockouts are permitted?*

Solution: Given: $\quad b = 12000, C_o = ₹\,12/\text{order}, C = 250$

$$C_c = 0.20 \times 250 = ₹\,50 \text{ and}$$

$$C_s = 0.25 \times 250 = ₹\,62.5$$

Economic order quantity, Q' if,

(1) Stockouts are not permitted,

$$Q' = \sqrt{\frac{2DC_o}{C_c}}$$

$$= \sqrt{\frac{2 \times 10000 \times 12}{50}}$$

$$= 75.89 \text{ units}$$

$$EOQ = 75.89 \text{ units}$$

(2) If stocksouts are permitted,

$$Q' = \sqrt{\frac{2DC_o}{C_c}\left(\frac{C_c + C_s}{C_c}\right)}$$

$$= \sqrt{\frac{2 \times 12000 \times 12}{50}\left(\frac{50 + 62.5}{62.5}\right)}$$

$$= 56.56 \text{ units}$$

$$EOQ = 56.56 \text{ units}$$

Example 6.18: *The following data is available for an item:*

(i) *Price per unit = ₹50*

(ii) *Number of orders per year = 20*

(iii) *Minimum total inventory cost = ₹8000*

(iv) *Inventory holding cost per unit per year = 48.*

Calculate:

(1) Annual demand of an item,

(2) Procurement cost/order,

(3) Inventory carrying cost as percentage of average inventory investment,

(4) Economic order quantity.

Solution: Given: $C = 50$, $n = 20$, $TVC' = 8000$, $C_c = ₹\ 8$

$$TVC = \sqrt{2D \times C_o \times C_c}$$

and $\quad 8000 = \sqrt{2 \times D \times C_o \times 8} \qquad\qquad \ldots (i)$

$\therefore \quad D \times C_o = 4000000$

$$Q' = \sqrt{\frac{2DC_o}{C_c}} + \sqrt{\frac{2 \times 4000000}{8}}$$

$\therefore \qquad Q' = 1000$

$\therefore$ Economic order quantity = 1000

We know that

$$N' = \frac{D}{Q'} = \frac{\text{Annual demand}}{\text{Optimal order quantity}}$$

$$D = N' \times Q'$$

$$= 20 \times 1000 = 20000 \text{ units}$$

$\therefore$ Annual demand = 20000 units

$\therefore \quad D \times C_o = 4000000$ from Eqn. (i)

$$\therefore \qquad C_o = \frac{4000000}{20000}$$

$$= ₹\ 200/\text{order}$$

Average inventory investment

$$= \sqrt{2DC_oC}$$

$$= \sqrt{2 \times 20000 \times 8 \times 200} = 8000$$

$$C_r = r \times \text{Average inventory investment}$$

$$8 = \frac{r \times 8000}{100} = r = 0.1\%$$

Example 6.19: *Compute EOQ for meetings the following data:*

Daily demand	*= 20 units*
Daily production	*= 25 units*
Cost of set-up	*= ₹2500*
Cost of holding inventory	*= ₹20/year*
No. of working days (Month)	*= 25*

Also find no. of production batches in a year.

Solution: $D = 20$ unit/day, $C_o = 2500$, $C_c = 20$/unit/year

$$EOQ = \sqrt{\frac{2DC_o}{C_c}\left(\frac{P}{P-d}\right)}$$

Annual demand $= 20 \times 25 \times 12 = 6000$/year

$$P = \text{Production rate} = 25 \times 25 \times 12$$

$$= 7500/\text{year}$$

Economic log size for each production run is given by

$$Q' = \sqrt{\frac{2DC_o}{C_c}\left(\frac{P}{P-d}\right)}$$

$$= \sqrt{\frac{2 \times 6000 \times 2500}{20}\left(\frac{7500}{7500-6000}\right)}$$

$$= 2738.6 \text{ units}$$

$$N' = \frac{Q}{Q'}$$

$$= \frac{6000}{2738.6}\ 1.19/\text{year}$$

Example 6.20: *Determine the EOQ for a product whose average consumption rate is 80 units/day. The use of each unit is ₹ 20/year. The cost of planning and receiving an order is ₹ 20. Assuming 300 total working days in a year. Obtain the annual inventory capital, if carrying cost is ₹10/order.*

Solution: Given: $D = 80$/day $= 80 \times 300 = 24000$ units/order, $C = 20$/year, $Co = ₹\ 20$/order, and $C_o = ₹\ 10$/order

We know that,

Total minimum inventory investment,

$$TVC' = \sqrt{2DC_cC_o} = \sqrt{2 \times 24000 \times 10 \times 20}$$

$$= 3098.38$$

Total cost $= TVC' + D \times C$

$$= 3098.38 + 24000 \times 24000 \times 20$$

$$= ₹\ 483098.38/\text{year}$$

Example 6.21: *The details of the component are as under:*

Annual requirement	*= 12000*
No. of working days	*= 300/ year*
Cost/unit	*= ₹5*
Order cost	*= ₹500*
Inventory carrying	*= ₹1.8/unit/year*
Break order cost	*= ₹4.2/unit/year.*

Calculate:

(1) Economic batch quantity

(2) Optimum back order quantity

(3) Optimum cycle time

(4) Maximum inventory level.

Solution: Given: $D = 12000$ units, $C = ₹\,5$/unit, $C_o = ₹\,500$,

$C_c = 1.8$/unit/year, $C_s = ₹\,4.2$/unit/year.

(1) $EOQ(Q)' = \sqrt{\dfrac{2DC_o}{C_c}\left(\dfrac{C_c + C_s}{C_s}\right)}$

$= \sqrt{\dfrac{2 \times 12000 \times 500}{1.8}\left(\dfrac{1.7 + 4.2}{4.2}\right)}$

$= 3086.066$

(2) Optimum batch to be back orderly,

$R' = Q'\left(\dfrac{C_c}{C_c + C_s}\right) = 3086.066\left(\dfrac{1.8}{1.8 + 4.2}\right)$

$= 962$ units

(3) Optimum cycle time,

$t' = \dfrac{Q'}{D} = \dfrac{3086.66}{12000}$

$= 0.257$ years

$= 0.257 \times 300$

$= 77.15$ (Working days)

(4) Maximum inventory level

I_{max} level $= Q' - R' = 3086.066 - 962$

$= 2124.066$ units

Example 6.22: *A company purchases a component for which it has a steady usage of 1000 price per year. The ordering cost is ₹ 12/order and the estimated cost of money invested in inventory is 20% year. The unit cost of three component:*

$₹10 = Per\ 0 < 9 \le 149$

$₹9.75 = 149 < 9 \le 498$

$₹9.50 = 9 \ge 500$

No shortages are allowed. What is the optimum weight size?

Solution: Given: $D = 1000$ units/year, $C_o = ₹$ order, $r - 20\%$

For first component, $C_c = 10 \times 0.2 = 2$

$Q' = \sqrt{\dfrac{2DC_o}{C_c}} = \sqrt{\dfrac{2 \times 10000 \times 12}{10 \times 0.2}}$

$= 109.54$

As it satisfies $0 < 9 \le 149$ this is feasible

For second component,

$Q' = \sqrt{\dfrac{2DC_o}{C_c}} = \sqrt{\dfrac{2 \times 10000 \times 12}{9.50 \times 0.2}}$

$= 110.94$

Not feasible as $149 < 9 \le 498$

For third component,

$Q' = \sqrt{\dfrac{2DC_o}{C_c}}$

$= \sqrt{\dfrac{2 \times 10000 \times 12}{9.50 \times 0.2}}$

$= 112.74$

Not feasible as $9 \ge 500$.

Total inventory cost:

(1) For first component, $C = 10$ and $Q = 109.54$

$TC = D \times C + \sqrt{DC_c\,C_o}$

$= 1000 \times 10 + \sqrt{2 + 10000 \times 10 \times 0.2 \times 12}$

$= ₹\,10219.08$/year

(2) For second component, $C = 9.75$ and $Q = 149$

$TC = D \times C + \dfrac{Q}{2}C_c + \dfrac{D}{Q}C_o$

$= 10000 \times 9.75 + \dfrac{149}{2} \times 9.75 \times 0.2$

$\quad + \dfrac{1000}{149} \times 12$

$= ₹\,9975.82$/year

(3) For third component, $Q = 500$ and $C = 9.50$

$TC = D \times C + \dfrac{Q}{2C_c} + \dfrac{D}{Q} \cdot C_o$

$= 10000 \times 9.50 + \dfrac{500}{2} \times 9.50 \times 0.2$

$\quad + \dfrac{1000}{500} \times 12$

$= ₹\,9999$/year

Example 6.23: *Find the optimal order quantity for a product for which the price breaks are as follows:*

Quantity (q)	Unit cost (₹)
$0 < q < 500$	₹ 10
$500 \le q < 750$	₹ 9.25
$750 \le q$	₹ 8.75

The monthly demand for the product is 200 units.

Storage cost is 2% of the unit cost and ordering cost is ₹100.   **[Dec. 2007]**

Solution:

EOQ for unit price of ₹ 8.75 $= \sqrt{\dfrac{2\,C_o D}{C_c}}$

$$= \sqrt{\dfrac{2 \times 100 \times 200}{8.75 \times 0.02}}$$

$$= 478 \text{ units (infeasible)}$$

EOQ for price of ₹ 9.25 $= \sqrt{\dfrac{2 \times 100 \times 200}{9.25 \times 0.02}}$

$$= 465 \text{ units (infeasible)}$$

EOQ for price of ₹ 10 $= \sqrt{\dfrac{2 \times 100 \times 200}{10 \times 0.02}}$

$$= 447 \text{ units (feasible)}$$

Total cost/month for order quantity of 447 units (optimal size)

$$= \sqrt{2 \cdot D \cdot C_o \cdot C_c} + C_D$$

$$= \sqrt{2 \times (10 \times 0.02) \times 100 \times 200}$$

$$+ (10 \times 200)$$

$$= ₹\ 2089.45$$

Total cost/month for order quantity of 500 units (non-optimal size)

$$= \dfrac{Q}{2} C_c + \dfrac{D}{Q} C_o + (D \times C)$$

$$= \left[\left(\dfrac{500}{2} \times 9.25 \times 0.02\right) + \left(\dfrac{200}{500} \times 100\right) + (9.25 + 200)\right]$$

$$= ₹\ 1936.25$$

Total cost/month for order quantity 750 units (non-optimal size)

$$= \dfrac{Q}{2} C_c + \dfrac{D}{Q} C_o + [D \times C]$$

$$= \left[\left(\dfrac{750}{2} \times 8.75 \times 0.02\right) + \left(\dfrac{200}{750} \times 100\right) + (8.75 + 200)\right]$$

$$= ₹\ 1842.30$$

QUESTIONS FOR PRACTICE

1. Discuss the importance of inventory control.

2. List and explain different types of inventory cost and sketch cost – quantity tread-off.

3. Explain the concept of economist order quantity.

4. Derive the formula for Economic lot size with constant demand.

5. Explain EOQ models with quantity discounts.

6. Write short notes on:

   (a) EOQ with finite rate of replenishment

   (b) ABC analysis

   (c) EOQ with price-breaks

   (d) Static inventory control model

   (e) Necessity of maintaining inventory

   (f) Classification of inventory model.

7. Discuss what is ABC analysis in inventory control.

8. A pharma company consumes annual 6000 kg of chemical costing ₹ 5/kg, ordering cost is ₹ 25 and carrying cost is 6% per year/kg of average inventory. Find EOQ and total inventory cost (including the cost of chemicals). If the supplier offers discount of 5% on the cost price for a single order of annual requirement should the factory accept it.   **[May 09]**

CHAPTER 7
PROJECT MANAGEMENT TECHNIQUES

7.1 INTRODUCTION TO CPM AND PERT

(Application and Basic Steps in PERT/CPM Technique)

- Any task which has definite and definable start and end which involves investment and use of one or more than one resources could be formed as project. A large project involves many departments doing different activities which are interrelated in such a logical sequence that some activities cannot start until others are completed. Hence, to co-ordinate activities of various departments/groups network scheduling technique is used.

- It minimizes delays, interruptions and bottlenecks. There are two main scheduling techniques, viz. Critical Path Method (CPM), Programme Evaluation and Review Techniques (PERT). These techniques were developed in 1956-58 by two different groups. CPM was developed by Walker and further developed by Mauchly Associates. PERT was developed by team of engineers on polar's Missile programme of U.S. Navy.

- Steps to be followed in CPM/PERT techniques. Scheduling by CPM/PERT consists of basic four steps :

  1. **Planning:** It is started with splitting the total project/task into small projects/tasks, which further divided into activities. These activities are analyzed by department.

  2. **Scheduling:** In this step, time chart showing start and finish time for each activity and their relationship is prepared. It indicates the criticality of the activity. In case of non-critical activities the schedule shows amount of float/slack time.

  3. **Allocation of Resources:** Resource is a physical variable, viz., labour, finance equipment and space which creates constraints for completion of project. Hence, resource allocation is generally a compromise based on the decisions made by the managers.

  4. **Controlling:** This is the final step in the project management. CPM applies the principle and management by expectation to identify areas that are critical to the completion of the project.

7.2 TERMS USED IN PROJECT MANAGEMENT (NETWORK MODEL)

- **Activity:** Any individual operation which consumes resources and has known starting and finishing time called as an activity. It is represented by an arrow and arrowhead represents the direction of progress. (Refer Fig. 7.1).

- **Event:** It represents start and end of an activity in terms of time. It is represented in the form of circle. In network, some of the activities results into or combines at single node called as **merging of event** and on the other hand when single activity gives birth to more than one activity called as **bursting of event**. (Refer Fig. 7.1).

- **Dummy Activity:** It does not consumes time and any resources but it expresses the dependency of one activity over other to maintain the logical sequence in completion of project. Generally it is indicated in the form of dotted line. (Refer Fig. 7.1). **[Dec. 2007]**

- **Predecessor and Successor Activities:** Before starting a particular activities, all other activities which have to be completed called as predecessor activities. And all those activities which have to follow the particular activity under consideration are called as its successor activities. (Refer Fig. 7.1).

Activity Representation	Description
(a) A ⟶ ◯ ⟶ B	Activity A must be completed before activity B can start.
(b) A ⟶ ◯ ⟶ B, C	Activity B and C can not start until activity A is completed [bursting of events].
(c) A, B ⟶ ◯ ⟶ C	Activity C can not start until both activities A and B are completed [merging of events].

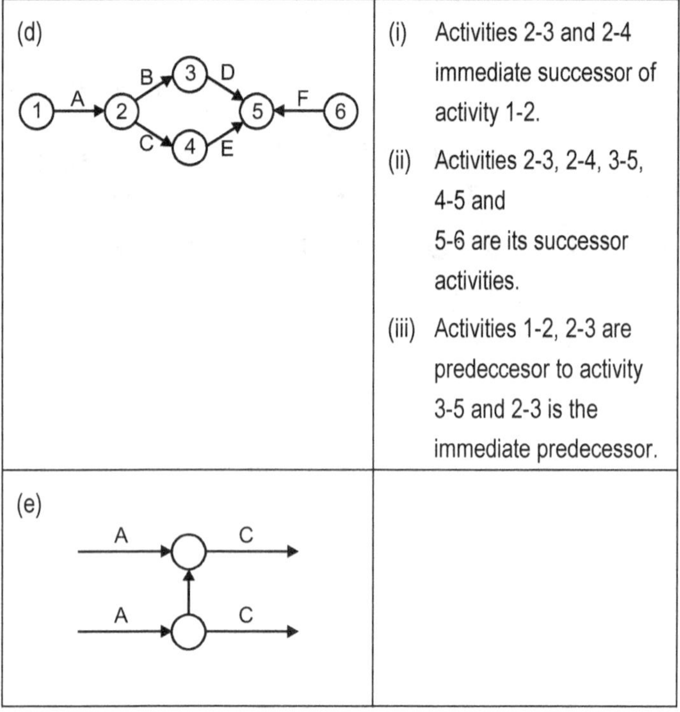

(d)

(i) Activities 2-3 and 2-4 immediate successor of activity 1-2.

(ii) Activities 2-3, 2-4, 3-5, 4-5 and 5-6 are its successor activities.

(iii) Activities 1-2, 2-3 are predeccesor to activity 3-5 and 2-3 is the immediate predecessor.

(e)

Fig. 7.1

7.3 NETWORK DIAGRAM WITH TIME ESTIMATES AND ANALYSIS

7.3.1 Fulkerson's Rule for Numbering Events in Network Diagram [May 2009]

- First network has to be drawn in a logical sequence, every event is assigned a number. Following are guiding rules to be followed for numbering the events:

  1. The initial event which has all outgoing arrows is numbered '1'.

  2. To convert some more nodes into initial events (which should be numbered 2, 3...) delete all the arrows coming out from node '1'.

  3. Delete all the arrows going out from these numbered events to create more initial events. Assign next numbers to these events.

  4. Continue until the final or terminal node which has arrows coming in, with no arrow going out is numbered.

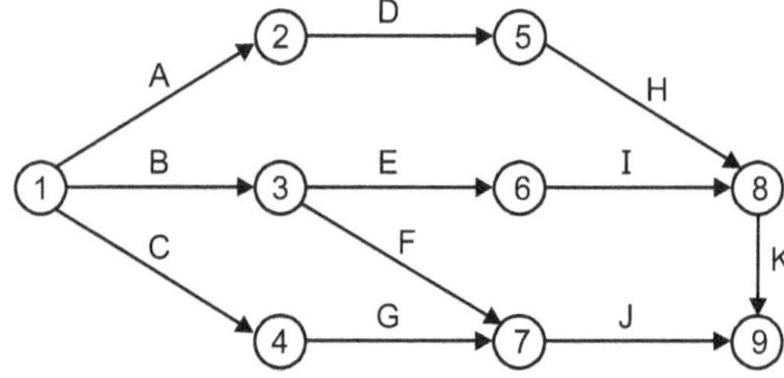

Fig. 7.2 : Illustrates numbering of the network

7.3.2 Critical Path Method (CPM)

- Horowitz describe the critical path method as a system for planning, scheduling and controlling a project. The operations involved in the project are represented graphically in a diagram called a **network**. The network has to be established considering operational sequence and interrelationship of various task.

- Then planner estimates the time it will take to do each operation. The **network** will usually consists of a number of parallel paths and some operations will be done simultaneously with others. The key to select the best network is in fact, a relatively small number of operations that to control the projects completion time. These operations are called 'critical operations', procedure of a path through the network is called the 'critical path'.

- **Projects Suitable for CPM:** In order to use CPM technique for scheduling effectively, a project must have a definite beginning and definite end. And it must also be capable of being divided into a sequence of operations. e.g.:

  ➤ Setting up a new department.
  ➤ Introducing a new product.
  ➤ Research and development projects.
  ➤ Engineering and architectural design.
  ➤ Assembling a large piece of machinery or aircraft.

- **Steps in CPM:**
  1. Analyze the project: Determine individual tasks or operations that are required.
  2. Show the sequence of operations on a network chart.
  3. Estimate the time it takes to perform each operation.
  4. Compute the critical path.
  5. Use this information to develop the most economical and efficient schedule.
  6. Use the schedule to control and monitor the job progress.
  7. Revise and update the schedule frequently throughout the execution of the project.

- The CPM approach can be extended to include the cost of each operation thus providing a means for determining an optimum time for completion of the project with a minimum total cost. The method relating operation times with costs is called Least cost scheduling.

- **Terms Used in CPM:** Forward pass consumption gives the earliest expected start and finish times for each activity. **[May 2009]**

1. **Earliest Start Time (ES):** It is the earliest event time of the tail end event.

$$ES_{ij} = E_i$$

where, ES_{ij} = Earliest start for an activity (i, j)

E_i = Earliest event occurrence time of event

2. **Earliest Finish Time (EF):** EF is the earliest time + Activity time

$$EF_{ij} = ES_{ij} + T_{ij}$$

where, T_{ij} = Time estimate of activity

3. **Earliest Event Time (EV):**

For event 'j', EV is the maximum of the earliest finish times of all activities ending into that event.

$$E_j = \text{Maximum of } (ES_{ij} + T_{ij})$$
$$= \text{Minimum } (L_i + T_{ij})$$

Backward pass consumptions specifies the latest event times by which all the activities must be completed without delaying the project completion time.

4. Latest finish time for an activity (i, j) equals to the latest event time of event,

$$j - LF_{ij} = L_{ij}$$

Latest starting time of activity (i, j) is $LS_{ij} = LF_{ij} = T_{ij}$

5. The slack of an event is the difference between latest and earliest event times.

6. **Floats :**   **[May 2009, Dec. 2009, May 2010]**

   (i) Total Float: It is the difference between the maximum time available and activity duration time.

   $$TF = LS_{ij} - ES_{ij} = (L_j - E_i) - T_{ij}$$

   where,

   E_i = Earliest expected completion time for tail event

   = Earliest starting time for an activity (i , j)

   L_i = Latest allowable completion time of head event

   = Latest finish time for activity (i, j)

   (ii) Free Float: Assuming all activities start as early as possible then the time available for an activity is called as free float.

   ∴ Free Float = $(E_j - E) - E_{ij}$

   It is the time period by which an activity can be delayed without affecting earliest start time of any immediately following activities assuming, the predecessor event occurs at its latest possible time and the successor event at its earliest possible time.

   Independent float = $(E_i - L_i) - E_{ij}$

7.3.3 Program Evaluation and Review Technique (PERT)

- Using CPM, it was assumed that the amount of time it took to perform an operation was known with some degree of certainty. For many projects that is realistic approach, viz., fixed maintenance schedules, planning for production etc. Experience and a tremendous amount of data of past projects is required to make accurate estimation of time.

- In some projects, a reliable time estimation is difficult. For example, research and development work, the design of new products, unusual construction projects and other instances. Where new or untried techniques or materials are being used involve operations for which realistic time estimates are not possible. PERT technique was designed to deal with this type of project. The main purpose in the analysis through PERT is to find out the completion for a particular event within specified date.

PERT requires estimating three times for performing an activity:

1. **The Optimistic time (T_o):** It is the shortest possible time in which the activity could be possibly be completed.

2. **The Most likely time (T_m):** It is the time the activity would take most often if it were repeated over and over again under the same conditions.

3. **Pessimistic time (T_p):** It represents the maximum time the activity could take. It takes in consideration that everything will go wrong.

- These three time estimates are used to determine an expected time (T_e) which is based on the recognition of the probabilistic nature of the activities involved.

$$T_e = \frac{T_o + 4T_m + T_p}{6}$$

- After identifying the critical path and the occurrence times of all activities, the important question is to be answered as what is the probability that a particular event will occur on or before the scheduled date ?

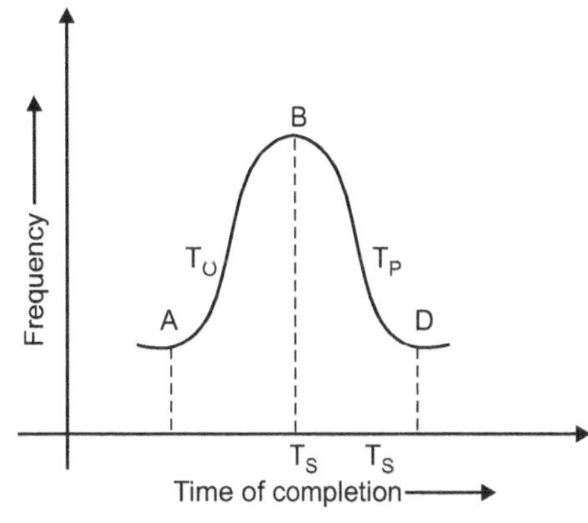

Fig. 7.3

- The probability of completing a project in scheduled time (T_s) is given by

$$P(T_s) = \frac{\text{Area under ABC}}{\text{Area under ABD}}$$

- The value of standard deviation for a network is calculated as:

Standard deviation for network = σ

where, variance for an activity

$$V = \sigma^2 = \frac{(T_p - T_o)^2}{6}$$

7.4 DIFFERENCE BETWEEN CPM AND PERT

[Dec. 2008, 2010, May 2011]

- Both CPM and PERT involve network diagrams, events, operations and critical path determination. The basic difference in the two, however, is the manner in which estimated times are determined. CPM is a deterministic model where activity is having single time, where PERT is probabilistic model, the expected time is calculated from three time estimates.

- Basically, CPM is activity oriented approach and PERT is event oriented approach. PERT is used where resources are always available but CPM is used where minimum overall costs are of prime importance. Hence, CPM is planning tool and PERT is controlling tool.

7.5 USES OF CPM AND PERT OR APPLICATION OF CPM AND PERT

[Dec. 2008, 2010, May 2011]

- These network techniques helps the production managers to achieve production target in minimum time and least cost (i.e. optimum utilization of resources).

- Network techniques helps to simulate the conditions to make necessary changes and improvements as and when required. And it provides number of check points.

- These techniques are widely applicable in projects of the following areas:

  - Civil construction
  - Research and development
  - Market survey
  - Maintenance planning
  - Inventory planning etc.

- A project consists of six activities. Draw the network diagram and calculate EST, LST, EFT, LFT and floats. Determine the critical path and find total project duration.

Table 7.1

Activity	Immediate Predecessor	Activity Duration
A	–	4
B	A	6
C	B	5
D	A	4
E	D	3
F	E, C	3

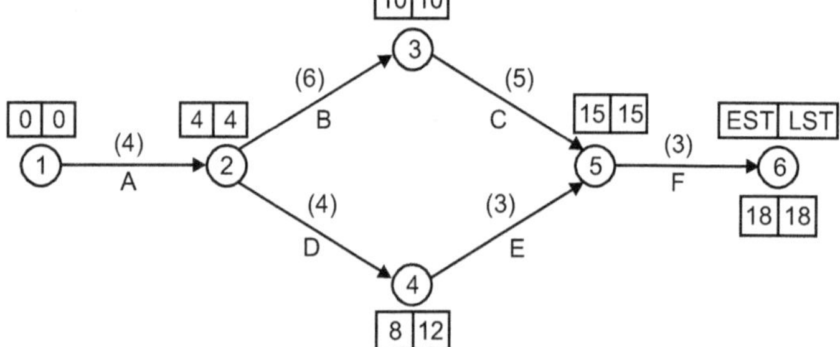

Fig. 7.4

The network diagram alongwith duration is as follows:

Table 7.2

Activity	Duration (Days)	EST	LST	EFT	LFT	Total Float	Free Float	Independent Float
A	4	0	0	4	4	0	0	0
B	6	4	4	10	10	0	0	0
C	5	10	10	15	15	0	0	0
D	4	4	8	8	12	4	0	0
E	3	8	12	11	15	4	4	0
F	3	15	15	18	18	0	0	0

SOLVED EXAMPLES

Example 7.1: *A established company has decided to add new product to its line. It will buy the product from manufacturing concern, package it and sell it to a number of distributions, selected on a geographical basis. Market research has indicated the volume expected and the size of the sales force required. The step shown in the following table 7.3 are to be planned.*

Table 7.3

Activity	Description	Time (Week)
a	Organise sale office	6
b	Hire salesman	4
c	Train salesman	7
d	Select advertise agency	2
e	Plan advertise campaign	4
f	Conduct advertise campaign	10
g	Design package	2
h	Set up packing facilities	10
i	Pack initial stock	6
j	Order stock from manufacturing	13
k	Select distributor	9
l	Sell to distributor	3
m	Ship stock	5

- The precedence relationships among these activities as shown in Fig. 7.5.

- The company can begin to organise to sales office, design the package and order the stock immediately. Also the stock must be ordered and the packaging facility must be set up before the initial stocks are packaged.

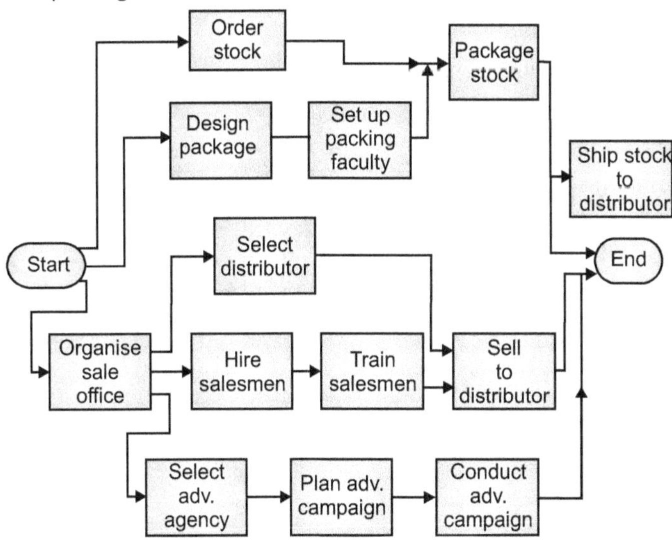

Fig. 7.5

(a) *Draw a arrow diagram for this project.*

(b) *Indicate the critical path.*

(c) *For each non-critical activity find the total and free float.*

Solution: Arrow diagram is shown in Fig. 7.6 with earliest start and finish, and latest start and finish.

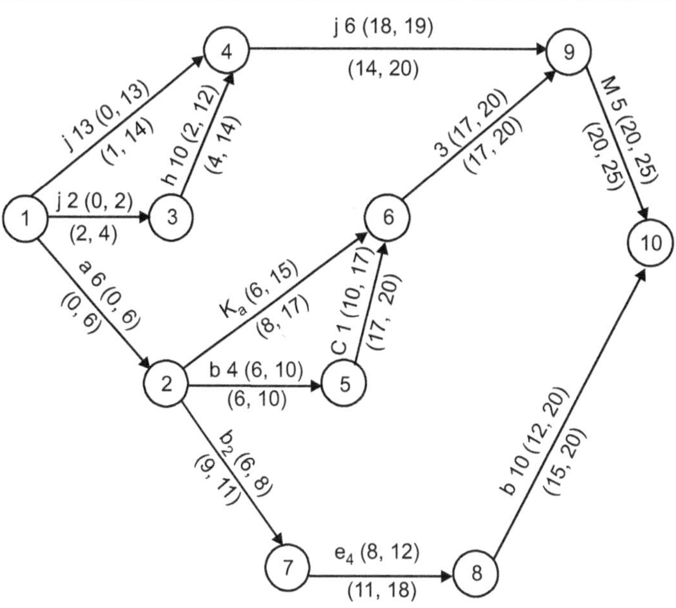

Fig. 7.6

The critical path of the project is 1-2-5-6-9-10 with a length of 25 weeks.

The critical activities of the project are a, b, c, d, and m.

The following table 7.4 shows, scheduling time, free and total float for the activities:

Table 7.4

	Activities												
	1-2	1-3	1-4	2-5	2-6	2-7	3-4	4-9	5-6	6-9	7-8	8-10	9-10
Time	6	2	13	4	9	2	10	6	7	3	4	10	5
ES	0	0	0	6	6	6	2	13	10	17	8	12	20
EF	6	2	13	10	15	8	12	19	17	20	12	22	25
LS	0	2	1	6	8	9	4	14	10	17	11	15	20
LF	6	4	14	10	17	11	14	20	17	20	15	25	25
TF	0	2	1	0	0	3	2	1	0	0	3	3	0
FF	0	0	0	0	2	0	1	1	0	0	0	3	0

In table : ES Earliest Start, EF – Earliest Finish.

LS : Latest start, LF Latest finish.

TF = Total float, LF – EF.   ... (7.1)

Free float : Earliest start of following activity - Earliest finish of the activity.

Example 7.2: *A small maintenance project consists of following 12 jobs. Draw the network of the project. Summarize CPM calculations in tabular form. Calculating the three types of floats for jobs and hence determine critical path.*

Table 7.5

Job	1-2	2-3	2-4	3-4	3-5	4-6	5-8	6-7	6-10	7-9	8-10	9-10
Duration	2	7	3	3	5	3	5	8	4	4	1	7

Solution: Network diagram:

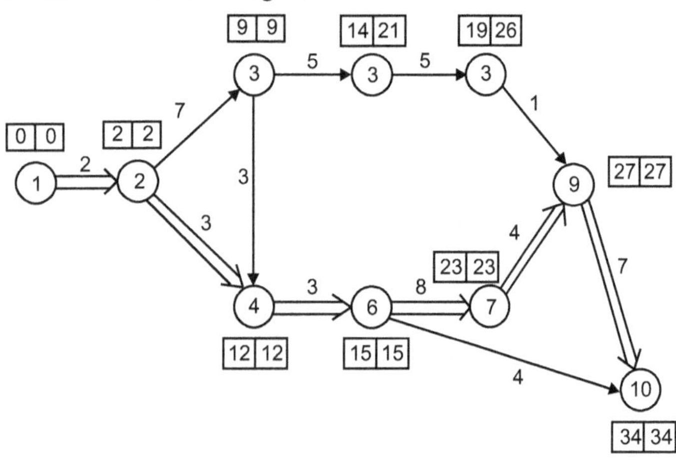

Fig. 7.7

Critical path = 1-2-4-6-7-9-10.

Table 7.6

Job	Duration	Total Float	Free Float	Independent Float
1-2	2	0	0	0
2-3	7	0	0	0
3-4	3	0	0	0
2-4	3	7	7	7
3-5	5	7	0	0
4-6	3	0	0	0
5-8	5	7	0	7
6-7	8	0	0	0
6-10	4	15	15	15
7-9	4	0	0	0
8-9	1	7	7	0
9-10	7	0	0	0

Example 7.3: *The following table 7.7 lists the jobs of a network alongwith their time estimates:*

(a) Draw the project network.

(b) Calculate the length and variance of the critical path.

(c) What is approximate probability that the job as per critical path will be completed in 24 hrs?

Table 7.7

Activity	T_o	T_m	T_p
1-2	1	3	5
2-3	2	5	6
2-4	4	6	7
2-5	8	10	12
3-5	0	0	0
3-6	4	8	9
4-7	5	7	14
5-7	7	10	16
6-7	0	0	0
6-8	6	9	12
7-9	1	3	7
8-9	3	5	7

Solution: Network Diagram:

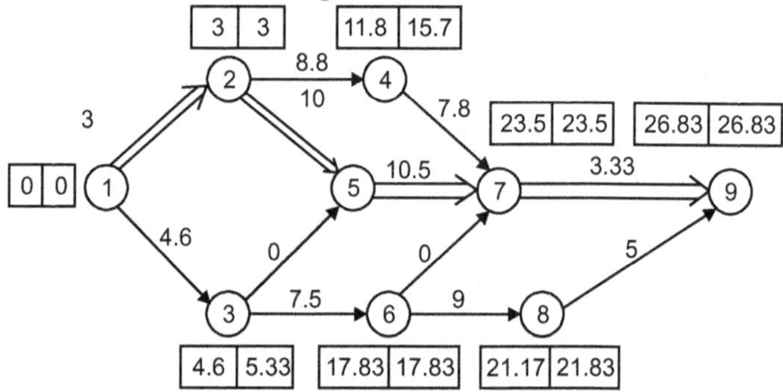

Fig. 7.8

Critical path = 1–2–5–7–9

Project duration,

(T_d) = 26.83 days.

Table 7.8

Activity	Expected time $\left(T_e = \dfrac{T_o + 4T_m + T_p}{6}\right)$	Variances $\sigma^2 = \left(T_e = \dfrac{T_p - T_o}{6}\right)^2$
1-2	3	0.44
2-3	4.66	0.44
2-4	5.83	0.25
2-5	10	0.44
3-5	0	0
3-6	7.5	0.634
4-7	7.83	2.25
5-7	10.15	2.25
6-7	0	0
6-8	9	1
7-9	3.34	1
8-9	5	0.44

Variance $= \sigma^2 = 0.44 + 0.44 + 2.25 + 1 = 4.13$

Standard deviation $= \sigma = \sqrt{4.13} = 2.032$

When,

due date $(T_s) = 24$ days $= 24 - 26.83$

$\therefore$ Probability $= 8.05\%$

Example 7.4: *A small project include seven major activities whose time destinations are listed in the table 7.9:*

Table 7.9

Steps	T_o	T_i	T_p	Activity
1-2	1	1	7	A
1-3	1	4	7	B
1-4	2	2	8	C
2-5	1	1	1	D
3-5	2	5	14	E
4-6	2	5	8	F
5-6	3	6	15	G

Draw network and find :

(a) Expected duration and variance of each activity.

(b) Expected project length.

(c) Variances and standard deviation of project length.

(d) Probability of getting project completed in 3 weeks earlier than expected. Maximum delay not more than 3 weeks.

(e) If project due date is 18 weeks, what is the probability of meeting the due date.

Solution: Network diagram :

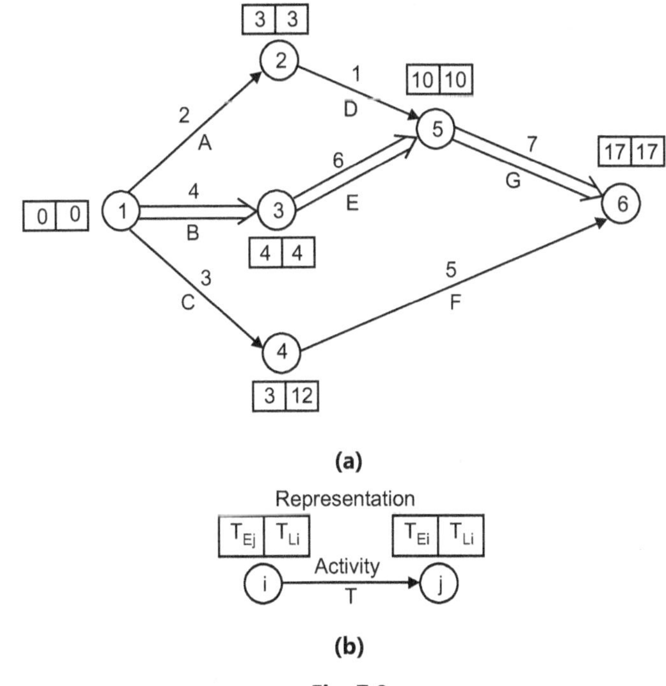

Fig. 7.9

where, $T_{E_i} =$ Earliest start time

$T_{E_j} =$ Earliest finish time

$T_{L_i} =$ Latest start time

$T_{ij} =$ Activity duration

Now,

Expected time $= T_e = \dfrac{T_o + 4T_i + T_p}{6}$

$\therefore$ Standard deviation $= \sigma = \dfrac{T_p - T_o}{6}$

and therefore,

Variance $= \sigma^2 = \left(\dfrac{T_p - T_o}{6}\right)^2$

The values are calculated and tabulated as follows:

Table 7.10

Activities		T_e	σ	σ^2
1-2	A	2	1	1
1-3	B	4	1	1
1-4	C	3	1	1
2-5	D	1	0	0
3-5	E	6	2	4
4-6	F	5	1	1
5-6	G	7	2	4

(a) Expected project length = 17 weeks

B - E - G is the critical path.

(b) Standard deviation of project length $= \sigma_B + \sigma_E + \sigma_G = 1 + 2 + 2 = 5$

(c) Variance of project length $= \sigma_B^2 + \sigma_E^2 + \sigma_G^2 = 1 + 4 + 4 = 9$

(d) Probability of getting completed in 3 weeks earlier than expected.

(e) Normal deviation.

$$Z = \frac{x - \bar{x}}{\sigma_p}$$

where, $x =$ Due duration = 3 weeks earlier

$y =$ Expected duration = 17 weeks

$\sigma_p =$ Probability of standard deviation

$$= \sqrt{\sigma_B^2 + \sigma_E^2 + \sigma_G^2} = \sqrt{1 + 4 + 4} = 1$$

$\therefore$ Normal deviation $= Z = \dfrac{14 - 17}{3} = 1$

[Ref. Table A from Appendix]

$\therefore$ Probability = 99%

(f) If due duration = 3 weeks

Then, $z = \dfrac{18 - 17}{3} = \dfrac{1}{3} = 0.33$

Example 7.5: *A small project is composed of scrap activities whose time estimates are listed below :*

Table 7.11

Activities		T_o	T_m	T_p
i	j			
1	2	3	6	15
1	6	2	5	14
2	3	6	12	30
2	4	2	5	8
3	5	5	11	17
4	6	3	6	15
6	7	3	9	27
5	8	1	4	7
7	8	4	19	28

(a) Draw network diagram.

(b) Calculate the length and variance of the critical path.

(c) What is the approximate probability that the job on critical path will be completed in 41 days ?

Solution: Network diagram is as follows:

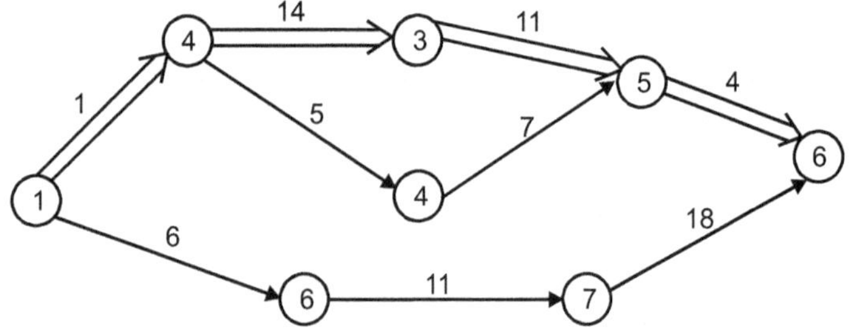

Fig. 7.10

Table 7.12

Activity (i – j)	T_e	σ	σ^2
1-2	7	2	4
1-6	6	2	4
2-3	14	4	16
2-4	5	1	1
3-5	11	2	4
4-5	7	2	4
6-7	11	4	16
5-8	4	1	1
7-8	18	4	16

(a) Expected project length = 36 days

Critical path : 1-2-3-5-8 :

(b) Standard deviation of project length = 2 + 4 + 2 + 1 = 9.

(c) Variance of project length = $\sigma_B^2 + \sigma_E^2 + 2_G^2 = 4 + 16 + 4 + 1 = 25$

(d) Probability of getting completed in 41 days.

Normal deviation = $z = \dfrac{x - \bar{x}}{\sigma_p}$

where,   x = Due duration = 41 days

$\bar{x}$ = Expected duration = 36 days

σ_p = Probability of standard deviation

$= \sqrt{\sigma_B^2 + \sigma_E^2 + \sigma_G^2} = 5$

$\therefore$ Normal deviation = $z = \dfrac{41 - 36}{5} = 1$

[Ref. Table A from Appendix]

$\therefore$ Probability = 84.13%

Example 7.6:

Table 7.13

Activity	1-2	1-3	2-4	2-6	3-4	4-5	4-6	5-7	6-7	7-8
NT	10	11	13	14	10	7	17	13	9	1

Find critical path, TF, FF, IF.

Solution:

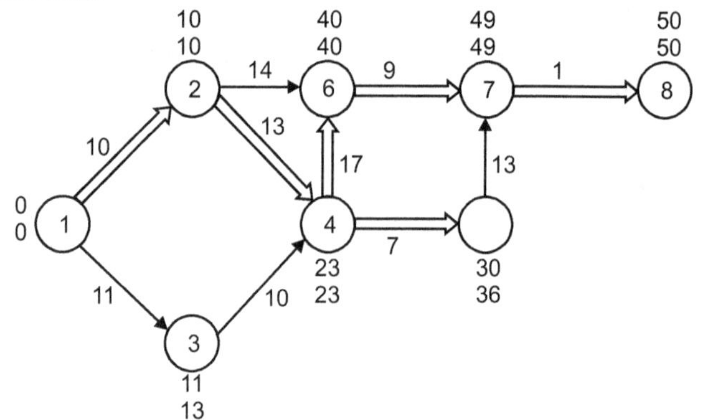

Fig. 7.11

$\therefore$ Critical path is $1 - 2 - 4 - 6 - 7 - 8$.

Table 7.14

Activity	Duration	TF	FF	IF	Earliest start		Latest start	
					Start	Finish	Start	Finish
1-2	10	0	0	0	0	10	0	10
1-3	11	2	2	2	0	11	2	13
2-4	13	0	0	0	10	23	10	23
2-6	14	16	16	16	10	24	26	40
3-4	10	2	0	2	11	21	13	23
4-5	7	6	6	6	23	30	29	36
4-6	17	0	0	0	23	40	23	40
5-7	13	0	0	6	30	43	36	49
6-7	9	0	0	0	40	49	40	49
7-8	1	0	0	0	49	50	49	50

Example 7.7: *Find critical path by drawing network, find the total duration for all projects. The activities predecessor and their delays are given below:*

Table 7.15

Activity	Predecessor	Duration
A	–	4
B	A	6
C	A	4
D	C	7
E	C	9
F	C	8
G	E	6
H	F	5
I	G, H	4
J	B	3
K	J	2

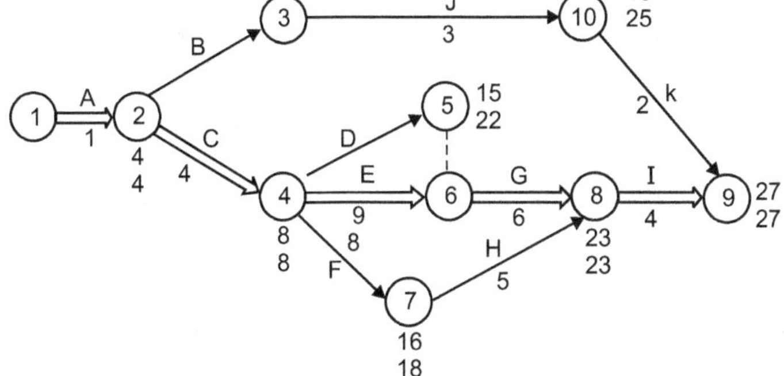

Fig. 7.12

Solution: Critical path = 2 – 4 – 6 – 8 – 9. Total project time = 27.

Example 7.8: *Determine optimal sequence, total float, independent float, free float and optimal path for project.*

Table 7.16

Activity	Predecessor
1-2	9
1-3	12
1-4	15
1-5	6
3-5	30
4-6	29
5-6	32

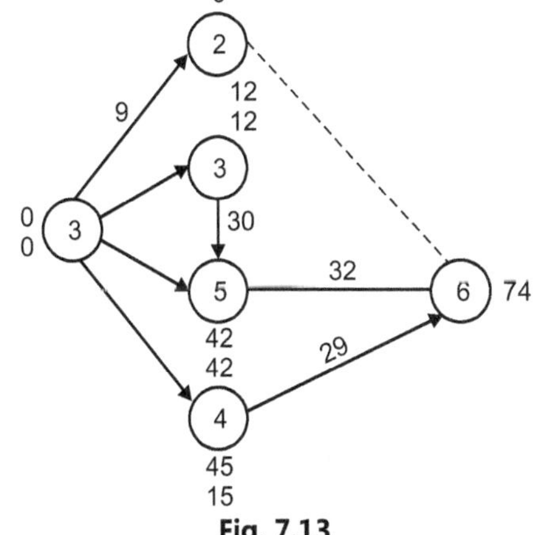

Fig. 7.13

Table 7.17

Activity	Duration	Earliest Start	Earliest Finish	Latest Start	Latest Finish	TF	IF	FF
1-2	9	0	9	65	74	0	0	0
1-3	12	0	12	0	12	0	0	0
1-4	15	0	15	30	15	0	30	30
1-5	6	0	6	36	42	34	36	36
3-5	30	12	12	12	42	0	0	0
4-6	29	15	44	45	74	0	30	0
5-6	32	42	74	42	74	0	0	0

Example 7.9: *Consider details of project as shown in Table 7.18, construct CPM table, find out project completion time and find out total float, independent float and free place.*

Table 7.18

Activity	Immediate Predecessor	Duration
A	–	4
B	–	8
C	–	5
D	A	4
E	A	5
F	B	7
G	B	4
H	C	8
I	C	3
J	D	6
K	E	5
L	F	4
M	G	12
N	H	7
O	I	10
P	J, K, L	5
Q	M, N, O	8

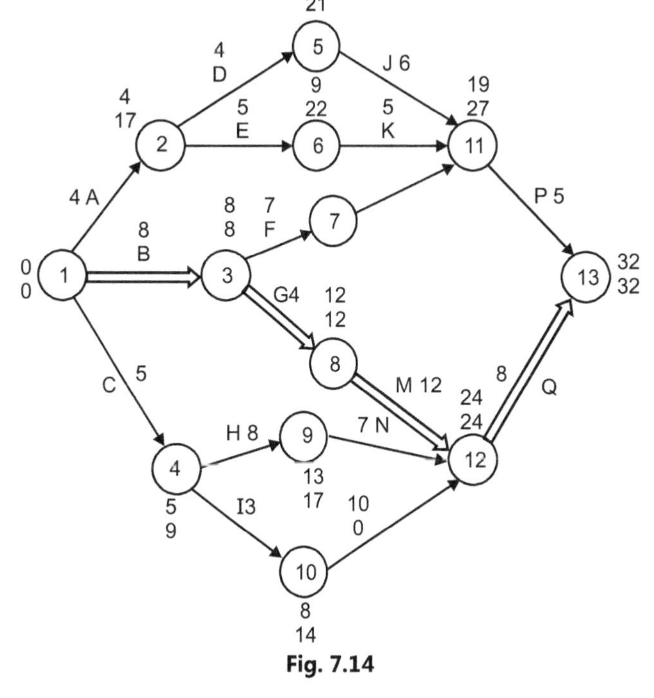

Fig. 7.14

Solution: Critical path = 1 – 3 – 8 – 12 – 13.

Table 7.19

Activity	Duration	Earliest		Latest		Float		
		Start	Finish	Start	Finish	Total	Independent	Free
A	4	0	4	13	17	13	0	0
B	8	0	8	0	8	4	0	0
C	5	0	5	4	9	4	0	0
D	4	4	8	17	21	13	0	0
E	5	4	9	17	22	13	0	0
F	7	8	15	16	23	8	0	0
G	4	8	12	8	12	0	0	0
H	8	5	13	9	17	4	0	0
I	3	5	8	11	14	6	0	0
J	6	8	14	21	27	13	0	5
K	5	9	14	22	27	13	0	5
L	4	15	19	23	27	8	0	0
M	12	12	24	12	24	0	0	0
N	7	13	20	17	24	4	0	4
O	10	8	18	14	24	6	0	6
P	5	19	24	27	32	8	0	8
Q	8	24	32	24	32	0	0	0

Example 7.10:

Table 7.20

Activity	Predecessor	Optimistic Time T_o	Most Likely Time T_m	Pessimistic T_p
A	–	6	7	8
B	–	1	2	9
C	–	1	4	7
D	A	1	2	3
E	A, B	1	2	9
F	C	1	5	9
G	C	2	2	8
H	E, F	4	4	4
I	E, F	4	4	0
J	D, H	2	5	14
K	I, J	2	2	8

Find idle time, C.P., σ and variance.

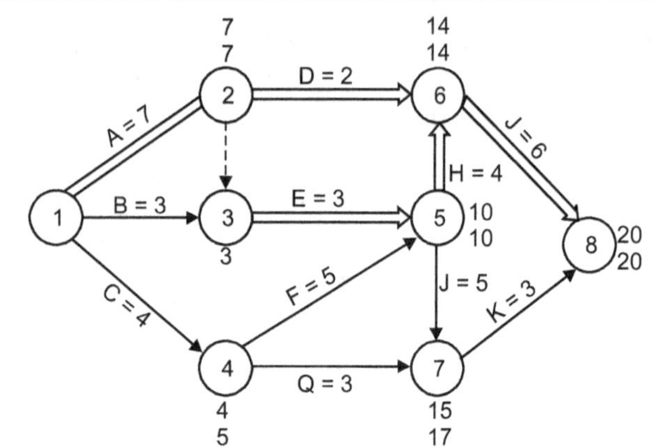

Fig. 7.15: Network diagram

Solution:

Table 7.21

Activity	T_o	T_m	T_p	Optimal duration (T_e)	Standard deviation (σ)	Variance $(\sigma)^2$
A	6	7	8	7	0.333	0.1089
B	1	2	9	3	1.33	1.7689
C	1	4	7	4	1	1
D	1	2	3	2	0.33	3.1089
E	1	2	9	3	1.33	1.7689
F	1	5	9	5	1.33	1.7689
G	2	2	8	3	1	1
H	4	4	4	4	0	0
I	4	4	0	5	1	1
J	2	5	14	6	2	4
K	2	2	8	3	1	1

Example 7.11: *A small project has the following data:*

[May 2011]

Table 7.22

Activity	1-2	1-3	1-4	2-5	3-5	4-6	5-6
Optimistic time	1	1	2	1	2	2	3
Most likely time	1	4	2	1	5	5	6
Pessimistic time	7	7	8	1	14	8	15

(i) *Draw network with critical path identification.*

(ii) *Calculate project duration, slack and floats.*

(iii) *What is the probability that the project will be completed in 4 weeks later than expected and 2 weeks earlier than expected?*

Given data:

Table 7.23

Z	– 0.5	– 1.00	–1.33	– 2.00	+ 1.65
P	0.30	0.25	0.16	0.09	0.95

Solution:

Table 7.24

Activity	T_o	T_m	T_p	T_e	σ $\dfrac{T_p - T_o}{6}$	V $\left(\dfrac{T_p - T_o}{6}\right)^2$
1 – 2	1	1	7	2	1	1
1 – 3	1	4	7	4	1	1
1 – 4	2	2	8	3	1	1
2 – 5	1	1	1	1	0	0
3 – 5	2	5	14	6	2	4
4 – 6	2	5	8	5	1	1
5 – 6	3	6	15	7	2	4

(i)

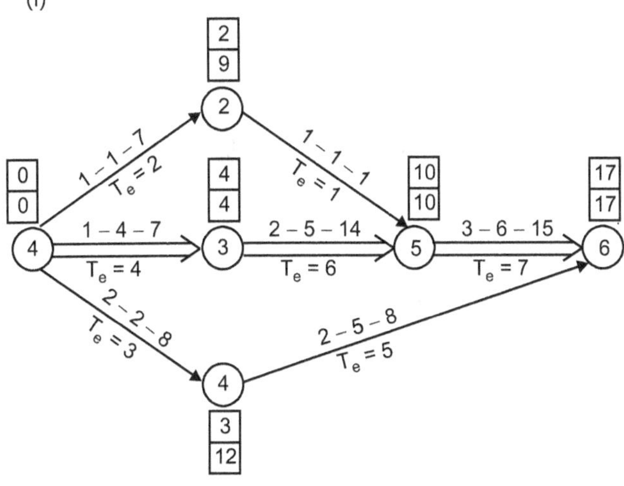

Fig. 7.16 : Various paths

(ii) Length of path $1 - 2 - 5 - 6 = 2 + 1 + 7 = 10$

$$1 - 2 - 5 - 6$$

Length of path $1 - 3 - 5 - 6 = 4 + 6 + 7 = 17$

Critical path $\rightarrow 1 - 3 - 5 - 6$

Length of path $1 - 4 - 6 = 3 + 5 = 8$

$$1 - 4 - 6$$

$\because$ $1 - 3 - 5 - 6$ has the longest duration, it is critical path of network.

$\therefore$ The expected project length = 17 weeks.

Variance:

$$V_{cp} = V_{1-3} + V_{3-5} + V_{5-6}$$
$$= 1 + 4 + 4$$
$$= 9$$
$$6 = 3 \text{ weeks}$$

(iii) Probability that project will be completed not more than 4 weeks later than expected time.

Expected time = 17 weeks

Scheduled time = 17 + 4 = 21 weeks

$$Z = \frac{21 - 17}{3} = 1.33$$
$$P = 0.9082 \text{ or } 90.82\%$$

Probability that project will be completed at least in 2 weeks earlier than expected time.

Expected time = 17 weeks

Scheduled time = 17 – 2 = 15 weeks

$$Z = \frac{15 - 17}{3} = -0.666$$

[Ref. Table A from Appendix]

$$P = 0.2899 \text{ or } 28.99\%$$

Example 7.12: *A project schedule has the following characteristics:* **[Dec. 2010]**

Table 7.25

Activity	T_o	T_m	T_p
1 – 2	1	2	3
2 – 3	1	2	3
2 – 4	1	3	5
3 – 5	3	4	5
4 – 5	2	3	4
4 – 6	3	5	7
5 – 7	4	5	6
6 – 7	6	7	8
7 – 8	2	4	6
7 – 9	4	6	8
8 – 10	1	2	3
9 – 10	3	5	7

(i) *Construct the network.*

(ii) *Find the variance for each activity.*

(iii) *Find the critical path and expected project completion time.*

Solution: (a) Variance calculation:

$$T_e = \frac{T_o + 3T_m + T_p}{6} \qquad V = \sigma^2 = \left[\frac{(T_p - T_o)}{6}\right]^2$$

Table 7.26

Activity	T_o	T_m	T_p	T_2	V
1 – 2	1	2	3	2	0.1089
2 – 3	1	2	3	2	0.1089
2 – 4	1	3	5	3	0.4356
3 – 5	3	4	5	4	0.1089
4 – 5	2	3	4	3	0.1089
4 – 6	3	5	7	5	0.4356
5 – 7	4	5	6	5	0.1089
6 – 7	6	7	8	7	0.1089
7 – 8	2	4	6	4	0.4356
7 – 9	4	6	8	6	0.1089
8 – 10	1	2	3	2	0.1089
9 – 10	3	5	7	5	0.4356

(b) Network diagram for given project:

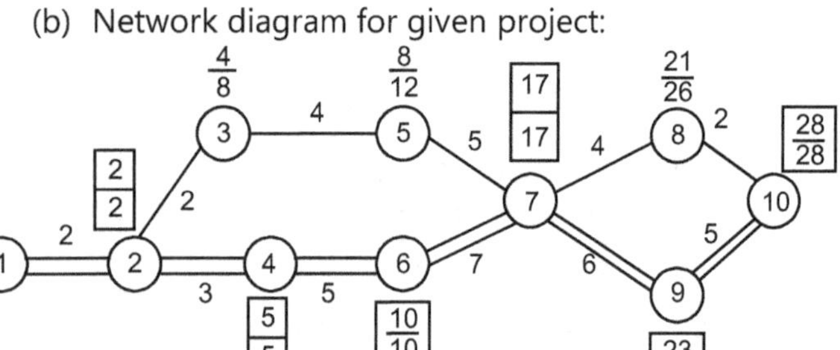

Fig. 7.17

Critical path: $1 - 2 - 4 - 6 - 7 - 9 - 10$.

(c) Expected project completion time = 28.

Example 7.13: *Find the optimal solution for the following network.* **[May 2010]**

Table 7.27

Activity	Succeeding Activity	Normal Duration	Crash Duration	Crash Cost	Normal Cost
A	B, C	8	8	500	500
B	D, E	4	3	1000	750
C	I	4	3	800	500
D	G	3	3	750	750
E	F	6	4	1500	800
F	H	9	6	2500	1600
G	–	5	4	500	400
H	–	7	5	800	600
I	F	8	5	3000	1500

Indirect cost = ₹150/- per day.

Solution:

(a) Activity network for the given problems is

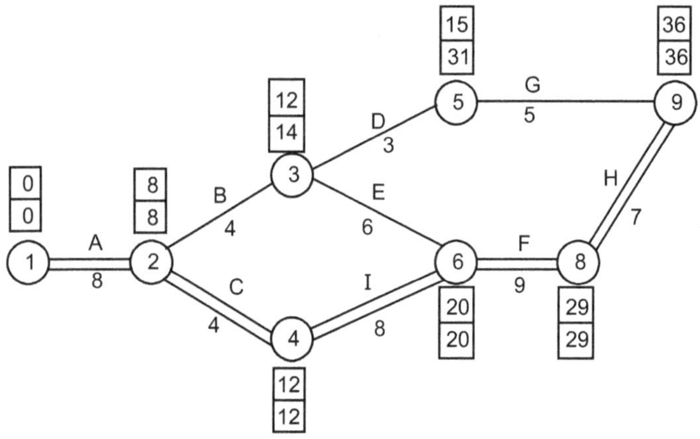

Fig. 7.18

Critical path: $A - C - I - F - H$

               $8 - 4 - 8 - 9 - 7$

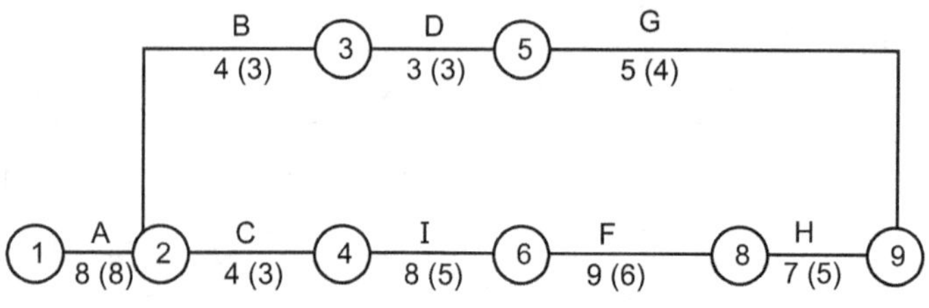

Fig. 7.19

(b) Project/Activity Cost – Time Analysis

Table 7.28

Activity	Normal Time	Normal Cost	Crash Time	Crash Cost	ΔC	ΔT	ΔC/ΔT
A	8	500	8	500	0	0	0
B	4	1000	3	750	250	1	250
C	4	800	3	500	300	1	300
D	3	750	3	750	0	0	0
E	6	1500	4	800	700	2	350
F	9	2500	6	1600	900	3	300
G	5	500	4	400	100	1	100
H	7	800	5	600	200	2	100
I	8	3000	5	1500	1500	3	500
Σ =		11350					

(c)   Project cost   = $11350 + (36 \times 150)$

                 =   16750

Table 7.29

Activity	Crash by	Cost slope	Increased cost	Saving	Net saving	New cost	New Duration
H	2	100	200	300	100	16650	34

Optimal solution for the given network is 34 days and cost is ₹ 16650.

7.6 DIFFERENCE BETWEEN PROJECT AND OTHER MANUFACTURING SYSTEMS

- Difference between project and other manufacturing systems, are as follows:

Project Management	Other Manufacturing System
Only during projects	Continuous process
Move from one stage to others	Single stage
Single clear objectives	No Single (clear)objectives
Definite start and end points	No Definite start and end points
More human resource related	Resources and machine related
Complex team building	Simple team building

7.7 SCOPE OF A PROJECT

7.7.1 Defining Scope of a Project

- A major decision at the outset of any project is to decide upon the organization and composition of the project team. In so doing, it is worth remembering that many members will have dual responsibilities of involvement in the project in addition to a commitment to other projects or management of a functional area on a day-to-day basis. It is at this stage that a project manager should be appointed and responsibilities made explicit for all members of the team.

- The selection of the team will be dependent upon the skill requirements of the project, and upon the matching of those skills to those possessed by individual members of the team. There may be a conflict here with hierarchical status.

- The project management team will, therefore, begin its task in advance of project proper so that a plan can be developed. An important first step is to set the objectives and then define the project, breaking it down into a set of activities and related costs. It is probably too early to determine exact resource implications at this stage, but expected requirements for people, supplies and equipment should at least be estimated during the planning stage.

7.7.2 Project Scheduling

- This phase is primarily concerned with attaching a timescale and sequence to the activities to be conducted within the project. Materials and people needed at each stage of the project are determined and the time each is to take will be set.

- A popular and easy to use technique for scheduling is the use of Gantt charts. Gantt charts reflect time estimates and can be easily understood. Horizontal bars are drawn against a time scale for each project activity, the length of which represent the time taken to complete. Letters or symbols can also be added to the left of each bar to show which other activities need to be completed before that one can begin

7.8 NECESSITY OF DIFFERENT PLANNING TECHNIQUES FOR PROJECT MANAGEMENTS AND USE OF NETWORKS FOR PLANNING OF A PROJECT

- As project scheduling is primarily concerned with attaching a timescale and sequence to the activities to be conducted within the project. Materials and people needed at each stage of the project are determined and the time each is to take will be set.

- A popular and easy to use technique for scheduling is the use of Gantt charts. Gantt charts reflect time estimates and can be easily understood. Horizontal bars are drawn against a time scale for each project activity, the length of which represent the time taken to complete. Letters or symbols can also be added to the left of each bar to show which other activities need to be completed before that one can begin.

- Using CPM, it was assumed that the amount of time it took to perform an operation was known with some degree of certainty.

- For many projects that is realistic approach, viz., fixed maintenance schedules, planning for production etc. Experience and a tremendous amount of data of past projects is required to make accurate estimation of time.

- In some projects, reliable time estimation is difficult. For example, research and development work, the design of new products, unusual construction projects and other instances.

- Where new or untried techniques or materials are being used involve operations for which realistic time estimates are not possible.

- PERT technique was designed to deal with this type of project.

- The main purpose in the analysis through PERT is to findout the completion for a particular event within specified date.

QUESTIONS FOR PRACTICE

1. Define and explain the terms:

   (i) CPM

   (iii) PERT (Refer Section 7.1)

2. Differentiate between CPM and PERT.

3. Discuss cost aspects and crashing of network.

4. Discuss the resource levelling in project management and use of histogram in it.

5. Write short notes on:

   (a) Updating the network

   (b) Types of floats

   (c) Role of statistical techniques in PERT.

6. Discuss the Fulkerson's rule used in network analysis.

7. A project has the following data of offer. **[May 2009]**

Head Activity	Tail Activity	Estimated Duration (Months)		
		Smaller	Middle	Highest
A	–	1	1	7
B	–	1	4	7
C	–	2	2	8
D	A	1	1	1
E	B	2	5	14
F	C	2	5	8
G	D, E	3	6	15
H	F, G	1	2	3

(a) Draw network and determine project duration.

(b) Calculate slacks.

(c) What duration of project will have 95% confidence on project completion?

◈ ◈ ◈

CHAPTER 8
TIME AND COST ANALYSIS

8.1 CRASHING (TIME-COST RELATIONSHIP TRADE-OFF) (TIME AND COST ESTIMATE)

[Dec. 2009, May 2011]

- Total project involve number of activities. Each activity of the project consumes some resources and hence cost associated with it. The overall total project cost consist of direct and indirect costs.

- Direct cost is the cost associated with the resources involved in the execution of individual activities, e.g., Cost of materials, machinery used, direct labour/sub-contracting payments etc.

- Indirect cost could be further classified as fixed cost e.g., insurance cost, administration expenses, taxes, excises etc. and variable indirect cost e.g., overheads, cost penalty for delay in work, depreciation etc.

- There will be specific activity duration for each activity for which direct cost will be minimum. If activity time requirement excides this limit then, more resources-ultimately more funds are required. Hence, crashing is the process by which the point could be achieved beyond which no further reducing in activity time will be possible irrespective of resources used. The time for the activity at minimum cost is called as **normal time** and the minimum time for the activity is called as **crash time**.

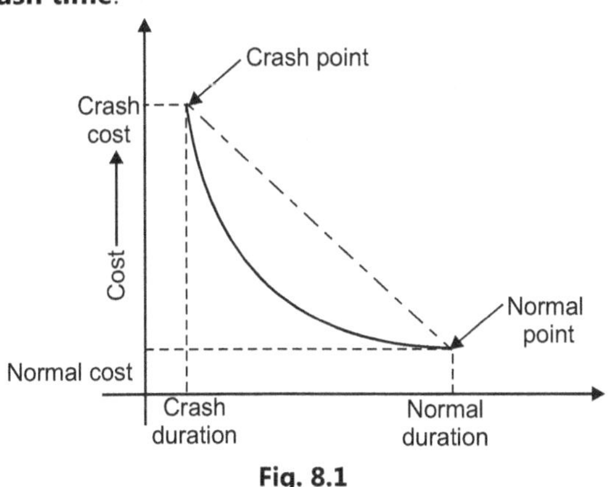

Fig. 8.1

Crashing Rules :

1. Construct network diagram and determine critical path.
2. Calculate the cost slope for different activities.
3. Those activities should be crashed first which have lower cost slope. And then calculate total cost of the project.

8.2 PROBABILISTIC TREATMENT OF PROJECT COMPLETION

- Using CPM, it was assumed that the amount of time it took to perform an operation was known with some degree of certainty.

- For many projects that is realistic approach, viz., fixed maintenance schedules, planning for production etc. Experience and a tremendous amount of data of past projects is required to make accurate estimation of time.

- In some projects, reliable time estimation is difficult. For example, research and development work, the design of new products, unusual construction projects and other instances.

- Where new or untried techniques or materials are being used involve operations for which realistic time estimates are not possible.

- PERT technique was designed to deal with this type of project.

- The main purpose in the analysis through PERT is to findout the completion for a particular event within specified date.

- PERT requires estimating three times for performing an activity:

The Optimistic Time (t_0)

- The shortest possible time required for the completion of an activity if all goes extremely well is known as optimistic time

The Pessimistic Time (t_p)

- The maximum possible time the activity will take if everything goes bad is known as pessimistic time.

The Most Likely Time (t_m)

- The time an activity will take if executed under normal condition is known as most likely time

- These three time estimates are used to determine an Expected time (TE) which is based on the recognition of the probabilistic nature of the activities involved.

$$T_e = \frac{T_o + 4T_m + T_p}{6}$$

- After identifying the critical path and the occurrence times of all activities, the important question is to be answered as what is the probability that a particular event will occur on or before the scheduled date ?

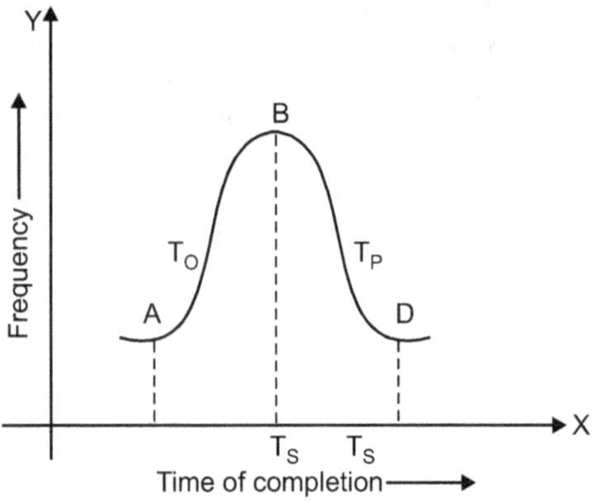

Fig. 8.2

- To find the probability of completion the project in time T, we calculate the standard normal variate(Z)

  It is expressed as

  $$Z = [T-T_{cp}]/y$$

  Where T = Project completion time or duration

  T_{cp} = Critical path duration or time, y=square root of sum of variance of critical path

The standard normal probability distribution table (Table in appendix) is used to find probability of completion of project. The probability is taken from table (read from the standard normal probability distribution table)

Variance (V) and Standard Deviation (σ) :

$$V = \text{variance for an activity is given as}$$

$$V = \sigma^2$$

Where $\sigma = [T_p-T_o]/6$

8.3 PROJECT UPDATE OR OPTIMIZATION (RESOURCE ALLOCATION AND RESOURCE LEVELING)

- Make adjustments in the PERT chart as the project progresses.

- As the project unfolds, the estimated times can be replaced with actual times.

- In cases where there are delays, additional resources may be needed to stay on schedule and the PERT chart may be modified to reflect the new situation.

Types of Cost or Elements of Cost:

- Total project involve number of activities.

- Each activity of the project consumes some resources and hence cost associated with it.

- The overall total project cost consist of direct and indirect costs

(i) Direct Cost

Direct cost is the cost associated with the resources involved in the execution of individual activities, e.g., Cost of materials, machinery used, direct labor/sub-contracting payments etc.

(ii) Indirect Cost

(a) Fixed Cost

Indirect cost could be further classified as fixed cost e.g., insurance cost, administration expenses, taxes, excises etc.

(b) Variable Cost

Variable indirect cost e.g., overheads, cost penalty for delay in work, depreciation etc.

8.3.1 Crashing

- Crashing refers to a particular variety of project schedule compression which is performed for the purposes of decreasing total period of time (also known as the total project schedule duration).

- The diminishing of the project duration typically take place after a careful and thorough analysis of all possible project duration minimization alternatives in which any and all methods to attain the maximum schedule duration for the least additional cost .

- The objective of crashing a network is to determine the optimum project schedule.

- Crashing may also be required to expedite the execution of a project, irrespective of the increase in cost.

- Each phase of the software design consumes some resources and hence has cost associated with it.

- In most of the cases cost will vary to some extent with the amount of time consumed by the design of each phase .The total cost of project, which is aggregate of the activities costs will also depends upon the project duration, can be cut down to some extent.

- The aim is always to strike a balance between the cost and time and to obtain an optimum software project schedule. An optimum minimum cost project schedule implies lowest possible cost and the associated time for the software project management

- There will be specific activity duration for each activity for which direct cost will be minimum. If activity time requirement excides this limit then, more resources-ultimately more funds are required.

- Hence, crashing is the process by which the point could be achieved beyond which no further reducing in activity time will be possible irrespective of resources used.

8.3.2 Crash Time

- The time for the activity at minimum cost is called as **Normal Time** and the minimum time for the activity is called as **Crash Time**.

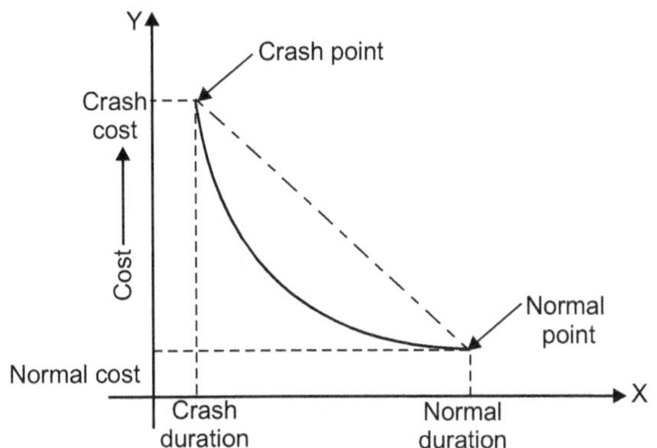

Fig. 8.3

- It is possible that some intermediate point may represent the ideal or optimal trade-off between time and cost for this activity.

8.3.3 Cost Slope

- The slope of the line connecting the normal point (lower point) and the crash point (upper point) is called the cost slope of the activity.

- The slope of this line can be calculated mathematically by knowing the coordinates of the normal and crash points:

$$\text{Cost slope} = \frac{\text{(crash cost - normal cost)}}{\text{(normal duration - crash duration)}} = \Delta C / \Delta T$$

$$= (N_c - C_c)/N_T - C_T$$

$$= \Delta C / \Delta T$$

- As the activity duration is reduced, there is an increase in direct cost. A simple case arises in the use of overtime work and premium wages to be paid for such overtime.

- Also overtime work is more prone to accidents and quality problems that must be corrected, so indirect costs may also increase. So, do not expect a linear relationship between duration and direct cost but convex function as shown in Fig. 8.3

8.3.4 Project Time-Cost Relationship (Time Cost Trade-Off Analysis)

- As total project costs include both direct costs and indirect costs of performing the activities of the project.

- If each activity of the project is scheduled for the duration that results in the minimum direct cost (normal duration) then the time to complete the entire project might be too long and substantial penalties associated with the late project completion might be incurred.

- At the other extreme, a manager might choose to complete the activity in the minimum possible time, called crash duration, but at a maximum cost.

- Thus, planners perform what is called time cost trade-off analysis to shorten the project duration.

- This can be done by selecting some activities on the critical path to shorten their duration.

- As the direct cost for the project equals the sum of the direct costs of its activities, then the project direct cost will increase by decreasing its duration.

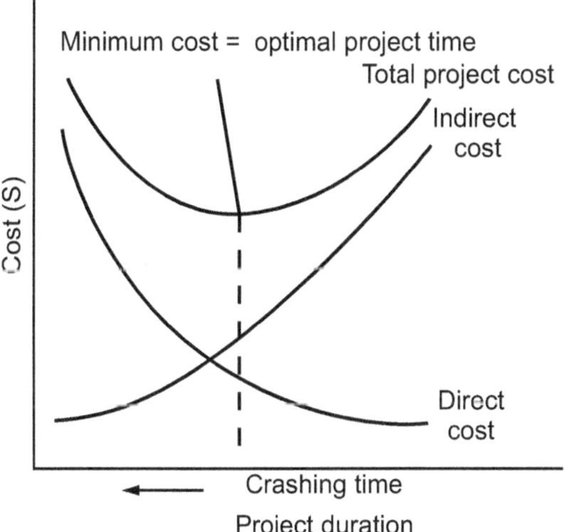

Fig. 8.4 : Project time-cost relationship

- On the other hand, the indirect cost will decrease by decreasing the project duration, as the indirect cost are almost a linear function with the project duration.

- Fig. 8.4 shows the direct and indirect cost relationships with the project duration. The project total time-cost relationship can be determined by adding up the direct cost and indirect cost values together.

Total project cost = direct cost + indirect cost

i.e Project Cost:

$$= ((\text{Project Direct Cost} + \text{Crashing cost of crashed activity}) + \text{Indirect Cost} \times \text{project duration}))$$

= ((Project total normal Cost
+ Crashing cost of crashed activity)
+ Indirect Cost per day x project
duration))

= ((Project total normal Cost + extra
Crashing cost of crashed activity)
+ Indirect Cost per day x project
duration))

where crashing cost of crashed activity = ΔT of crash activity x slope =ΔC

Note:

(i) Extra crash cost of critical activity is taken as cumulative addition

(ii) Write results of project cost calculation in tabular form

- The optimum project duration can be determined as the project duration that results in the least project total cost.

8.3.5 Crashing Rules

Important Steps :

1. Construct network diagram and determine critical path.
2. Calculate the cost slope for different activities.
3. Those activities should be crashed first which have lower cost slope. And then calculate total cost of the project

Detailed Steps:

Step 1: Find Earliest time estimates for all the activities, it is denoted as TE (ES)

Step 2: Find latest time estimates for all the activities, it is denoted as TL (LT)

Step 3: Determine the Critical Path.

Step 4: Compute the cost slope (i.e., cost per unit time) for each activity according to the following formula:

Cost slope = (crash cost-normal cost)/ (normal
duration -crash duration)
= $\Delta C / \Delta T$

Step 5: Among the critical path identify the activity with the minimum cost slope, and crash the activity by 1 day(or by 1 week or by 1 month or by 1 year as per duration specified).

Step 6: Calculate the project cost. Identify new critical path.

Total project cost = direct cost + indirect cost

i.e Project Cost:

= ((Project Direct Cost + Crashing cost of
crashed activity) + (Indirect Cost
x project duration))

= ((Project total normal Cost
+ Crashing cost of crashed activity)
+ (Indirect Cost per day x project
duration))

= ((Project total normal Cost + extra
Crashing cost of crashed activity)
+ Indirect Cost per day x project
duration))

where crashing cost of crashed activity = ΔT of crash activity x slope = ΔCIteration Step:

Step 7: Now in the new critical path select the activity with the next minimum cost slope, and crash by one day

Step 8: Repeat this process until all the activities in the critical path have been crashed by 1 day.

Step 9: Once all the activities along the critical path are crashed by one day, Repeat the process again i.e. goes to step5.

Step 10: Find the minimum project cost and identify the activities which do not lie along the critical path

Step 11: Now see if the project cost can be further reduced without affecting the project duration time then performuncrashing. i.e uncrash the activities which do not lie along the critical path.

- Uncrashing should start with an activity having the maximum cost slope. An activity is to be expanded only to the extent that it itself may become critical, but should not affect the original critical path.

8.3.6 Criticality Theorem

- Helps in selecting which activities to crash, especially when dealing with large project network
- Checks condition where crash duration of critical path is greater than normal duration of non-critical path(s), in which case do not need to consider activities on non-critical path

i.e. crash duration of critical path> normal duration of non-critical path

Note:

- While crashing a critical activity, only those non critical activities will require crashing which start from the tail event of critical activity under consideration or earlier and finish at the head event of the critical activity or later than that.

- Only those activity are crashed having cost slope (or crash cost) less than or equal to indirect cost i.e cost slope $\leq$ indirect cost

- Crash critical activity such that more than one parallel critical path may be possible

8.3.7 Recourses Leveling V/s Resource Allocation

Resource Leveling :

It is used for:

- How to smooth resource demands?
- No problem with time or resources.
- Strategy ?
- Method of moments?
- Method of double moments?
- Multi-Resources?
- Desired (best) profiles ?

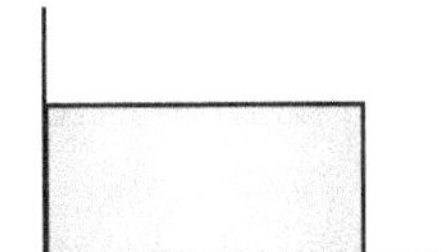 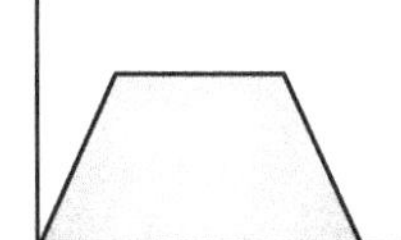

Fig. 8.5

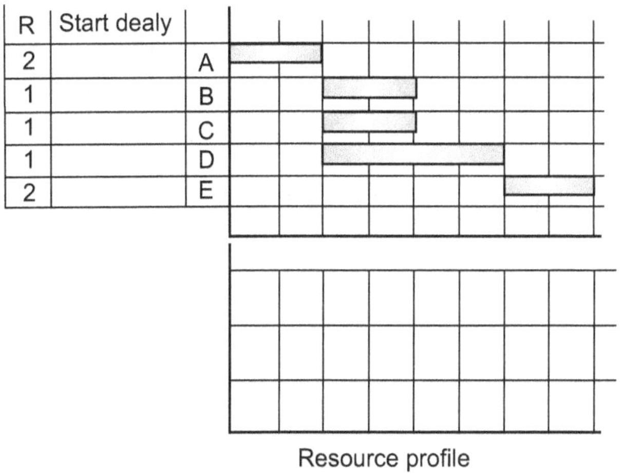

R	Start dealy									
2		A								
1		B								
1		C								
1		D								
2		E								

Resource profile

Fig. 8.6

Strategy : _______________________________

For Example : *Two schedule alternatives have associated resource profiles as shown below, which alternative would you choose and why? Also calculate the total worker-weeks needed for both cases :*

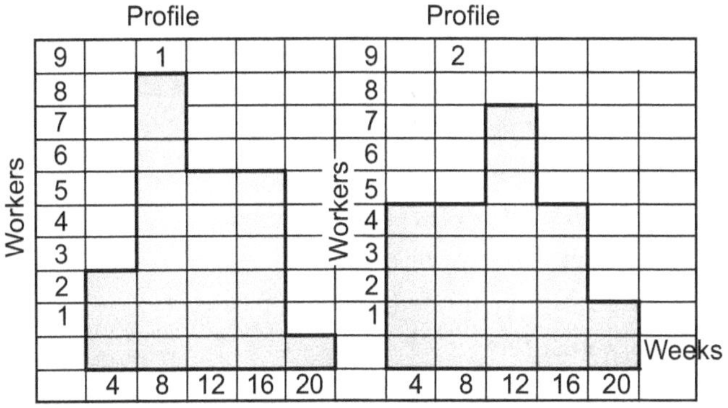

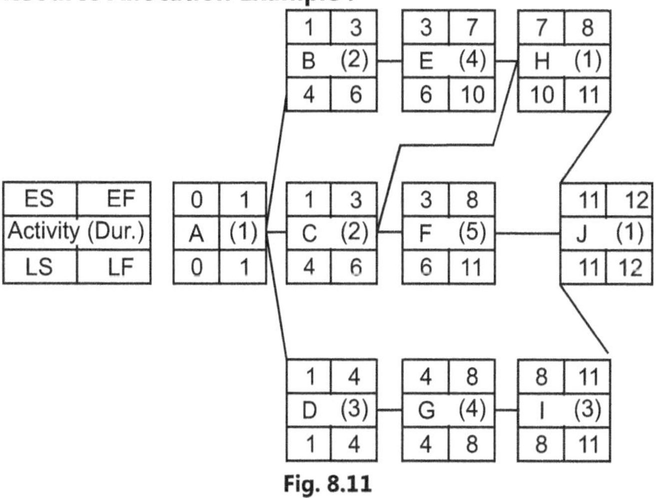

Fig. 8.7

Mx =

My =

Resource Allocation :

It is used for :

- Allocate limited resources to top-priority activities.
- Strategy ?
- Heuristic rules
- Inconsistency among existing software
- Excel Implementation
- EasyPlan Optimization

Resource limit = 2/day

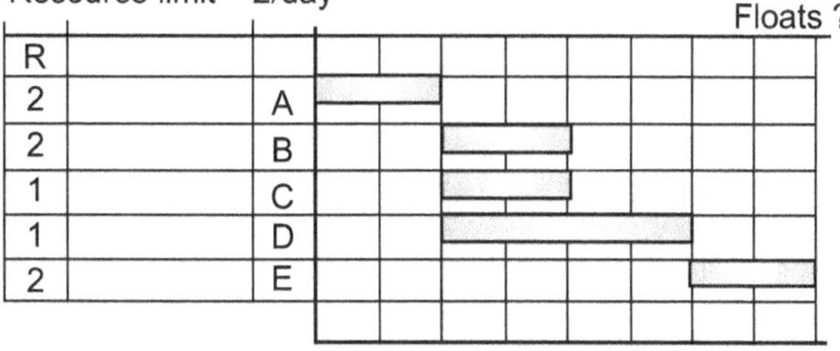

Fig. 8.8

Solution :

R	Start dealy								
2		A							
2		B							
1		C							
1		D							
2		E							

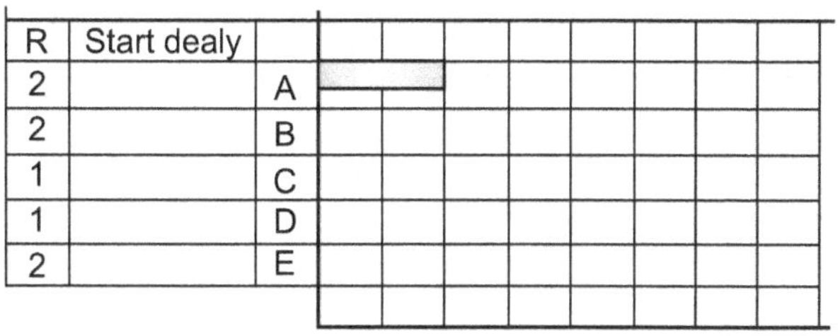

Fig. 8.9

Strategy : _______________________________

Another Solution :

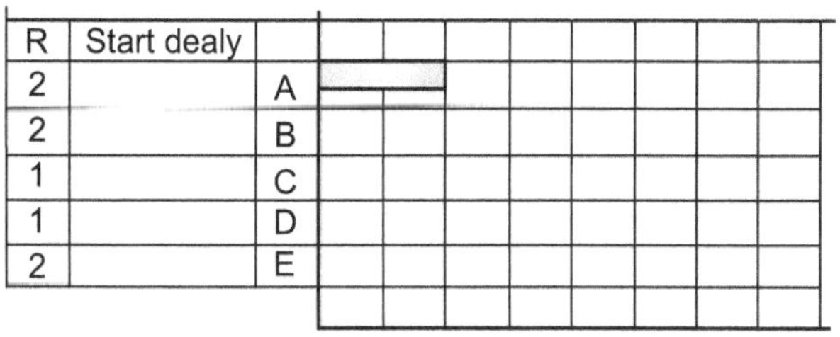

R	Start dealy								
2		A							
2		B							
1		C							
1		D							
2		E							

Fig. 8.10

Priority Rules :

Resource Allocation Example :

Fig. 8.11

Table 8.1

Time	Eligible Activities	Resources (limit = 2)	Duration	Rule (ELS)	Decision	Finish Time

Microsoft Project Solution ?

Case of Multi-Resources? Case of Multi-Resources Multi-Skills?

Meeting Deadline :

- Activity time-cost relationship? Linear V/s Discrete (Cheap & Slow versus Fast & Expensive)
- Cost Slope?
- Project time-cost relationship?
- Strategy to meet deadline?

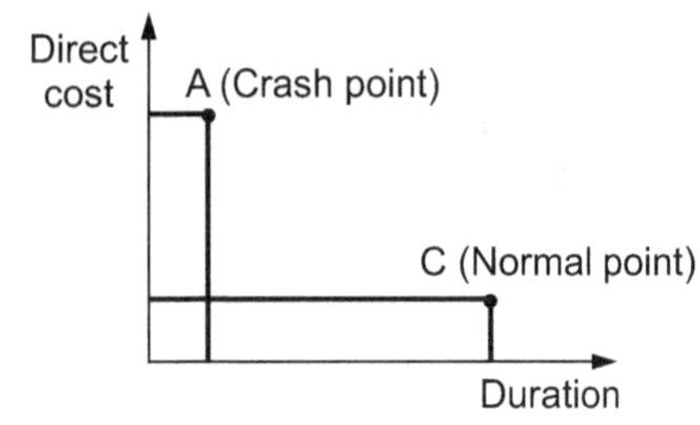

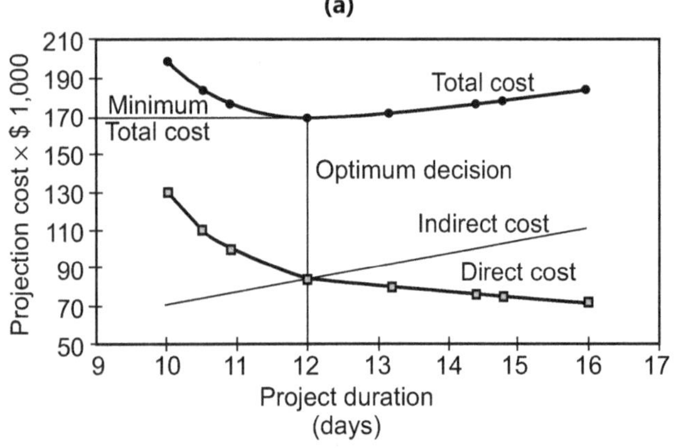

Fig. 8.12

For Example : Durations and cost slopes are shown. We need to meet a 12-day deadline.

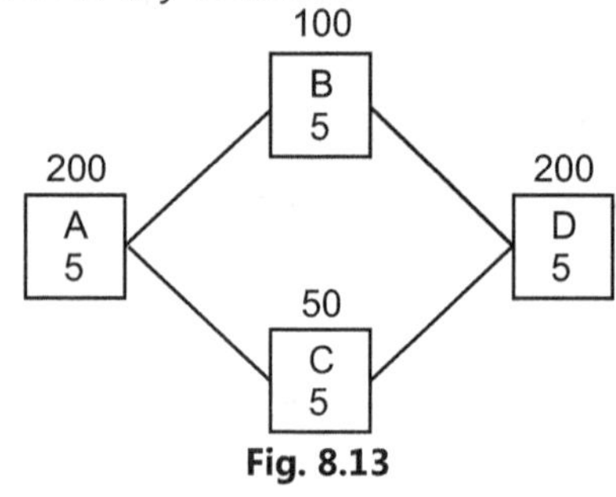

Fig. 8.13

For Example : Normal and crash data for the tasks are shown. What is the optimum project duration? How can the project be finished in 20 days?

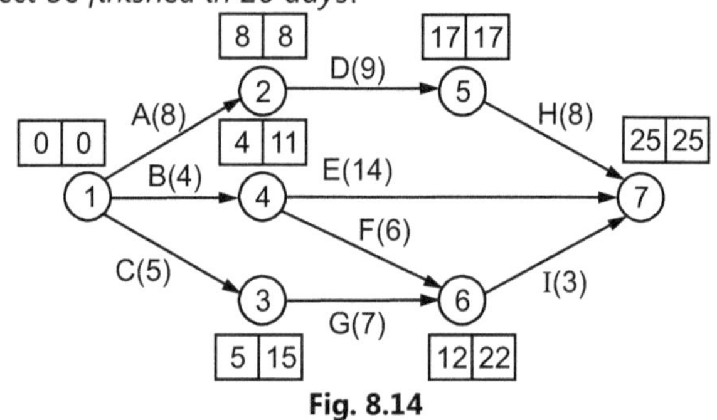

Fig. 8.14

Table 8.2

Activity	Normal Duration	Normal Cost	Crash Duration	Crash Cost	Critical	Crash Cost/Day
A	8	16000	6	19000		
B	4		No crashing			
C	5					
D	9	18000	7	19000		
E	14		No crashing			
F	6					
G	7		No crashing			
H	8	16000	6	18000		
I	3		No crashing			

Notes : Total "Normal" cost of all other tasks = $70,000
Daily indirect cost is $1,000/day.

Cash Flow Analysis :

- **Factors Affecting Interest Payment :**
  - Credit from Suppliers
  - Reporting Periods (payment frequency)
  - Interest rate
  - Subcontracting
  - Mobilization payment
  - Better scheduling
  - Hold back percentage
  - Bid unbalancing

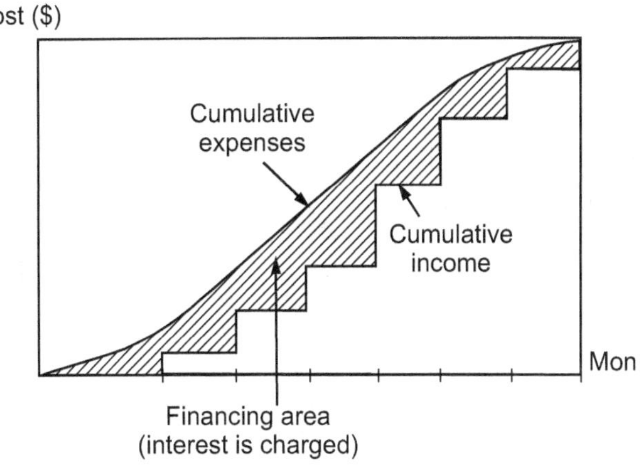

Fig. 8.15

Terms Used Every Period :

$$\text{Cost} = (\text{Direct} + \text{Indirect}) \text{ estimates}$$

$$\text{Expenses} = \text{Cost (if suppliers can give you a credit)}$$

$$\text{Budget} = \text{Cost} + \text{Markup} \times \text{Cost}$$

$$= \text{Cost} (1 + \text{Markup})$$

$$\text{Income} = \text{Owner payment} = \text{Budget} - \text{Holdback}$$

If Everything Goes Well :

$$\text{Expenses} = \text{Cost}$$

$$\text{Income} = \text{Budget}$$

$$\text{Profit} = \text{Cost} \times \text{Markup}$$

For Example : Overdraft Calculations :

Data :

- Direct costs (evenly distributed) on the bar chart;
- Indirect cost = \$5000 per month;
- Contractors markup = 5%;
- Reporting period = monthly;

- Owner retainage = 10% (no retainage after work is 50% complete);
- Payments made 30 days after invoice made; and
- Interest rate is 1% monthly.

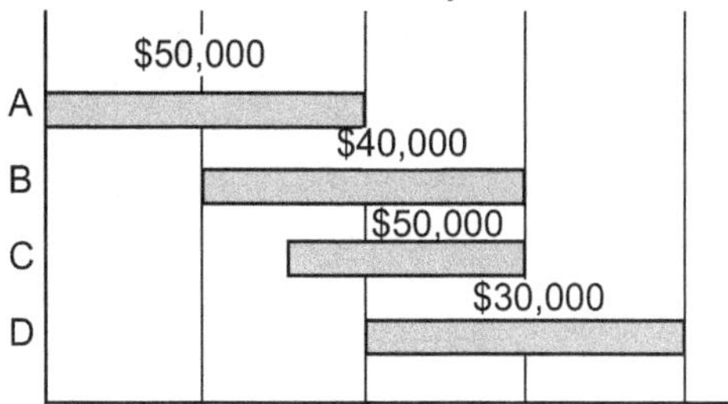

Fig. 8.16

Table 8.3

Monthly direct cost	\$25000	\$65000	\$75000	\$15000
Monthly indirect cost	\$ 5000	\$ 5000	\$ 5000	\$ 5000
Total monthly costs	\$30000	\$70000	\$80000	\$70000
Cumulative monthly costs	\$30000	\$100000	\$180000	\$200000

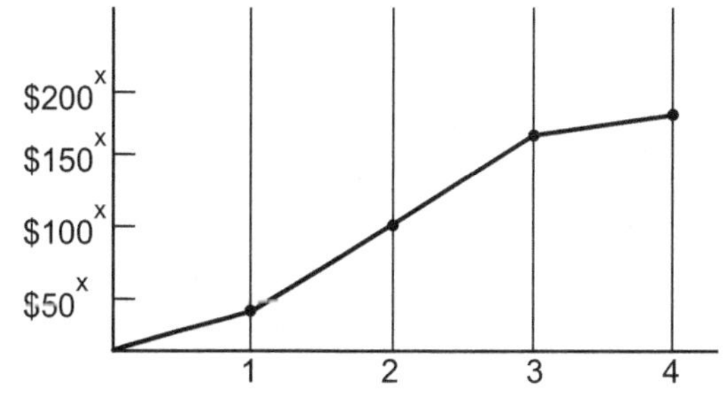

Fig. 8.17

Table 8.4
Overdraft Calculations

Month	1	2		3		4				
Direct cost	\$25000	\$65000		\$75000		\$15000				
Indirect cost	5000	5000		5000		5000				
Subtotal	30000	70000		80000		20000				
Markup	1500	3500		4000		1000				
Total billed	31500	73500		84000		21000				
Retainage with held	3150	7350		0		0				
Payment received			\$28350		\$66150		\$84000		\$31500	
Total cost to date	30000	100000		18000		200000		200000		
Total amount billed to date	31500	105000		189000		210000		210000		
Total paid to date			28350		94500		178500		210000	
Overdraft end of month	30000		100300		152953		108333		25416	(–)5830
Interest of overdraft balance	300		1003		1530		1083		254	0
Total amount financed	\$30300		\$101303		\$154483		\$109416		\$25670	

Note : A simple illustration only. Most lenders would calculate interest charges more ?? on the amount time involved employing daily interest factors.

Managing Both : Deadline and Resources :

Subcontract
Deadline = 10 days and resource limit = 2/day

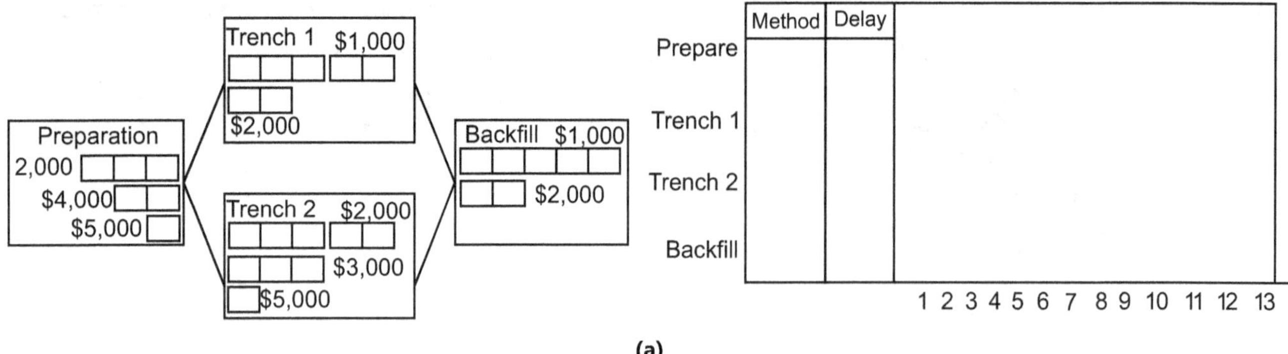

(a)

If markup = 10%, Retention = 5% and indirect cost = $100/day, Draw the Cash Flow Chart. What is the effect of bid unbalancing?

Table 8.5

Start delay	Method	Bat chart	Direct $	Ind.$	Budget (Markup = 10%)	Adjust.	Budget (Markup = 10%)
				$100 ser day			
		1 2 3 4 5 6 7 8 9 10	Sum:	Sum:	Sum:		Sum:

	9
	8
	7
	6
	5
	4
	3
	2
	1
	0
	1 2 3 4 5 6 7 8 9 10

(b)

Fig. 8.18

SOLVED EXAMPLES

Example 8.1 : *A project consists of six activities. Draw the network diagram and Determine the critical path*

Table 8.6

Activity	Immediate Predecessor	Activity Duration (week)
A	–	4
B	A	6
C	B	5
D	A	4
E	D	3
F	E, C	3

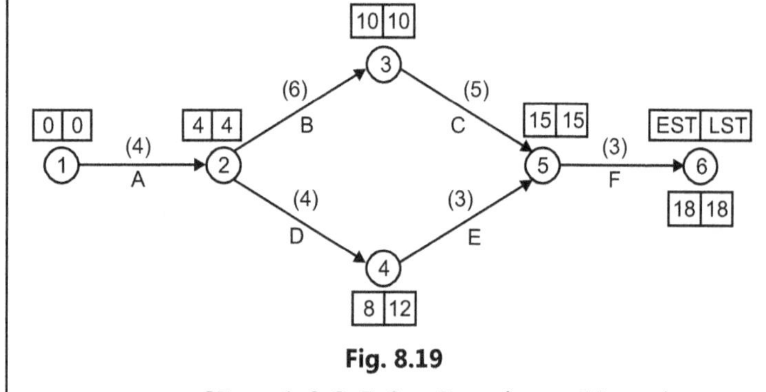

CP = 1-2-3-5-6,   Duration = 18 week

Example 8.2 : *A small project consists of 13 activities. Their precedence relationships and duration in days is given in table.*

Table 8.7

Activity	Predecessor	Duration (Days)
A	–	6
B	A	4
C	B	7
D	A	2
E	D	4
F	E	10
G	–	2
H	G	10
I	J, H	6
J	–	13
K	A	9
L	C, K	3
M	I, L	5

(i)   Construct the project network.

(ii) Find the critical path.

(iii) Find total completion time of the project.

Solution:

Network diagram for the given project is

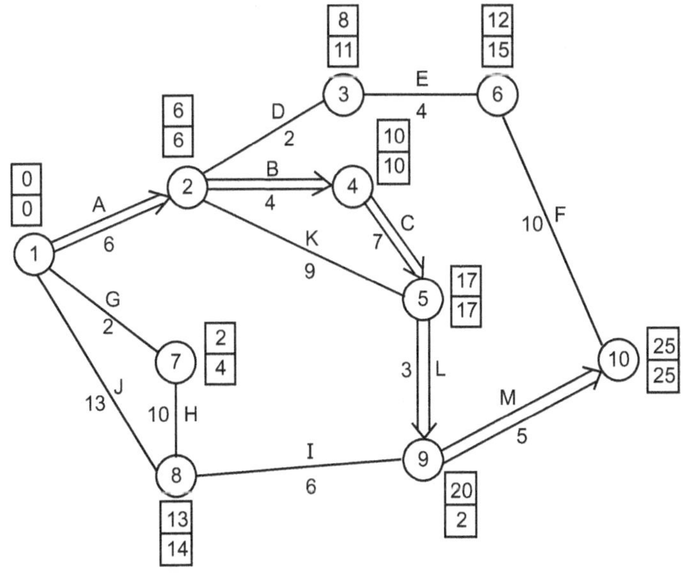

Fig. 8.20

Critical path: A – B – C – L – m

Total completion time of project = 25.

Example 8.3 : *Find the optimal solution for the following network.*

Table 8.8

Activity	Succeeding Activity	Normal Duration
A	B, C	8
B	D, E	4
C	I	4
D	G	3
E	F	6
F	H	9
G	–	5
H	–	7
I	F	8

Solution:

(a)   Activity network for the given problems is

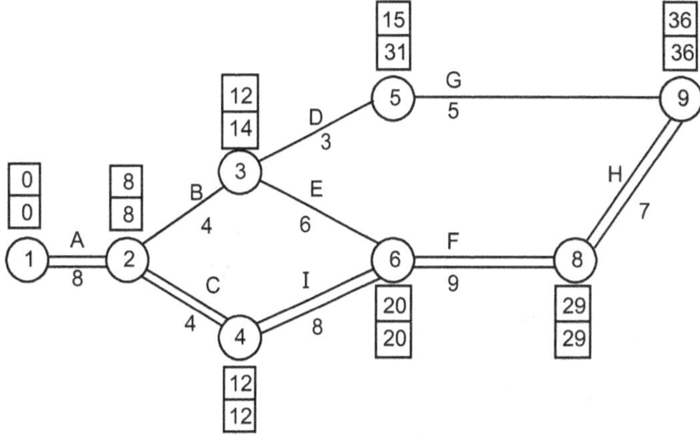

Fig. 8.21

Critical path: A – C – I – F – H

8 – 4 – 8 – 9 – 7

Example 8.4 : *A small maintenance project consist of following 12 jobs. Draw the network of the project. Summarize CPM calculations in tabular form. Calculating the floats for jobs and hence determine critical path.*

Table 8.9

Job	1-2	2-3	2-4	3-4	3-5	4-6	3-8	6-7	7-9	6-10	8-9	9-10
Duration	2	7	3	3	5	3	5	8	4	4	1	7

Solution: Network diagram:

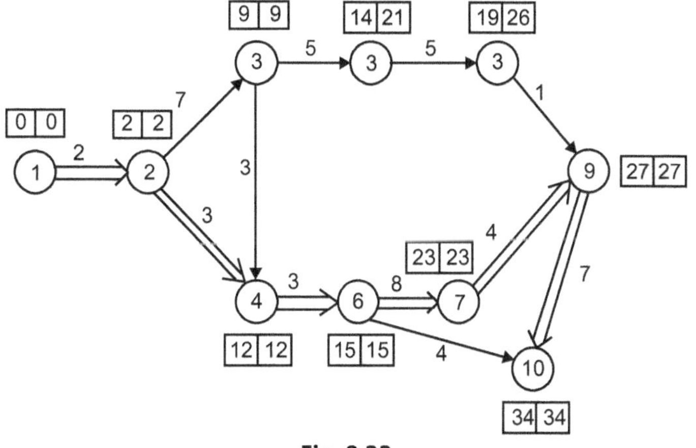

Fig. 8.22

Critical path= 1-2-4-6-7-9-10.

Table 8.10

Job	Duration	Float
1-2	2	0
2-3	7	0
3-4	3	0
2-4	3	7
3-5	5	7
4-6	3	0
5-8	5	7
6-7	8	0
6-10	4	15
7-9	4	0
8-9	1	7
9-10	7	0

Example 8.5 : *A project consists of six activities. Draw the network diagram and calculate EST, LST, EFT, LFT and float(idle time). Determine the critical path and find total project duration.*

Table 8.11

Activity	Immediate Predecessor	Activity Duration (week)
A	–	4
B	A	6
C	B	5
D	A	4
E	D	3
F	E, C	3

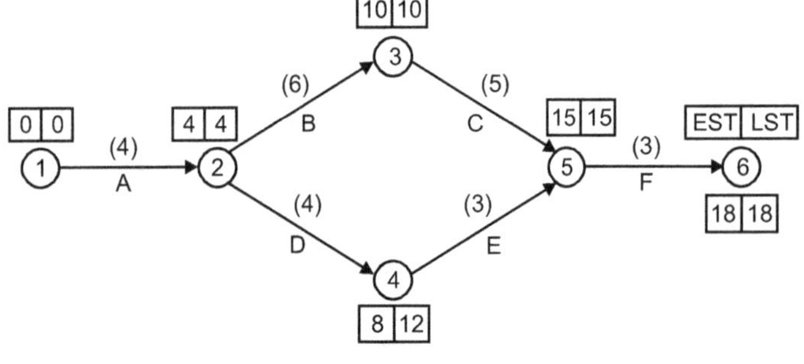

Fig. 8.23

CP=1-2-3-5-6, Duration =18 week

The network diagram alongwith duration as follows:

Table 8.12

Activity	Duration (Days)	EST	LST	EFT	LFT	Float
A	4	0	0	4	4	0
B	6	4	4	10	10	0
C	5	10	10	15	15	0
D	4	4	4	8	12	4
E	3	8	12	11	15	4
F	3	15	15	18	18	0

EST(from tail event)

$$LST = LFT - T$$

(from tail event)

$$EFT = EST + T$$

LFT (LST from lead event)

$$Float = LFT - EFT$$

Numerical Practice (for PERT)

Example 8.6 : *Find (i) C.P., (ii) σ and (iii) variance*

Table 8.13

Activity	Predecessor	Optimistic Time T_o	Most Likely Time T_m	Pessimist T_p
A	–	6	7	8
B	–	1	2	9
C	–	1	4	7
D	A	1	2	3
E	A, B	1	2	9
F	C	1	5	9
G	C	2	2	8
H	E, F	4	4	4
I	E, F	4	4	10
J	D, H	2	5	14
K	I, J	2	2	8

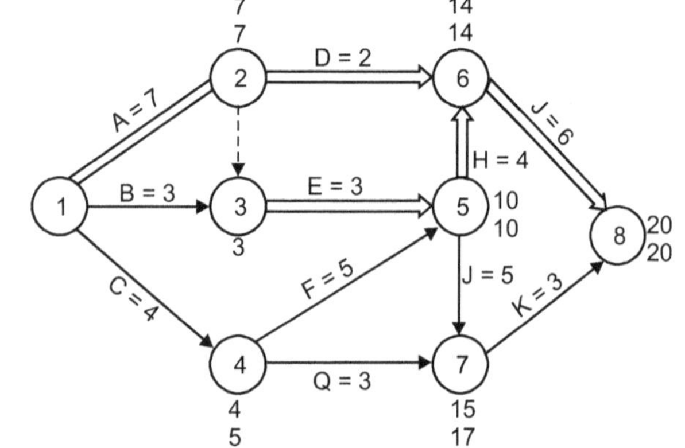

Fig. 8.24 : Network diagram

Solution:

Table 8.14

Activity	T_o	T_m	T_p	Optimal Duration (T_e)	Standard Deviation (σ)	Variance (σ^2)
A	6	7	8	7	0.333	0.11088
B	1	2	9	3	1.33	1.7689
C	1	4	7	4	1	1
D	1	2	3	2	0.33	0.11088
E	1	2	9	3	1.33	1.7689
F	1	5	9	5	1.33	1.7689
G	2	2	8	3	1	1
H	4	4	4	4	0	0
I	4	4	10	5	1	1
J	2	5	14	6	2	4
K	2	2	8	3	1	1

Example 8.7 : *For small project given below. Draw network and calculate expected duration of project. What is completion of probability that project completed within 30 days? What is duration of 90.7% probability of completion ?*

Table 8.15

Activity	T_o	T_m	T_p	Optimal Duration	σ	v
1-2	2	5	14	6	2	4
1-3	9	12	15	12	1	1
2 4	5	14	17	13	2	4
3-4	2	5	8	5	1	1
3-5	6	6	12	7	1	1
4-5	8	17	20	16	2	4

Solution:

Optimal duration = T_e

$$= \frac{T_o + T_p + 4T_m}{6}$$

Network diagram:

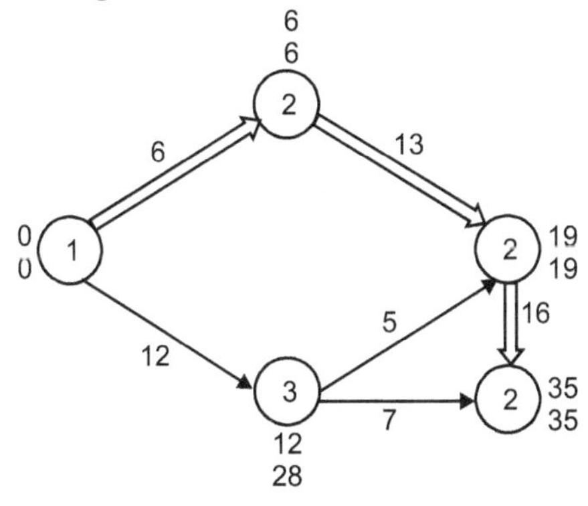

Fig. 8.25

Critical path = $1 - 2 - 4 - 5$

Contains

$$1 - 2 \rightarrow 6$$
$$2 - 4 \rightarrow 13$$
$$4 - 5 \rightarrow 16$$
$$\underline{\qquad\ 35 \text{ days}}$$

Scheduled duration = 30 days < 35 days

$$= z = \frac{30 - 35}{\sqrt{12}}$$

$$= 1.44 \approx 1.4$$

$\therefore$ $\qquad Z = 1.4$

$\therefore$ $\qquad$ at Z = 1.4, P = = 91.9%

Now for P = 90.7 %, Z = 1.32

$\therefore$ $\qquad 1.32 = \dfrac{x - 35}{\sqrt{12}}$

$$x = 39.57 \approx 40 \text{ days are required}$$

Example 8.8 : *Construct project network, find the expected duration variance. Find out the critical path and expected completion time. Also find the probability of complete project on or before 30 weeks. If probability of completion = 90% then find the project completion time*

Solution:

Table 8.16

Activity	Predecessor	T_o	T_m	T_p Time	Optimal	σ
A	–	3	5	8	5.167	0.833
B		6	7	9	7.167	0.5
C	A	4	5	9	5.5	0.833
D	B	3	5	8	5.167	0.833
E	A	4	6	9	6.167	0.833
F	C, D	5	8	11	8	1
G	C, D, E	3	6	9	6	1
H	F	1	2	9	3	1.33

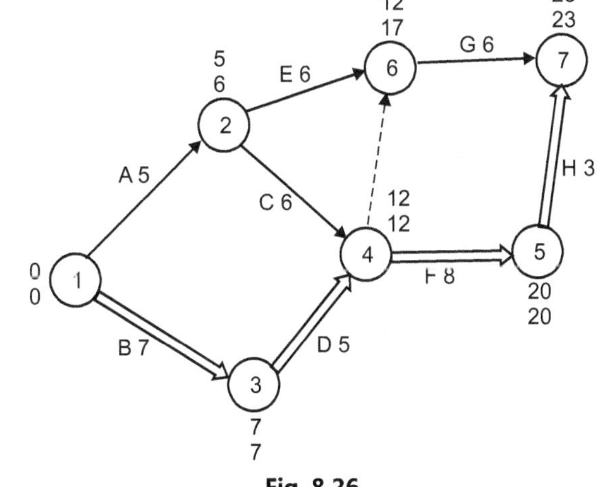

**Fig. 8.26

Critical path = $1 - 3 - 4 - 5 - 7$

Expected completion time = 23 weeks

Now standard deviation for network is

or　　　　$= \sqrt{0.25 + 0.694 + 1 + 1.77}$

Now probability for completion of work in 30 weeks

　　　　$z = 1.88$

∴ From table probability for completion of task in 30 weeks is 97.1%.

If probability of completion = 90% then the project completion time is

For 90% probability, $z = 1.3$

∴ $\dfrac{T_{exp} - 23}{(1.927)^2} = 1.3$

　　　　$T_{exp} = 23 + 4.8 = 28$ days

Example 8.9 : *The following table gives data on normal time and cost and crash time & cost for a project*

Table 8.17

Activity	Normal		Crash	
	Time (Weeks) (TN)	COST (₹) (CN)	Time (Weeks) (TC)	Cost (₹) (CN)
1-2	3	300	2	400
2-3	3	30	3	30
2-4	7	420	5	580
2-5	9	720	7	810
3-5	5	250	4	300
4-5	0	0	0	0
5-6	6	320	4	410
6-7	4	400	3	470
6-8	13	780	10	900
7-8	10	1000	9	1200

Total cost = ₹ 4220

The indirect cost per week is ₹ 50.

1. *Draw the network for the project and critical path.*

2. *Find optimum time and optimum cost.*

3. *Determine minimum total time and corresponding cost.*

Solution:

Step 1: Find Earliest time estimates for all the activities, it is denoted as TE (ES)

Step 2: Find latest time estimates for all the activities, it is denoted as TL (LT)

Step 3: Determine the Critical Path.

Step 4: Compute the cost slope (i.e., cost per unit time) for each activity according to the following formula:

　　Cost slope = (crash cost-normal cost)/ (normal duration -crash duration) $= \Delta C / \Delta T$

Step 5: Among the critical path identify the activity with the minimum cost slope, and crash the activity by 1 day(or by I week or by I month or by I year as per duration specified).

Step 6: Calculate the project cost. Identify new critical path.

Project Cost:

Total project cost =direct cost + indirect cost

i.e. Project Cost:

　　=((Project Direct Cost + Crashing cost of crashed activity) +(Indirect Cost × project duration))

　　=((Project total normal Cost + Crashing cost of crashed activity) + (Indirect Cost per day x project duration))

　　= ((Project total normal Cost + extra Crashing cost of crashed activity) +(Indirect Cost per day × project duration))

where crashing cost of crashed activity = ΔT of crash activity x slope $= \Delta C0$ Iteration Step:

Step 7: Now in the new critical path select the activity with the next minimum cost slope, and crash by one day

Step 8: Repeat this process until all the activities in the critical path have been crashed by 1 day.

Step 9: Once all the activities along the critical path are crashed by one day, Repeat the process again i.e. goes to step5.

Step 10: Find the minimum project cost and identify the activities which do not lie along the critical path

Step 10: Now see if the project cost can be further reduced without affecting the project duration time then perform uncrashing. i.e. uncrash the activities which do not lie along the critical path.

Uncrashing should start with an activity having the maximum cost slope. An activity is to be expanded only to the extent that it itself may become critical, but should not affect the original critical path.

Example 8.10 : *Indirect cost ₹ 50 per week*

Table 8.18

Activity	Normal		Crash		ΔT	ΔC	Δc/ΔT
	Time (Week) (TN)	Cost (₹) (CN)	Time (Weeks) (TC)	Cost (₹) (CN)			
1-2	3	300	2	400	1	100	100
2-3	3	30	3	30	0	0	0
2-4	7	420	5	480	2	160	80
2-5	9	720	7	810	2	90	45
3-5	5	250	4	300	1	50	50
4-5	0	0	0	0	0	0	0
5-6	6	320	4	410	2	90	45
6-7	4	400	3	470	1	70	70
6-8	13	780	10	900	3	120	40
7-8	10	1000	9	1200	1	200	200

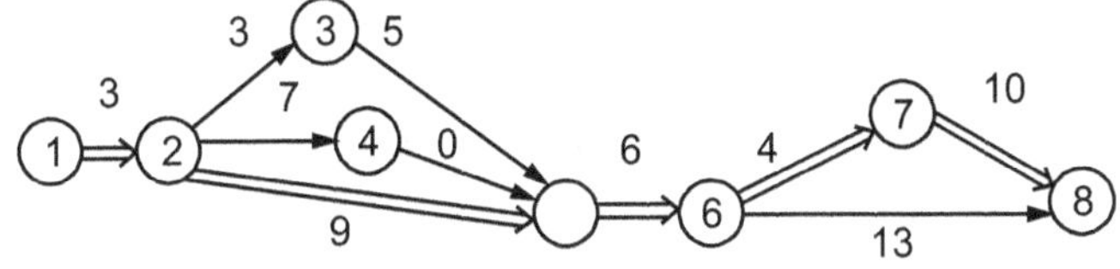

Fig. 8.27 : Project network

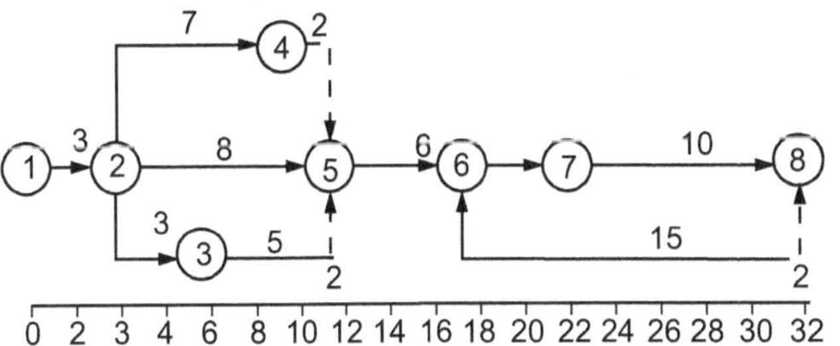

Fig. 8.28 : Time scale diagram of critical path 1-2-5-6-7-8

Crashing in activity 2-5 in 1 week (as this activity has least slope=45)

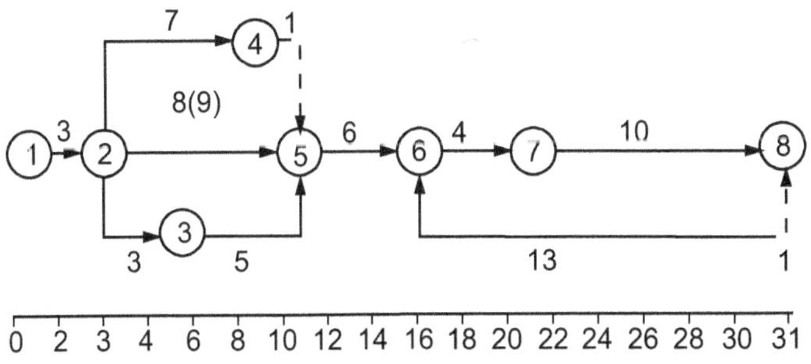

Fig. 8.29

Crashing in activity 5-6 in 2 week

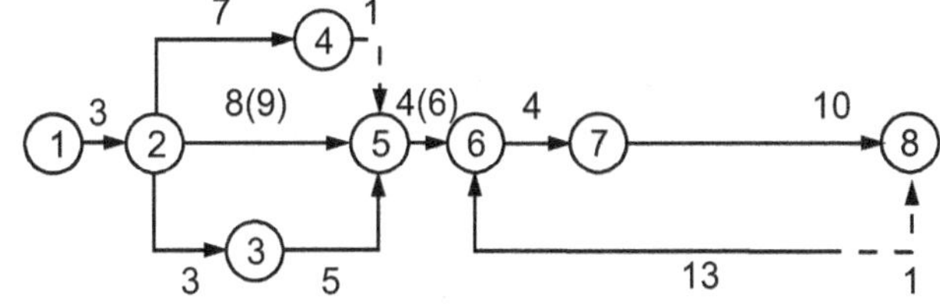

Fig. 8.30

Crashing in activity 6-7 in 1 week

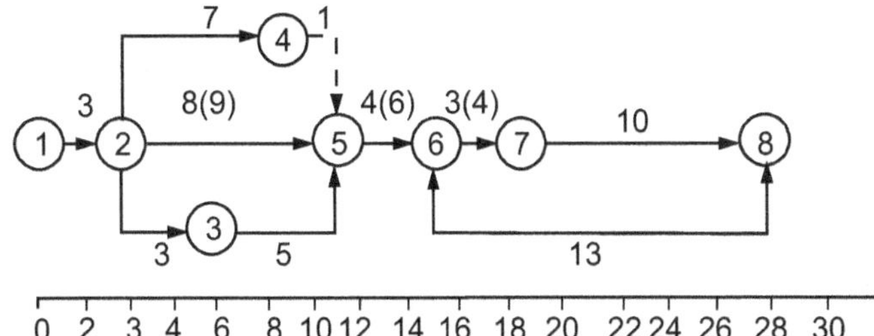

Fig. 8.31

Crashing in activity 1-2 in 1 week

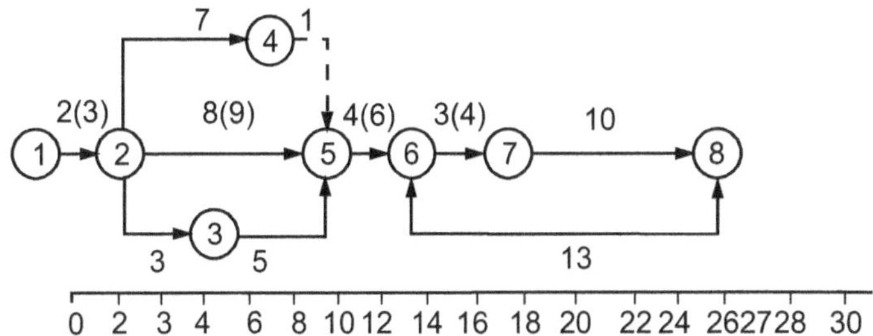

Fig. 8.32

Crashing in activity 6-8 in 2 week

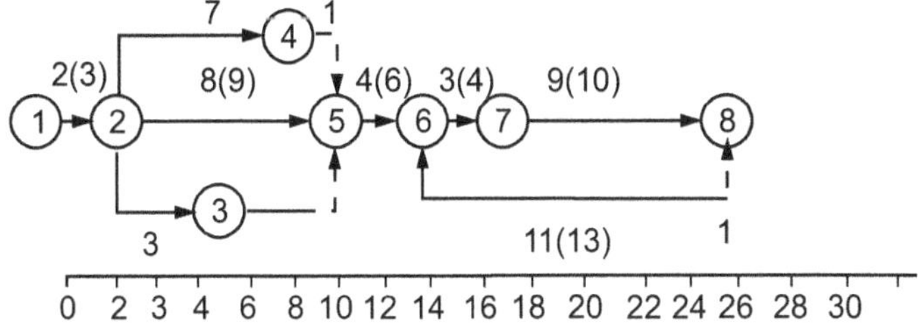

Fig. 8.33

Crashing in activity 7-8 in 1 week

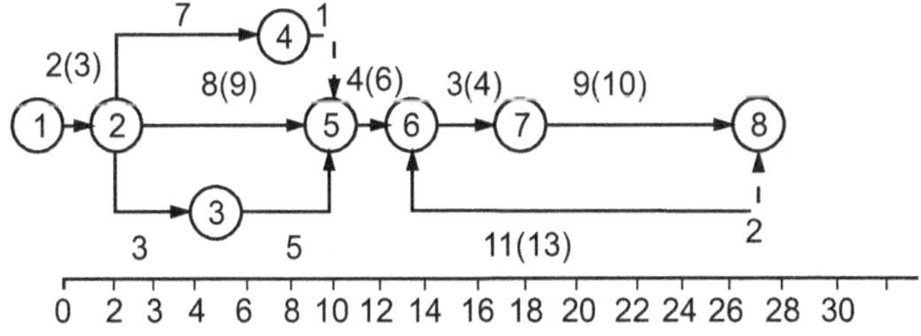

Fig. 8.34

Crashing in activity 3-5 in 1 week

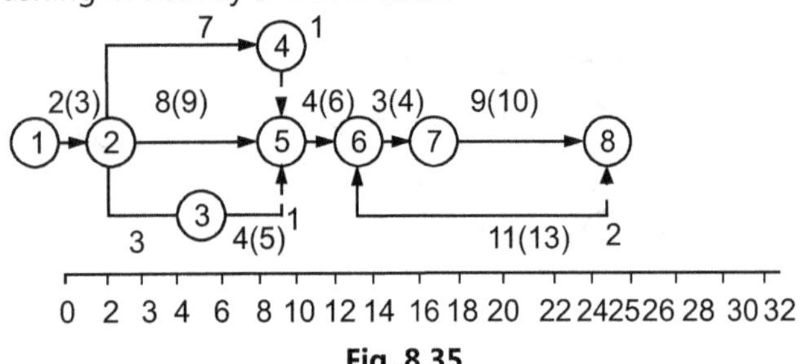

Fig. 8.35

Again Crashing in activity 2-5 in 1 week

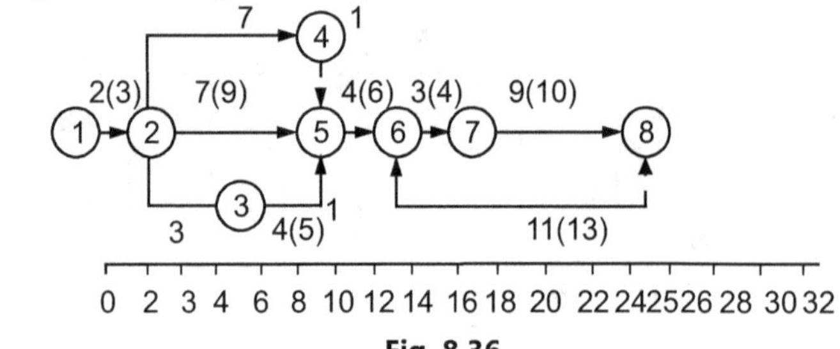

Fig. 8.36

The above results are summarized in tables below:

Table 8.20

Activity Crashed	No. of Week Crash	Represent Fig.	Weeks Saved in project	Project duration (Week)	Normal Direct Cost (DC) ₹	Indirect Cost (IC) ₹	Crash Cost (CC) ₹	Total Cost = DC + IC + CC ₹
CPM	-	Fig. 8.28	–	32	4220	$32 \times 50 = 1600$	0	5820
2-5	1	Fig. 8.29	1	31	4220	$31 \times 50 = 1550$	$1 \times 45 = 45$	5815
5-6	2	Fig. 8.30	2	29	4220	$29 \times 50 = 1450$	$45 + 2 \times 45 = 135$	5805 Least
6-7	1	Fig. 8.31	1	28	4220	$28 \times 50 = 1400$	$135 + 1 \times 70 = 205$	5825
1-2	1	Fig. 8.32	1	27	4220	$27 \times 50 = 1350$	$305 + 2 \times 40 = 385$	5955
6-8	2	Fig. 8.33	-	27	4220	$27 \times 50 = 1350$	$305 + 2 \times 40 = 385$	5900
7-8	1	Fig. 8.34	1	26	4220	$26 \times 50 = 1300$	$385 + 1 \times 200 = 585$	6105
3-5	1	Fig. 8.35	-	26	4220	$26 \times 50 = 1300$	$585 + 1 \times 50 = 635$	6155
2-5	1	Fig. 8.36	1	25	4220	$25 \times 50 = 1250$	$635 + 1 \times 45 = 680$	6150

Indirect cost = indirect cost per week x project duration

The above table indicates the optimum cost (₹ 5805 least) and optimum duration (29 weeks)

Alternative Approach or Method :

Example 8.11 : *The following Table 8.18 shows activity of project along N_T, C_T, N_C, C_C, indirect cost ₹60 per day .*

Table 8.21

Activity	N_T (day)	C_T (day)	C_C	N_C
1 – 2	20	17	720	600
1 – 3	25	25	300	300
2 – 3	10	8	440	300
2 – 4	12	6	700	100
3 – 4	5	2	350	200
4 – 5	10	5	650	350

(i)   Draw network and identify C.P.

(ii) Calculate N_T and corresponding cost.

(iii) Crash the activities and determine optimal duration of project and corresponding cost.

Solution:

Critical path = $1 - 2 - 3 - 4 - 5$

T_C = Direct cost   + indirect cost

T_C = total normal cost + indirect cost per week × project duration

T_C = $1850 + 45 \times 60 = ₹ 4550$

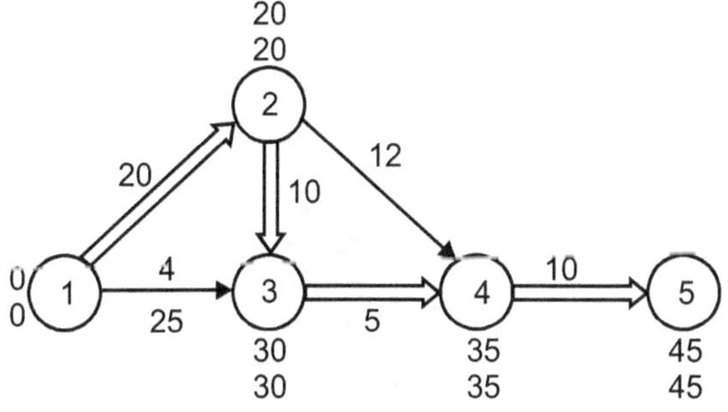

Fig. 8.37

For Crashing:

Table 8.22

Activity	N_T	N_C	C_T	C_C	ΔC ($C_C - N_C$)	ΔT ($N_T - C_T$)	$\dfrac{\Delta C}{\Delta T}$
1 – 2	20	600	17	720	120	3	40
1 – 3	25	300	25	300	0	0	0
2 – 3	10	300	8	440	140	2	70
2 – 4	12	100	6	700	600	6	100
3 – 4	15	200	2	350	150	3	50
4 – 5	10	350	5	650	300	5	60
		$\Sigma = 1850$					

Activity $1 - 2$ has less cost slope. Hence, crash activity $1 - 2$ by 3 days.

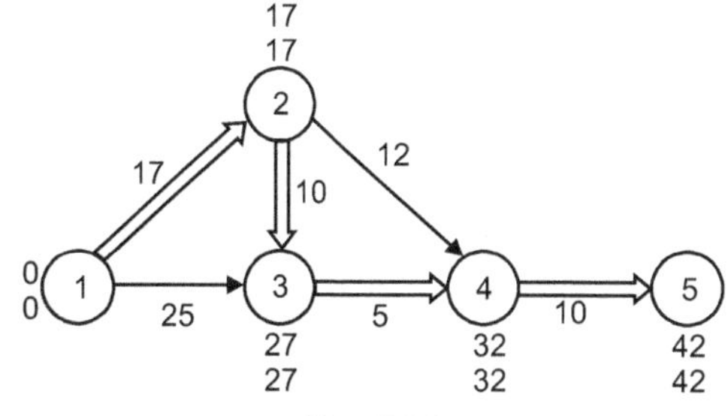

Fig. 8.38

New Critical path = $1 - 2 - 3 - 4 - 5$

Direct cost = Normal cost + crash cost

where crash cost

= ΔT of crash activity x slope $= \Delta C$

Direct cost = $1850 + 3(40) = ₹ 1970$

Indirect cost=indirect cost per day x project duration

Indirect cost = $60 (42) = ₹ 2520$

Total project cost = direct cost + indirect cost

∴ Total project cost = $1970 + 2520 = ₹ 4490$

Now activity $3 - 4$ has less cost slope.

Hence, crash $3 - 4$ by 3 days.

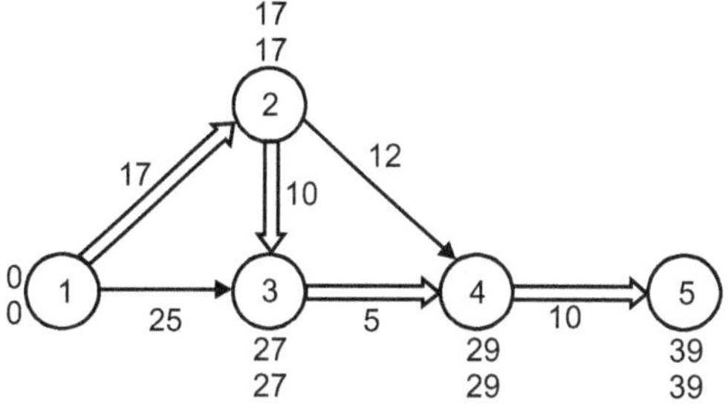

Fig. 8.39

New Critical path = $1 - 2 - 3 - 4 - 5$

Direct cost=Previous direct cost + crash cost

Direct cost = $1970 + 3(50)$

= $₹ 2120$

Indirect cost = $39 (60) = ₹ 2340$

Total cost = $₹ 4460$

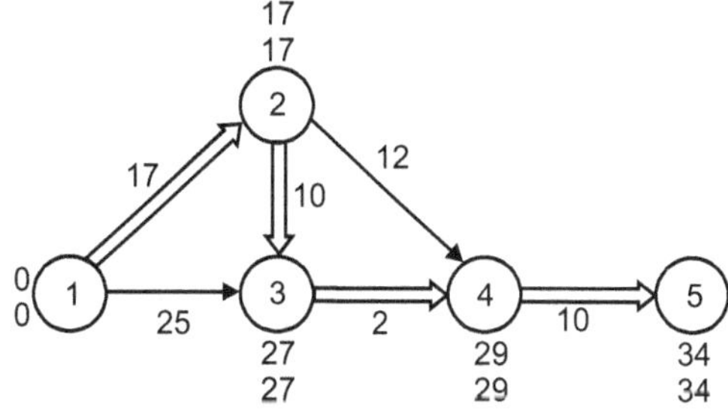

Fig. 8.40

Now, crash activity $4 - 5$ by 5 days.

New C_p = $1 - 2 - 3 - 4 - 5$

Direct cost = $2120 + 5 (60)$

Indirect cost = $60 (34) = ₹ 2040$

Total cost = ₹ 4460

Now crash Activity 2 – 3 has left crash 2 – 3 by 2 days.

Critical path = 1 – 2 – 4 – 5

Direct cost = 1970 (direct cost of previous activity 1-2) + 2 (70)

= ₹ 2110

Indirect cost = 60 (34) = ₹ 2040

∴ Total cost = ₹ 4150

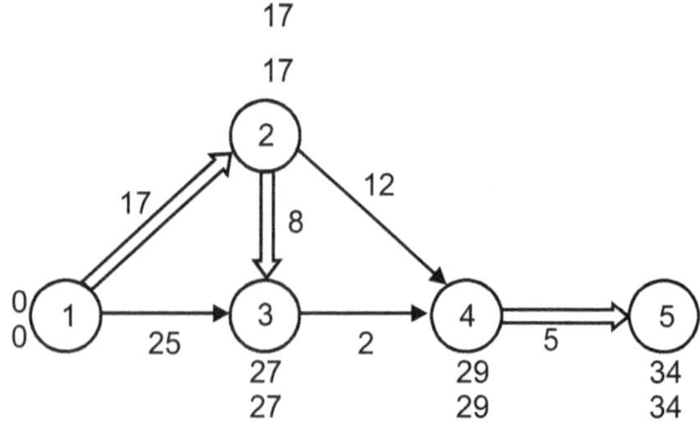

Fig. 8.41

Crash activity 2 – 4 by 6 days (this is non critical activity but has next slope (Greater than previous crashed critical activity slope)

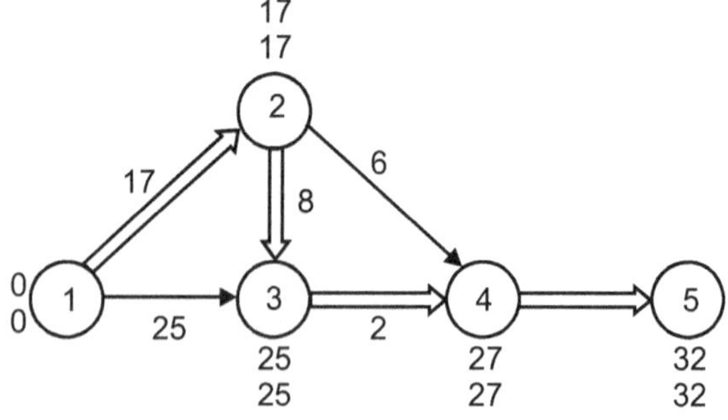

Fig. 8.42

New C_P = 1 – 2 – 3 – 4 – 5

Direct cost = 2110

(direct cost of previous activity 2-3) + 6 (100) = ₹ 2710

Indirect cost = 60 (32) = ₹ 1920

∴ Total cost = ₹ 4630

Table 8.23

Activity	Duration	Total cost	Remark
1 – 2	42	₹ 4490	
1 – 3	-	-	0 slope
2 – 3	34	₹ 4150	optimum
2 – 4	32	₹ 4630	
3 – 4	39	₹ 4460	
4 – 5	34	₹ 4460	

The above table indicates the optimum cost (₹ 4150 least) and optimum duration (34 weeks)

Also shows the maximum crashing after project duration is 34 weeks

Example 8.12 : *Consider the data of project as shown in the Table 8.24.*

Table 8.24

Activity	N_T	C_T	N_C	C_C
1 – 2	8	5	800	950
1 – 3	5	3	500	700
1 – 4	9	6	600	1250
2 – 5	10	8	900	1300
3 – 5	5	3	700	1100
3 – 6	6	5	1200	1500
4 – 6	7	5	1300	1400
5 – 7	2	1	400	500
6 – 7	4	2	500	900

If the indirect cost/week is ₹ 300 find the optimal crash project completion time.

Solution:

Table 8.25

Activity	N_T	C_T	N_C	C_C	ΔT	ΔC	$\Delta C/\Delta T$
1 – 2	8	5	800	950	3	150	50
1 – 3	5	3	500	700	2	200	100
1 – 4	9	6	600	1250	3	650	216.67
2 – 5	10	8	900	1300	2	400	200
3 – 5	5	3	700	1100	2	400	200
3 – 6	6	5	1200	1500	1	300	300
4 – 6	7	5	1300	1400	2	100	50
5 – 7	2	1	400	500	1	100	100
6 – 7	4	2	500	900	2	400	200

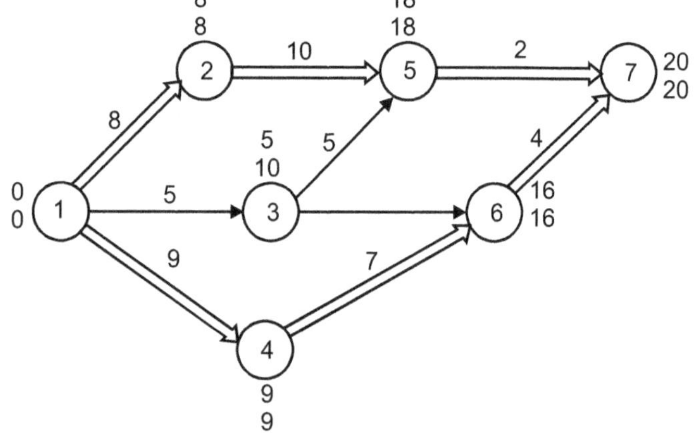

**Fig. 8.43

Example 8.13 : *Find the optimal solution for the following network.*

Table 8.26

Activity	Succeeding Activity	Normal Duration	Crash Duration	Crash Cost	Normal Cost
A	B, C	8	8	500	500
B	D, E	4	3	1000	750
C	I	4	3	800	500
D	G	3	3	750	750
E	F	6	4	1500	800
F	H	9	6	2500	1600
G	–	5	4	500	400
H	–	7	5	800	600
I	F	8	5	3000	1500

Indirect cost = ₹RS 150/- per day.

Solution:

(a) Activity network for the given problems is

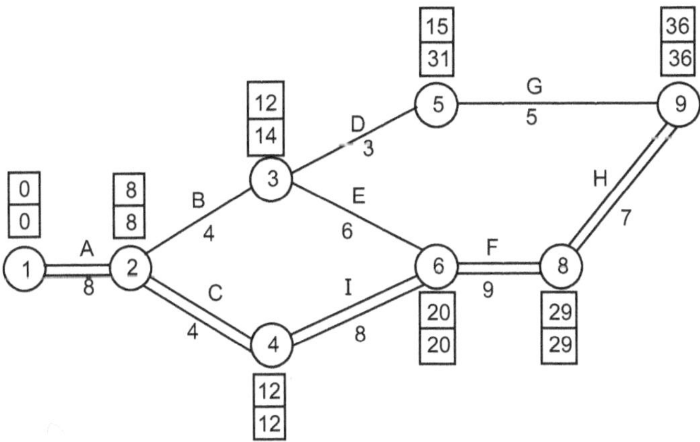

Fig. 8.44

Critical path: A – C – I – F – H

      8 – 4 – 8 – 9 – 7

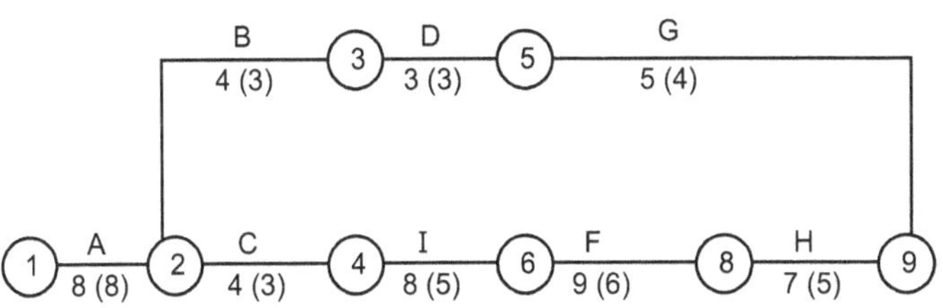

Fig. 8.45

(b) Project/Activity Cost – Time Analysis

Table 8.27

Activity	Normal		Crash		ΔC	ΔT	ΔC/ΔT
	Time	Cost	Time	Cost			
A	8	500	8	500	0	0	0
B	4	1000	3	750	250	1	250
C	4	800	3	500	300	1	300
D	3	750	3	750	0	0	0
E	6	1500	4	800	700	2	350
F	9	2500	6	1600	900	3	300
G	5	500	4	400	100	1	100
H	7	800	5	600	200	2	100
I	8	3000	5	1500	1500	3	500
Σ =		11350					

Optimal solution for the given network is 34 days and cost is ₹ 16650.

Example 8.14 : *A project consist of 6 activities from A to F. The time cost trade of data given in the table. Indirect cost per day is ₹ 60 and direct cost ₹ 1000.*

Table 8.28

Steps	Activity	Normal Duration	Minimum Duration	Cost of Crashing
1-2	A	9	6	20
1-3	B	8	5	25
1-4	C	15	10	30
2-4	D	5	3	10
3-4	E	10	6	15
4-5	F	2	1	10

Draw network diagram, determine CPM and find,

(i) Optimum duration of project and cost,

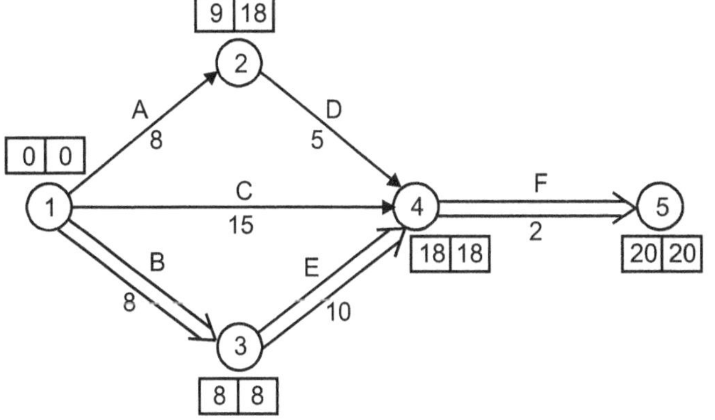

Fig. 8.46

Solution : Optimum cost = ₹ 2030, Optimum duration = 15 days.

Example 8.15 : *The following table shows the details of a project consider overhead cost = ₹ 600.*

Table 8.29

Activity	Predecessor	Time in days		Difference	Cost	
		Normal	Crash	Δt	Normal	Crash
A	–	3	2	1	800	1600
B	–	6	4	2	2200	3200
C	–	5	3	2	2000	2600
D	A	5	3	2	3000	4000
E	B	4	2	2	2400	3800
F	B	6	4	2	3400	4200
G	B	3	2	1	1600	2400
H	C, G	4	3	1	1200	1600
I	D, E	8	6	2	4400	5600
J	E	4	2	2	2000	2800
K	C, G, F, J	6	4	2	2600	3600
L	C, G, F, J	5	4	1	1200	1600
M	I, K	3	2	1	1000	1400
N	H, L	7	5	2	3800	5200

(i) Draw the network showing critical path

(ii) Draw time scale network

(iii) Find optimum duration and cost.

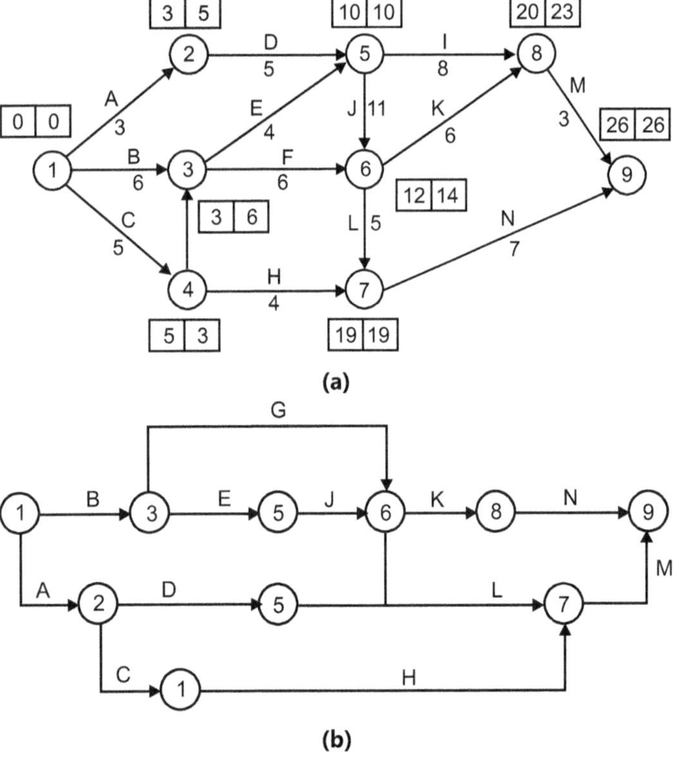

(a)

(b)

Fig. 8.47

Optimum Project duration = 22 roman days and

Optimum Total cost = ₹ 46500

Example 8.16 : *A local company intends to carry out a capital labour. The following table shows for each activity needed, normal time N_T, shortest time N_C, the activity completed and cost per day, induced cost of completing the eight activities N_T is ₹RS 174000 (fixed cost or normal cost) excluding side overhead. Overhead cost of general side activity is RS 4800/day. Calculate normal duration of project cost and critical path, calculate optimal duration and corresponding cost, shortest time and associated cost.*

Table 8.30

Activity	N_T	N_C	Cost of Reduction/ Day (ΔC)	Slope ΔC/ ΔT
1 – 2	6	4	2400	1200
1 – 3	8	4	2700	675
1 – 4	5	3	1500	750
2 – 4	3	3	–	–
2 – 5	5	3	1200	600
3 – 6	12	8	6000	1500
4 – 6	8	5	1500	500
5 – 6	8	6	–	–

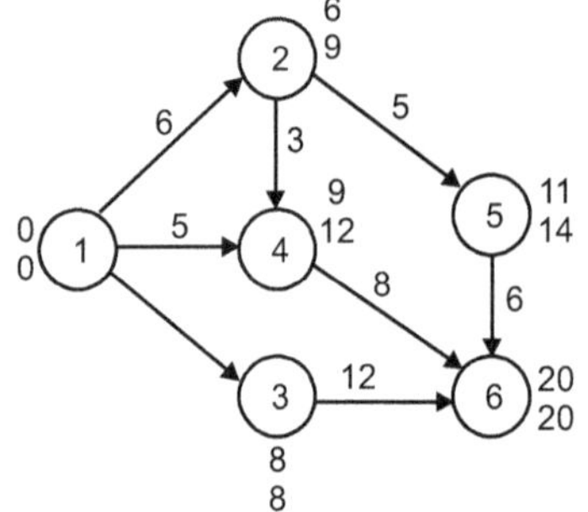

Fig. 8.48

Example 8.17 : *To explain the process of crashing a network to reach the optimum project schedule, let us consider the network shown in Fig. 8.49. With each activity is associated normal direct cost and crash direct cost, the normal duration time and crash duration time. The complete data is given in table 8.31. The network has been drawn for normal conditions and the times shown along the arrows are normal duration times*

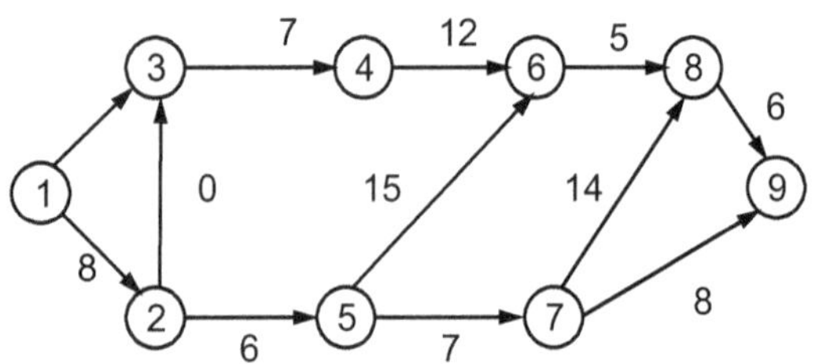

Fig. 8.49

Table 8.31

Activity	Normal		Crash		ΔT	Δc	Δc/Δt
	Time	Cost	Time	Cost			
1-2	8	7000	3	10000	5	3000	600
1-3	4	6000	2	8000	2	2000	1000
2-3	0	0	0	0	0	0	0
2-5	6	9000	1	11500	5	2500	500
3-4	7	2500	5	3000	2	500	250
4-6	12	10000	8	16000	4	6000	1500
5-6	15	12000	10	16000	5	4000	800
5-7	7	12000	6	14000	1	2000	2000
6-8	5	10000	5	10000	0	0	0
7-8	14	6000	7	7400	7	1400	200
7-9	8	6000	5	12000	3	6000	2000
8-9	6	6000	4	7800	2	1800	900

Table 8.32 : Result of Calculations Based on Unit Crashing

Activity Crashed	Weeks Saved	Project Duration	Direct Cost	Indirect Cost	Total Cost
Nil	0	41	86500	41000	127500
7-8	1	40	87500	40000	127500
2-5	1	39	87200	39000	126200
1-2	1	38	87800	38000	125800
8-9	1	37	88700	37000	125700
5-6	0	37	89500	37000	126500
7-8	1	36	89700	36000	125700
1-2	1	33	91800	33000	124800
3-4	0	32	92650	32000	124850
7-8	0	32	92850	32000	124850
2-5	1	31	93350	31000	124350
3-4	0	31	93600	31000	124600
2-5	1	30	94100	30000	124100
2-5	0	30	94600	30000	124600
1-2	0	30	94700	30000	124700
1-3	1	29	95700	29000	124700
2-5	0	29	98200	28000	126200
4-6	1	27	99700	27000	126700
5-6	0	27	100500	27000	127500
4-6	1	26	102000	26000	128000
7-8	1	25	104500	25000	129500
Uncrashing		30	93400	30000	123400

- In Table 8.32 the results shows how the total cost of the project is reduced as the total duration is crashed. Before uncrashing the minimum cost of project is ₹ 124100 for the project duration of 30 days and after uncrashing the minimum cost will be ₹ 123400 for the project duration of 30 days.

- The following Fig. 8.50 depicts the results obtained

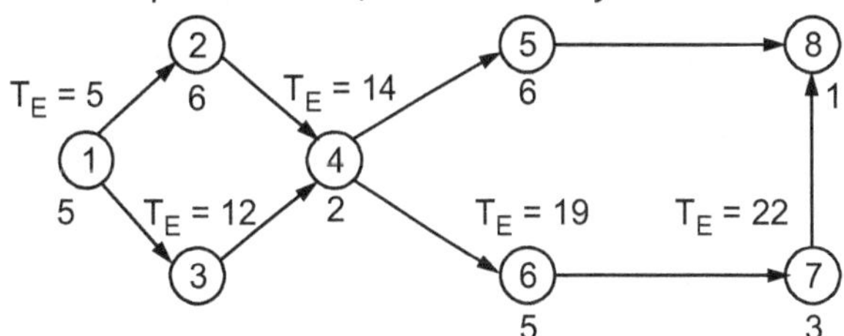

Fig. 8.50 : Project Duration vs Cost Analysis

- In this the algorithm proposed for unit crashing reduces the cost of project. When activities are crashed by one day then only the crashing cost corresponding to one day is increased thereby reducing the project duration as well as cost.

- A C++ program is may be developed to achieve the above results. This approach is well suitable for places where cost is of major consideration.

Example 8.18 : *Using information from the table, indicate expected completion time for each activity.*

Fig. 8.51 : CPM Chartwith TE

- Calculate earliest expected completion time for each activity (TE) and the entire project.

- The earliest expected completion time for a given activity is determined by summing the expected completion time of this activity and the earliest expected completion time of the immediate predecessor.

Rule: if two or more activities precede an activity, the one with the largest TE is used in calculation (e.g., for activity 4, we will use TE of activity 3 but not 2 since 12 > 11). the critical path 1,3,4,6,7,8

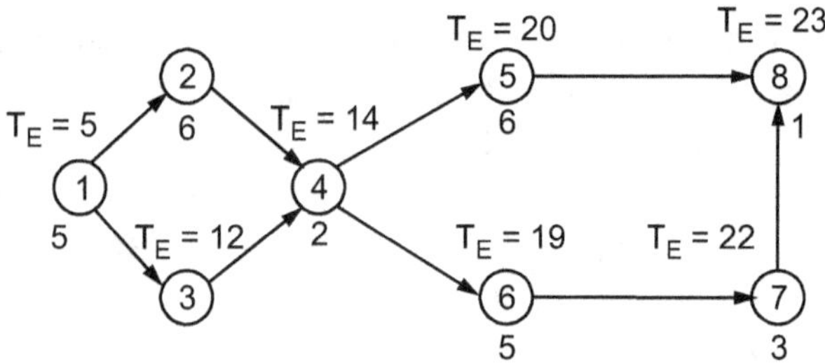

Fig. 8.52 : Critical path

The critical path represents the shortest time, in which a project can be completed. Any activity on the critical path that is delayed in completion delays the entire project. Activities not on the critical path contain slack time and allow the project manager some flexibility in scheduling.

PERT

Table 8.33

Activity	Immed. Predec.	Optimistic Time (Hr.)	Most Likely Time (Hr.)	Pessimistic Time (Hr.)
B	--	1	4.5	5
C	A	3	3	3
D	A	4	5	6
E	A	0.5	1	1.5
F	B,C	3	4	5
G	B,C	1	1.5	5
H	E,F	5	6	7
I	E,F	2	5	8
J	D,H	2.5	2.75	4.5
K	G,I	3	5	7

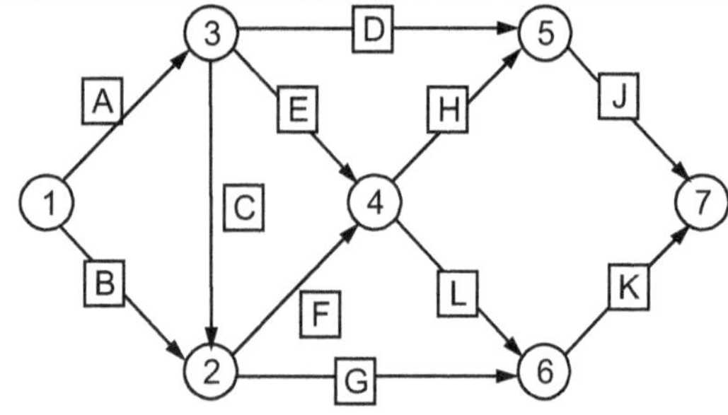

Fig. 8.53 : PERT Network

Activity	Expected Time	Variance
A	6	4/9
B	4	4/9
C	3	0
D	5	1/9
E	1	1/36
F	4	1/9
G	2	4/9
H	6	1/9
I	5	1
J	3	1/9
K	5	4/9

Table 8.34

Activity	ES	EF	LS	LF	Slack	
A	0	6	0	6	0	*critical
B	0	4	5	9	5	
C	6	9	6	9	0*	
D	6	11	15	20	9	
E	6	7	12	13	6	
F	9	13	9	13	0 *	
G	9	11	16	18	7	
H	13	19	14	20	1	
I	13	18	13	18	0*	
J	19	22	20	23	1	
K	18	23	18	23	0*	

The estimated project completion time is the Max EF at node 7 = 23.

$$s^2 = s^2_A + s^2_C + s^2_F + s^2_H + s^2_K$$
$$= 4/9 + 0 + 1/9 + 1 + 4/9$$
$$= 2$$

$$\sigma\text{path} = 1.414$$

$$z = (24 - 23)/ = \sigma(24\text{-}23)/1.414 = .71$$

From the Standard Normal Distribution table:

$$P(z \le .71) = .5 + .2612 = .7612$$
$$P(T \le 24) = 0.7612$$

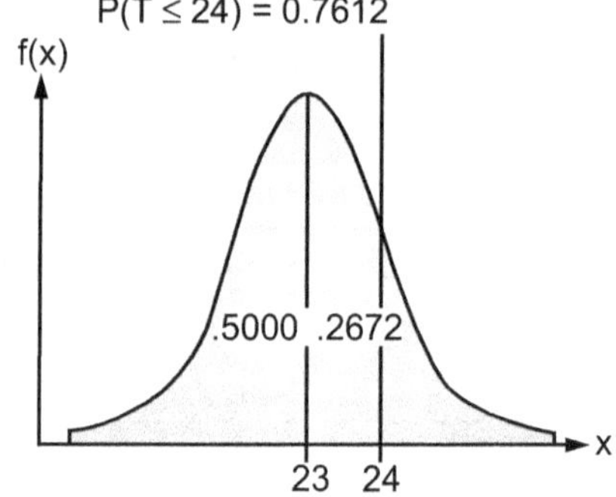

Fig. 8.54 : Probability the project will be completed within 24 hrs

PERT/COST

Table 8.35 : Time Cost data

Activity	Normal Time	Normal Cost RS	Crash Time	Crash Cost	Allowable Crash Time	Slope
1	12	3000	7	5000	5	400
2	8	2000	5	3500	3	500
3	4	4000	3	7000	1	3000
4	12	50000	9	71000	3	7000
5	4	500	1	1100	3	200
6	4	500	1	1100	3	200
7	4	1500	3	22000	1	7000
		75000		110700		

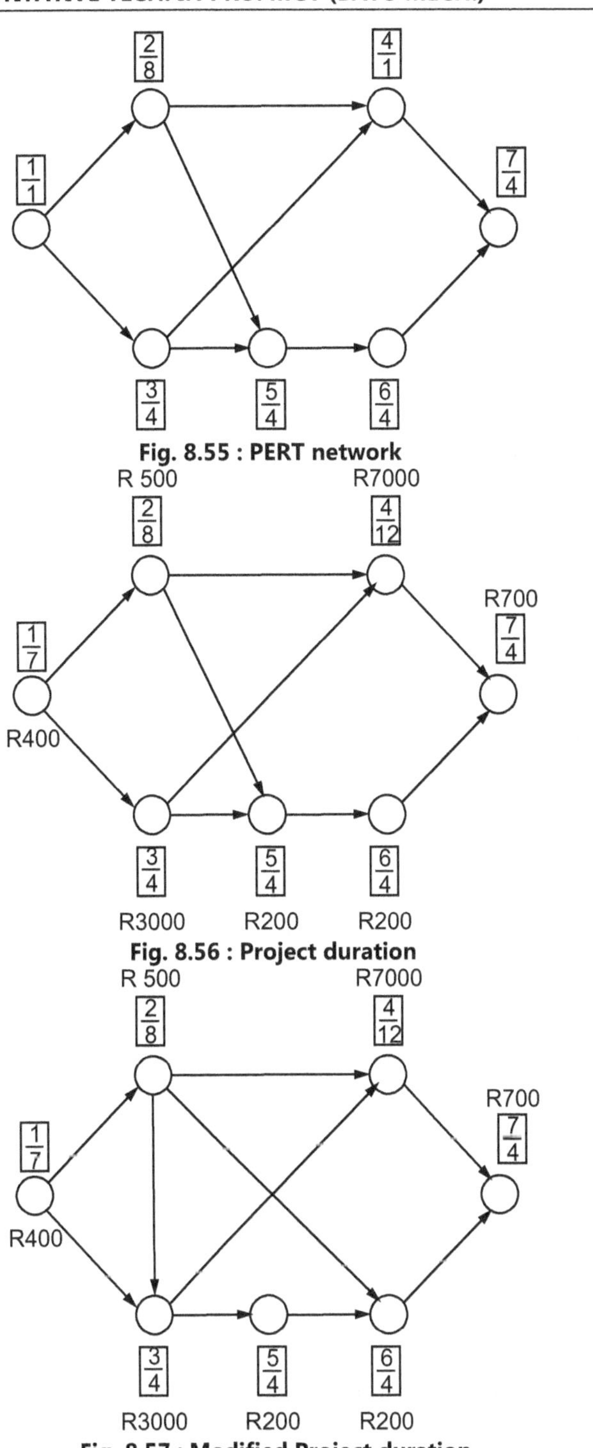

Fig. 8.55 : PERT network

Fig. 8.56 : Project duration

Fig. 8.57 : Modified Project duration

Example 8.19 : *Oil Credit Card Operation Project*

Table 8.36

Activity	Description	Immediate Predecessors	Time	Resources
A	Select office site	–		
B	Create Organizational and Financial Plan	–		
C	Determine personnel Requirements	B		

D	Design Facility	A, C		
E	Construct Interior	D		
F	Select Personnel to move	C		
G	Hire New Employees	F		
H	Move Records, Key Personnel, etc.	F		
I	Make Financial Arrangements with Institutions in Des Moines	B		
J	Train New Personnel	H, E, G		

Gas Storage Tank Construction Project

Table 8.37 : Descriptions, Durations and Precedence Relations of Work Components

Activity	Duration	Amount of Labor Requirement	Immediate Predecessors
Foundation Works			
1.1 Sub base leveling	3 days	6 workers	-
1.2 Lean Concrete	2 days	4 workers	1.1
1.3 Formwork to Foundation	5 days	3 workers	1.2
1.4 Rebar installation of Foundation	10 days	20 workers	1.2
1.5 Heating conduit and inclinometer installation	6 days	4 workers	1.2
1.6 Anchor plates installation	4 days	4 workers	1.4, 1.5
1.7 Pouring concrete to foundation	1 days	20 workers	1.3, 1.6
1st Lift			
2.1 Rebar Installation to Wall	4 days	18 workers	1.7
2.2 Formwork to Inner Tank	2 days	18 workers	2.3
2.4 Rebar Installation of Outer Wall	2 days	18 workers	1.7

Conti...

2.5 Formwork to Buttress and Outer Tank	10 days	8 workers	2.6	
2.6 Post-Tensioning Duct Installation	5 days	3 workers	2.3, 2.4	
2.7 Pouring concrete	1 days	12 workers	2.2, 2.5	
2nd Lift				
3.1 Rebar Installation to Wall	3 days	18 workers	2.7	
3.2 Formwork to Inner Tank	1.5 days	15 workers	3.3	
3.3 Embedment Installation	3 days	4 workers	3.1	
3.4 Rebar installation of Outer Wall	1.5 days	15 workers	2.7	
3.5 Formwork to Buttress and Outer Tank	8 days	10 workers	3.6	
3.6 Post-Tensioning Duct Installation + Anchor	3 days	5 workers	3.3, 3.4	
3.7 Pouring Concrete	1 days	12 workers	3.2, 3.5	

Indirect cost = ₹ 500 per day

Table 8.38

N_T	C_T	N_c	C_c	Diff in Cost	Diff in Time	Slope	Activity
3	3	36000	36000	0	0	0	A= 1.1
2	2	16000	16000	0	0	0	B =1.2
5	4	30000	34000	4000	1	4000	C= 1.3
10	7	400000	409000	9000	3	3000	D =1.4
6	4	48000	50000	2000	2	1000	E =1.5
4	3	32000	32500	500	1	500	F =1.6
1	1	40000	40000	0	0	0	G= 1.7
4	3	144000	148000	4000	1	4000	H =2.1
2	1	72000	73500	1500	1	1500	I =2.2
3	3	36000	36000	0	0	0	J= 2.3
2	1	72000	77000	5000	1	5000	K =2.4
10	6	160000	170000	10000	4	2500	L =2.5
5	3	30000	42000	12000	2	6000	M =2.6
1	1	24000	24000	0	0	0	N =2.7
3	2	108000	115000	7000	1	7000	O= 3.1
1.5	1.5	45000	45000	0	0	0	P =3.2
3	2	24000	28500	4500	1	4500	Q= 3.3
1.5	1.5	45000	45000	0	0	0	R= 3.4
8	6	160000	165500	5500	2	2750	S =3.5
3	2	30000	38000	8000	1	8000	T =3.6
1	1	24000	24000	0	0	0	U =3.2
		Total 1576000					

Table 8.39

Activity Crashed	No Day Crashed	Represent Fig.	No Days Saved	Project Duration	Total Direct Normal Cost	Crash Cost	Indirect Cost	Total Proj. Cost
CPM				61	1576000		30500	1606500
F	1		1	60	1576000	500	30000	1606500 least
L	4			56	1576000	10500	28000	1614500
S	2			54	1576000	16000	27000	1619000
D	3			51	1576000	25000	25500	1626500
H	1			50	1576000	29000	25000	1630000
Q	1			49	1576000	33500	25500	1634000
M	2			47	1576000	45500	23500	1645000
O	1			46	1576000	52500	23000	1655500
T	1			45	1576000	60500	23500	1659000
K	1			44	1576000	65500	22000	1663500

Ans. :

Optimum duration = 60 days

Optimum cost = ₹ 1606500)

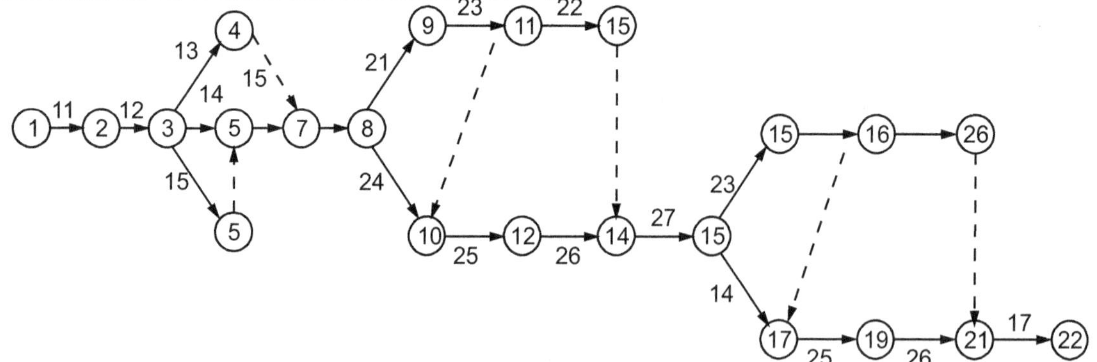

Fig. 8.58

$$CP = \text{A-B-D-F-G-H-J-J'-M-L-N-O-Q-Q'-T-S-U (61 DAYS)},$$

$$\text{float} = 45 \text{ days}$$

$$CP = \text{1-2-3-5-7-8-9-10-11-12-14-15-16-18-17-19-21-22 (61 DAYS)},$$

$$\text{float} = 45 \text{ days}$$

Example 8.20 : *The following table 8.40 shows the details of a project consider overhead cost = ₹600.*

Table 8.40

Activity	Predecessor	Time in days		Difference	Cost	
		Normal	Crash	Δt	Normal	Crash
A	–	3	2	1	800	1600
B	–	6	4	2	2200	3200
C	–	5	3	2	2000	2600
D	A	5	3	2	3000	4000
E	B	4	2	2	2400	3800
F	B	6	4	2	3400	4200
G	B	3	2	1	1600	2400
H	C, G	4	3	1	1200	1600
I	D, E	8	6	2	4400	5600
J	E	4	2	2	2000	2800
K	C, G, F, J	6	4	2	2600	3600
L	C, G, F, J	5	4	1	1200	1600
M	I, K	3	2	1	1000	1400
N	H, L	7	5	2	3800	5200

(i) Draw the network showing critical path.

(ii) Draw time scale network.

(iii) Find optimum duration and cost.

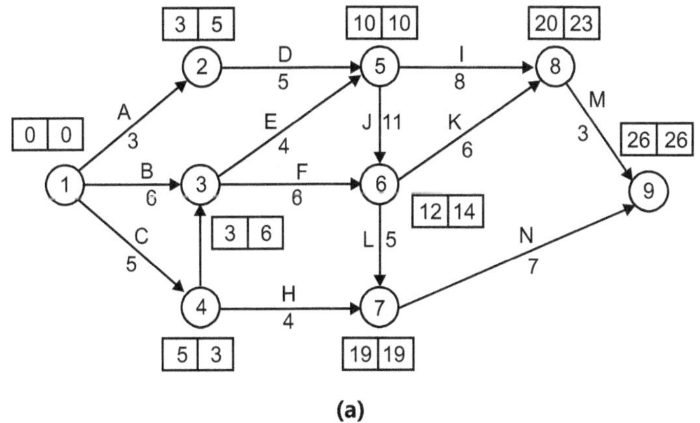

(a)

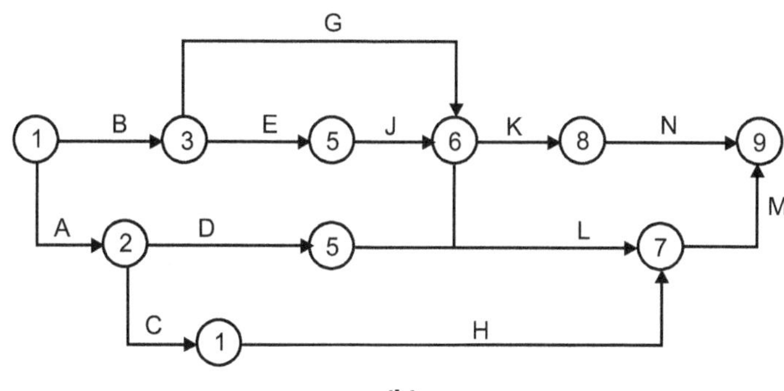

(b)

Fig. 8.59

Solution:

Table 8.41

Iteration	Activity	Slope	Δt	Project duration	Direct cost	Indirect cost	Total cost
0	–	–	–	26	31600	15600	47200
1	L	400	1	25	32000	15000	47000
2	J	400	2	23	32800	13800	46600
3	B	500	2	22	33300	13200	46500

(i) It is decided to crash 'B' activity, therefore E, J, L, N are also get crashed.

(ii) But crashing activities B and N results into increase in total cost as slope will become greater than ₹ 600.

Hence, crash only L, J, B.

∴ Project duration = 22 roman days and

Total cost = ₹ 46500

Example 8.21: *A project consist of 6 activities from A to F. The time cost trade of data given in the table. Indirect cost per day is ₹60 and direct cost ₹1000.*

Table 8.42

Steps	Activity	Normal Duration	Minimum Duration	Cost of Crashing
1-2	A	9	6	20
1-3	B	8	5	25
1-4	C	15	10	30
2-4	D	5	3	10
3-4	E	10	6	15
4-5	F	2	1	10

Draw network diagram, determine CPM and find,

(i) *Normal duration of project and cost,*

(ii) *Optimum duration of project and cost,*

(iii) *Minimum duration of project and cost.*

Solution: Network diagram:

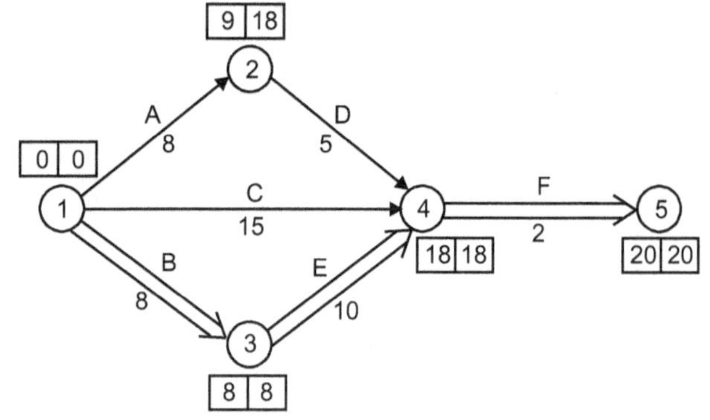

Fig. 8.60

Critical path = B – E – F

Project time duration = 20 days

$$\therefore \text{ Total cost } = 1000 + (20 \times 60)$$
$$= ₹\ 2200.$$

Table 8.43

Activity	ΔT	ΔC/T (Given)
A	3	20
B	3	25
C	5	30
D	2	10
E	4	15
F	1	40

Crashing Rules :

(i) Critical activities are to be crashed first i.e., B, E and F.

(ii) Cost slope i.e., ΔC/ΔT should be less than or equal to indirect cost per day.

(iii) Minimum cost-slope activity is to be crashed first.

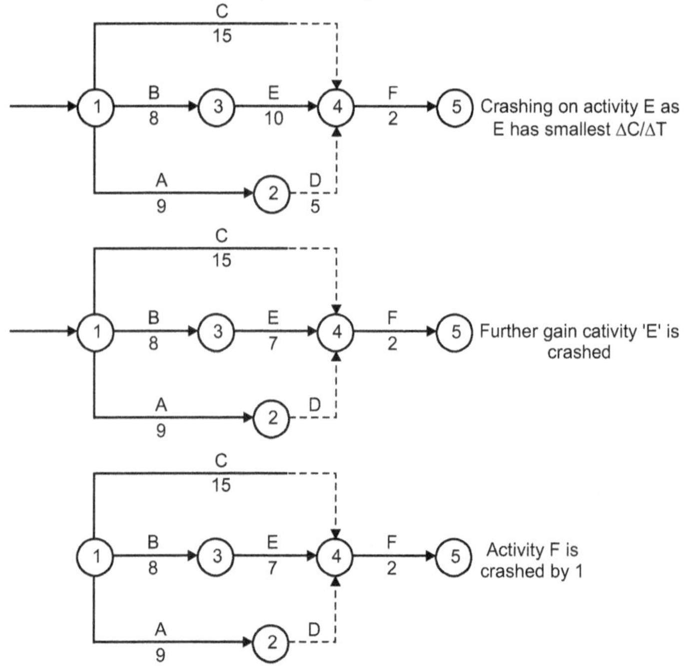

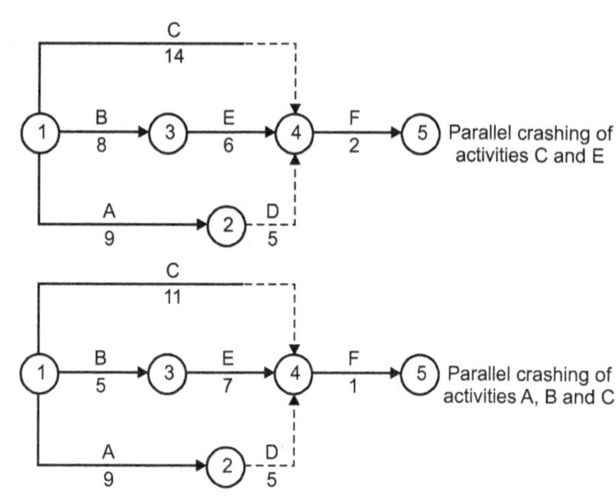

Fig. 8.61

Crashing Calculations:

Table 8.44

Activity	Days Crashed	Cost Slope	Increasing Cost	Saving	Net Saving	New Duration	New Cost
E	3	15	45	180	13	17	2065
F	1	40	40	60	20	16	2045
C	1	(15 + 30)	45	60	15	5	2030
A, B, C	3	(25 + 20 + 30)	225	180	(−45)	12	2075
					Increased cost		

(i) Normal cost = ₹ 200, Normal duration = 20 days.

(ii) Optimum cost = ₹ 2030, Optimum duration = 15 days.

(iii) Maximum cost = ₹ 2075, Minimum duration = 12 days.

Example 8.22: *An electronics firm has signed a contract to install an instrument landing device at the local airport. The complete installation can be broken down into fourteen separate activities. Each activity (labelled A through N), its predecessor activities, normal times and cost, crash time and cost are given below. The contract specifies that the installation will be completed within 18 days. There is a penalty of ₹ 10000 per day beyond the specified complete time.*

Table 8.45

Activity	Predecessor	Normal Time	Normal Cost	Crash Time	Crash Cost
A	–	3	32000	2	36000
B	–	5	55000	4	50000
C	–	6	57500	4	70000
D	A	7	75000	5	85000
E	A	4	42000	3	49000
F	B, D	2	8000	2	18000
G	C	4	42500	4	48500
H	A	8	85000	5	10600
I	C	5	57500	4	53500
J	C	7	67000	5	73500
K	E, F, G	4	40000	3	44600
L	H, I	6	65000	4	75000
M	L	3	28000	2	33500
N	J, K	5	52500	4	57500

(i)   *What is the normal time to complete the installation?*

(ii) *What is the shortest possible time for completion of the installation?*

(iii) *What is the most economical period of time in which to schedule the installation E?*

(iv) *What is the minimum total cost (installation plus penalty)?*

Solution: Arrow dig with normal time is as shown in Fig. 8.62. Then critical path can be obtained.

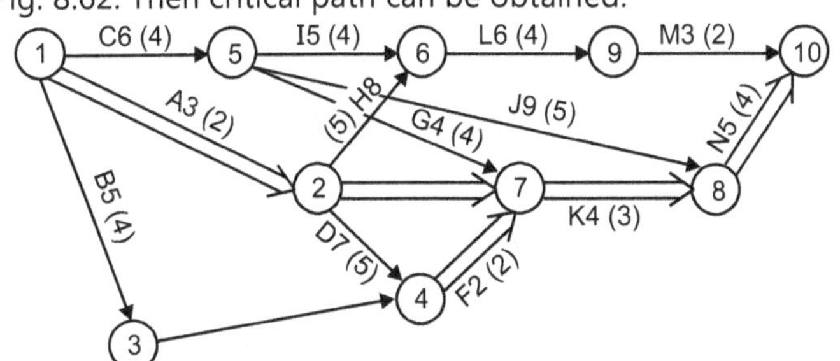

Fig. 8.62

Table 8.46

Sr. No.	Activity Path or Activity	Duration
1.	C, I, L, M	20
2.	C, J, N	18
3.	C, J, K, N	19
4.	A, E, K, N	16
5.	A, D, F, K, N	21

So the critical path is A, D, F, K, N with normal duration of 21 days and minimum project length is 17 days.

The shortest possible time for completion of installation is 16 days.

The schedule take more time as per critical path. So it can be reduced by crashing some of the activities. As activity laying on critical path control the duration of the installation, therefore the duration of some activities laying on critical path is reduced.

Table 8.47

Path	Duration	Duration after crashing			
		(1)	(2)	(3)	(4)
C, I, L, M	20	20	19	18	17
C, J, N	18	18	18	18	17
C, G, K, N	19	19	19	18	17
A, E, K, N	16	16	16	16	16
A, D, F, K, N	21	20	19	18	17

Crashing cost/day is calculated as:

$$(\Delta C) \text{ Crashing cost} = \frac{\text{Crash cost} - \text{Normal cost}}{\text{Normal time} - \text{Crash time}}$$

Table 8.48

Activity	A	B	C	D	E	F	G
C.C.	4000	– 5000	6250	5000	7000	–	–
Activity	H	I	J	K	L	M	N
C.C.	7000	– 4000	3000	4000	5000	5500	5000

Table 8.49

Crashing	Critical Path	Options	Crashing Cost	Deasion	Duration After Crash
1.	A. D, F, K, N	A, D, K, N	As in table	Crash (A)	20
2.	A, D, F, K, N, C, I, L, M	D, K, N C, I, L, M Combine option (R + I)	K + I 4000 + (– 4000) = 0 = 0	Crash K + I	19
3.	A, D, F, K, N C, I, L, M C, G, K, N	D, N C, L, M, C, N Combine D + C N + L N + M	 11250 10000 10500	Crash N + L	18
4.	A, D, F, K, N C, I, L, M C, G, K, N C, J, N	D C, L, M, C C, F Combine	 11250 (C + D)	Crash C + D	17

Table 8.50

Duration	Normal Cost	Crash Cost	Overheads	Total Cost
21	715000	–	10000 × 3	747000
20	715000	4000	10000 × 2	741500
19	715000	4000	10000 × 1	731500
18	715000	14000	0	731500
17	715000	25250	0	742750

From the table we get the total cost of installation of project is 73100 for 18 days. After 18 days if project is crashed the total cost is increased (as in table).

So the most economical period of time in which to schedule installation is within 8 days.

The minimum total cost of installation is 731500.

Example 8.23: *For small project given below. Draw network and calculate expected duration of project. What is completion of probability that project completed within 30 days? What due data has about 90% change of being meet?*

Table 8.51

Activity	T_o	T_m	T_p	Optimal Duration	$\sigma = \left(\dfrac{T_p - T_o}{\sigma}\right)$	$\left(\dfrac{T_p - T_o}{\sigma}\right)^2$
1-2	2	5	14	6	2	4
1-3	9	12	15	12	1	1
2-4	5	14	17	13	2	4
3-4	2	5	8	5	1	1
3-5	6	6	12	7	1	1
4-5	8	17	20	16	2	4

Solution:

Table 8.52

Z	%
0.6	72.6
0.8	78.6
1	84.4
1.5	88.5
1.4	91.9
1.6	94.6

Optimal duration = T_c

$$= \frac{T_o + T_p + 4T_m}{6}$$

Network diagram:

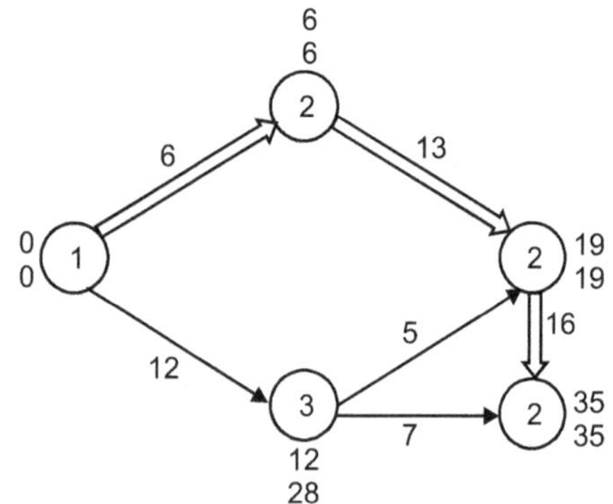

Fig. 8.63

Critical path $= 1 - 2 - 4 - 5$

Contains $1 - 2 \rightarrow 6$

$2 - 4 \rightarrow 13$

$4 - 5 \rightarrow \underline{16}$

35 days

Scheduled duration = 30 days < 35 days

$$z = \frac{T_{sche} - T_{exp}}{\sqrt{\sigma\phi^2}}$$

$$= \frac{30 - 35}{\sqrt{12}} = 1.44 \approx 1.4$$

$\therefore$ $z = 1.4$

$\therefore$ % = 91.9 **[Ref. Table A from Appendix]**

$\therefore$ $1.4 = \dfrac{x - 35}{\sqrt{12}}$

$x = 39.85 \approx 40$ days are required

Example 8.24: *Consider the following project activities, predecessor, optimistic, most likely and pessimistic.*

Table 8.53

Activity	Predecessor	Optimistic	Most likely	Pessimistic
A	–	3	6	9
B	–	2	5	8
C	A	2	4	6
D	B	2	3	10
E	B	1	3	11
F	C, D	4	6	8
G	E	1	5	16

Draw network diagram. Determine critical path calculate variance and standard deviation of project length. What is probability that project will be completed in 16 weeks?

Table 8.54

Z	0	0.5	1	1.5	2
P	0.5	0.692	0.841	0.933	0.972

Solution:

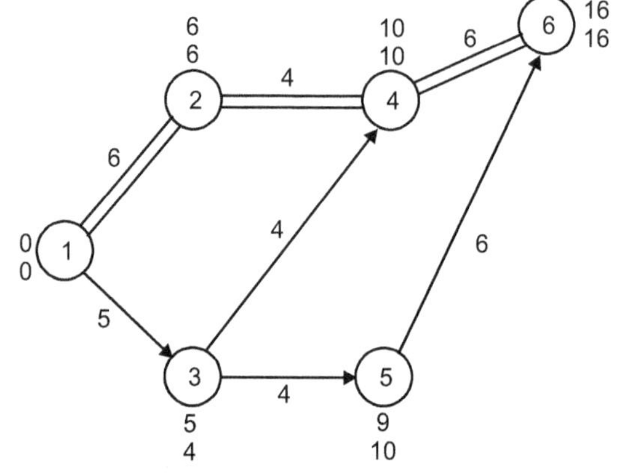

Fig. 8.64

Table 8.55

T_o	T_m	T_p	Averate time $\dfrac{T_o + 4T_m + T_p}{6}$	$\sigma = \dfrac{T_p - T_o}{6}$	σ^2
3	6	9	6	1	1
2	5	8	5	1	1
2	4	6	5	0.66	0.4356
2	3	10	4	1.33	1.7689
1	3	11	4	1.66	2.7556
4	6	8	6	0.66	0.4356
1	5	16	6	2.33	5.4289

Critical path $= 1 - 2 - 4 - 6$

Path	V	σ^2
1 – 2	1	1
2 – 4	0.66	0.4356
4 – 6	0.66	0.4356

$$Z = \frac{16 - 18}{\sqrt{\sigma^2_{cr}}} = \frac{2}{\sqrt{1.871}} = 1.46$$

[Ref. Table A from Appendix]

Probability is 0.933.

Example 8.25: *Consider the following data of project:*

Table 8.56

Activity	Predecessor	T_o	T_m	T_p
A	–	3	5	8
B	–	6	7	9
C	A	4	5	9
D	B	3	5	8
E	B	4	6	9
F	C, D	5	8	11
G	C, D, E	3	6	9
H	F	1	2	9

Construct project network, find the expected duration variance. Find out the critical path and expected completion time. Also find the probability of complete project on or before 30 weeks.

Solution:

Table 8.57

Activity	Predecessor	T_o	T_m	T_p Time	Optimal	σ
A	–	3	5	8	5.167	0.833
B	–	6	7	9	7.167	0.5
C	A	4	5	9	5.5	0.833
D	B	3	5	8	5.167	0.833
E	A	4	6	9	6.167	0.833
F	C, D	5	8	11	8	1
G	C, D, E	3	6	9	6	1
H	F	1	2	9	3	1.33

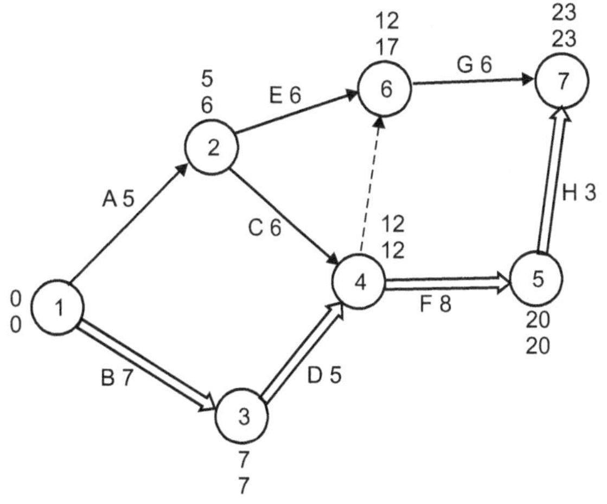

Fig. 8.65

Critical path $= 1 - 3 - 4 - 5 - 7$

Expected completion time $= 23$ weeks

Now standard deviation for network is

or　　　$= \sqrt{0.25 + 0.694 + 1 + 1.77}$

σ for network $= 1.927$

Now probability for completion of work in 30 weeks

$$= \frac{30 - 23}{(1.927)^2} = 1.88$$

∴　From table probability for completion of task in 30 weeks is 97.1%.

If probability of completion $= 90\%$ then the project completion time is

For 90% probability, N.D. $= 1.3$

∴　$\dfrac{T_{exp} - 23}{(1.927)^2} = 1.3$

$$T_{exp} = 23 + 4.8 = 28$$

Example 8.26: *A local company intends to carry out a capital labour. The following table shows for each activity needed, normal time, shortest time, the activity completed and cost per day, for reducing time per day remain same irrespective of number of days induced cost of completing the eight activities N_T is ₹ 174000 excluding side overhead. Overhead cost of general side activity is 4800/day. Calculate normal duration of project cost and critical path, calculate optimal duration and corresponding cost, shortest time and associated cost.*

Table 8.58

Activity	N_T	N_C	Cost of Reduction/ Day	Slope $\dfrac{a}{b-c}$
1 – 2	6	4	2400	1200
1 – 3	8	4	2700	675
1 – 4	5	3	1500	750
2 – 4	3	3	–	–
2 – 5	5	3	1200	600
3 – 6	12	8	6000	1500
4 – 6	8	5	1500	500
5 – 6	8	6	–	–

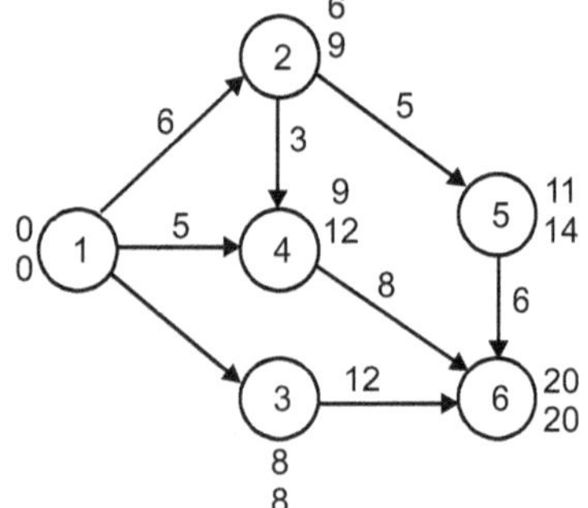

Fig. 8.66

Solution:

Critical path = $1 - 3 - 6$

Total cost = $174000 + (20 \times 4800) = 270000$ units

Slope of $1 - 3$ is less than that of $3 - 6$

Crashing activity $1 - 3$ by 4.

Network:

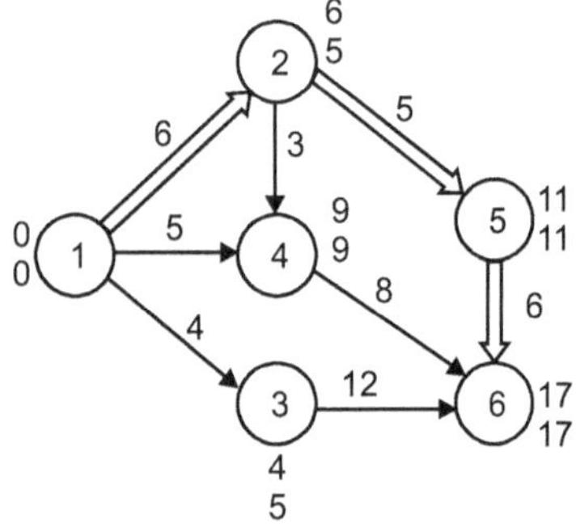

Fig. 8.67

Critical path = $1 - 2 - 5 - 6$

Total cost = $174000 + (17 \times 4800) + 4 (2700)$

= 258500 units

Slope of activity $2 - 5$ is less.

Now, crashing activity $2 - 5$ by 2.

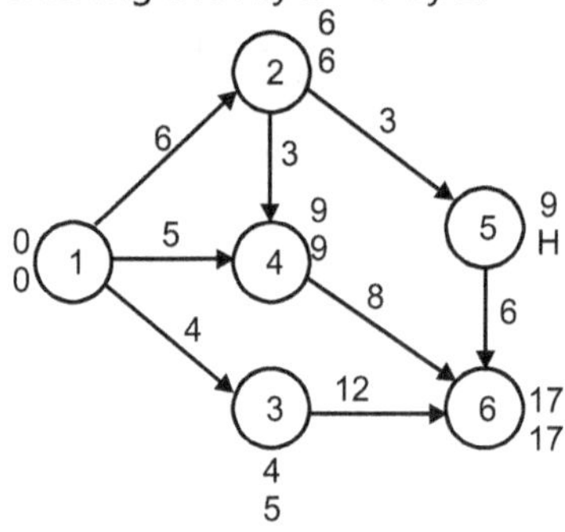

Fig. 8.68

Critical path = $1 - 2 - 4 - 6$

Total cost = $174000 + 17 \times 4800 + 4 \times \dfrac{2700}{4}$

$+ \dfrac{1200}{2} \times 2$

= 259500 units

Crashing activity $4 - 6$ by 3.

Network:

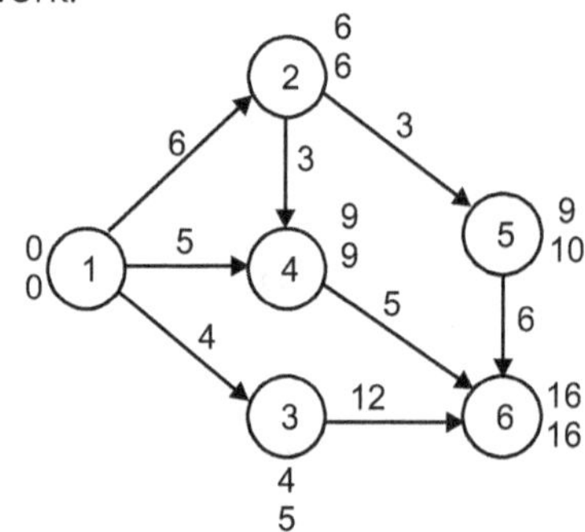

Fig. 8.69

Critical path = $1 - 3 - 6$

Total cost = $174000 + 16 \times 4800 + 4 \times \dfrac{2700}{4}$

$+ \dfrac{1200}{2} \times 2 + \dfrac{1500}{3} \times 3$

= 256200 units

Now, crashing activity $3 - 6$ by 4.

Network:

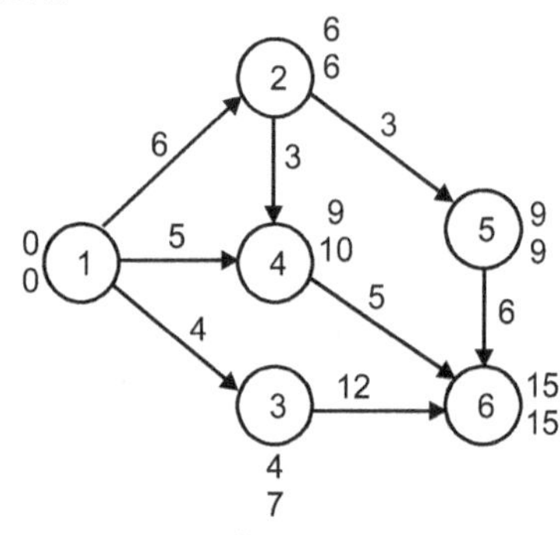

Fig. 8.70

Critical path $= 1 - 2 - 3 - 6$

$$\text{Total cost} = 174000 + 15 \times 4800 + 4 \times \frac{2700}{4}$$

$$+ \frac{1200}{2} \times 2 + \frac{1500}{3} \times 3$$

$$+ \frac{6000}{4} \times 4 = 257400 \text{ units}$$

So, after crashing activity 3 - 6, total cost increases. So crashing of activities is upto crashing activity 4 - 6 by 3. The cost is 256200 units. The shortest time required is 16 units.

Example 8.27: *The following table shows activities, their normal and crash duration and crash slope. Their overhead cost is ₹160.*

Table 8.59

Activities	Normal Duration	Crash Duration	Crash Slope
1 – 2	6	4	80
1 – 3	8	4	90
1 – 4	5	3	30
2 – 4	3	3	–
2 – 5	5	3	40
3 – 6	12	8	200
4 – 6	8	5	50
5 – 6	6	6	–

Calculate normal duration, normal cost of project. Find lowest cost duration company is interested in finding out total cost of project if all activities are crashed.

Solution:

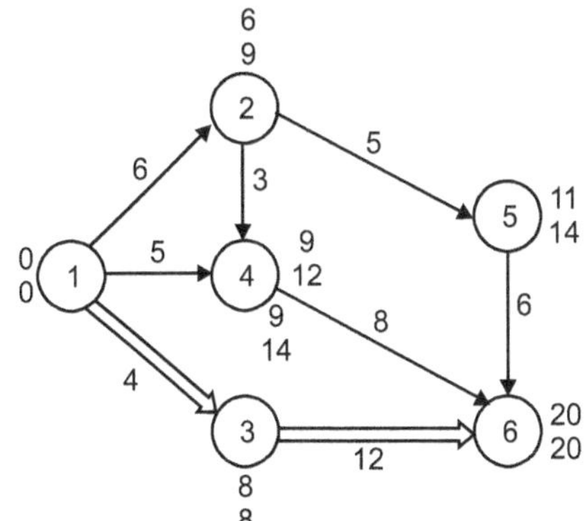

Fig. 8.71

Critical path $= 1 - 3 - 6$

Total cost $= 160 \times 20 = 320$

Activity to be crashed is 1 – 3:

$2 - 4 = 4$ weeks to be crashed

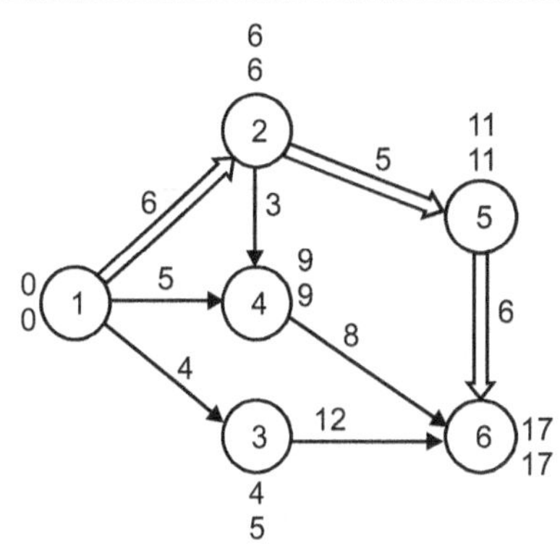

Fig. 8.72

Two critical paths are observed:

(i) $1 - 2 - 5 - 6$

(ii) $1 - 2 - 4 - 6$

Activity to be crashed is $2 - 5$

i.e. $6 - 4 = 2$ weeks

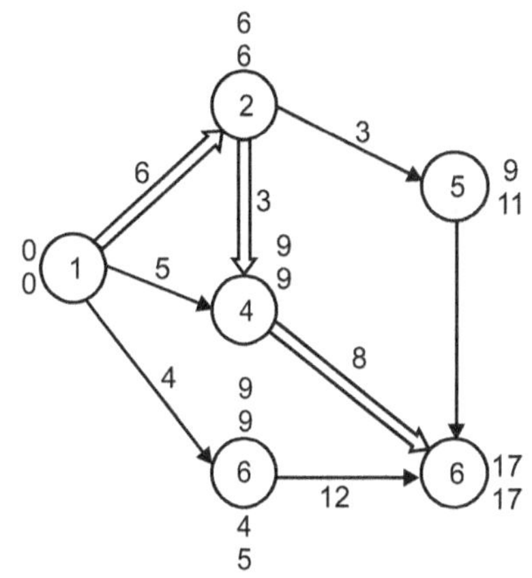

Fig. 8.73

Total cost $= 160 \times 17 + 4 \times 40 = 2880$

Critical path $= 1 - 2 - 4 - 6$

Total cost $= 160 \times 17 + 3 \times 40 + 4 \times 40 = 3000$

Activity to be crashed is $4 - 6$:

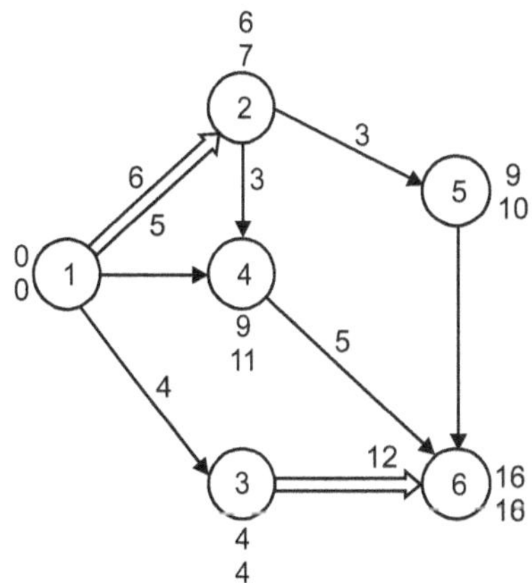

Fig. 8.74

Critical path is $3 - 6$:

$$= 160 \times 16 + 3 \times 40 + 5 \times 50 + 4 \times 40$$

$$= 3090$$

Activity to be crashed is 3 − 6:

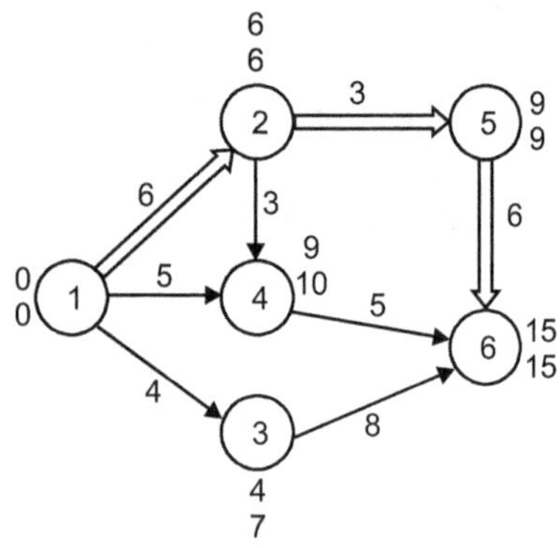

Fig. 8.75

Critical path is 1 − 2 − 5 − 6:

Total cost = $160 \times 15 + 3 \times 40 + 5 \times 50 + 8 \times 200$

$+ 4 \times 40$

= 4530

Example 8.28: *Consider the data of project as shown in the table 8.60.*

Table 8.60

Activity	N_T	C_T	N_C	C_C
1 − 2	8	5	800	950
1 − 3	5	3	500	700
1 − 4	9	6	600	1250
2 − 5	10	8	900	1300
3 − 5	5	3	700	1100
3 − 6	6	5	1200	1500
4 − 6	7	5	1300	1400
5 − 7	2	1	400	500
6 − 7	4	2	500	900

If the fixed cost/week is ₹ 300 find the optimal crash project completion time.

Solution:

Table 8.61

Activity	N_T	C_T	N_C	C_C	ΔT	ΔC	$\Delta C/\Delta T$
1 − 2	8	5	800	950	3	150	50
1 − 3	5	3	500	700	2	200	100
1 − 4	9	6	600	1250	3	650	216.67
2 − 5	10	8	900	1300	2	400	200
3 − 5	5	3	700	1100	2	400	200
3 − 6	6	5	1200	1500	1	300	300
4 − 6	7	5	1300	1400	2	100	50
5 − 7	2	1	400	500	1	100	100
6 − 7	4	2	500	900	2	400	200

Table 8.62

Network No.	Critical Path	Activity Crashed	Crash Days	Slope	T_C
1.	1 − 2 − 5 − 7	−	−	−	12900
	1 − 4 − 6 − 7				
2.	1 − 4 − 6 − 7	1 − 2	3	50	13050
3.	1 − 4 − 6 − 7	4 − 6	2	50	12550
4.	1 − 2 − 5 − 7	6 − 7	2	200	12650
5.	1 − 2 − 5 − 7	5 − 7	1	100	12450
	1 − 4 − 6 − 7				
6.	1 − 4 − 6 − 7	2 − 5	2	200	12850
7.	1 − 2 − 5 − 7	1 − 4	3	216.67	12900

The optimum cost is ₹ 12450 by crashing upto K weeks, then the cost increases.

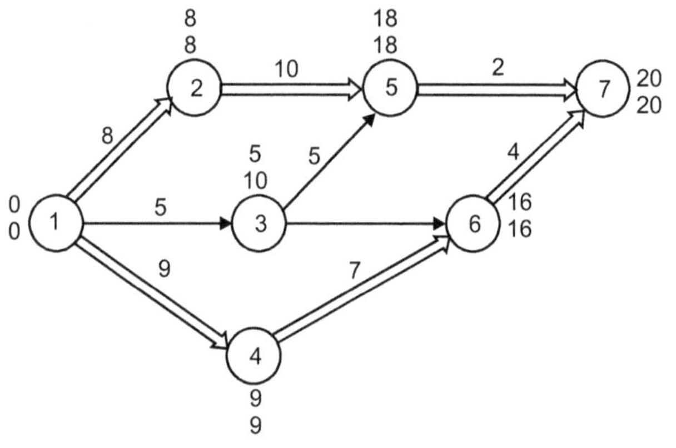

Fig. 8.76

Critical path:

1. $1 − 2 − 5 − 7$

2. $1 − 4 − 6 − 7$

1. $T_C = 5900 + 20 + 300 = 12900$

2. Consider critical path $1 − 2 − 5 − 7$:

As $1 − 2$ has minimum slope (50), we crash activity $1 − 2$ by 3 days.

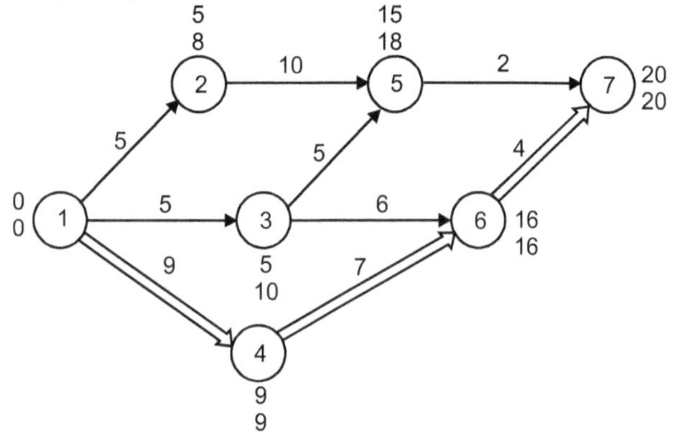

Fig. 8.77

$T_C = 6900 + 20 \times 300 + 3 \times 50 = 13050$

3. As activity 4 – 6 has minimum slope so crash activity by 2 days.

    Critical path: $1 - 4 - 6 - 7$

    $$T_C = 6900 + 18 \times 300 + 3 \times 50 + 2 \times 50$$
    $$= 12550$$

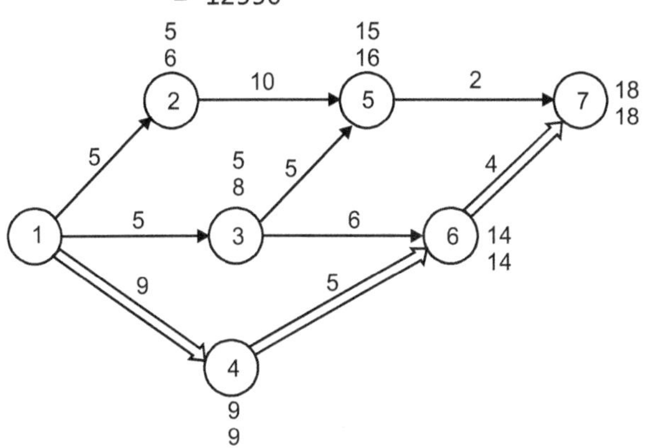

Fig. 8.78

4. As activity 6 – 7 has minimum slope crash it by 2 days.

    Critical path: $1 - 2 - 5 - 7$

    $$T_C = 6900 + 17 \times 300 + 3 \times 50 + 2 \times 50$$
    $$+ 2 \times 200 = 12650$$

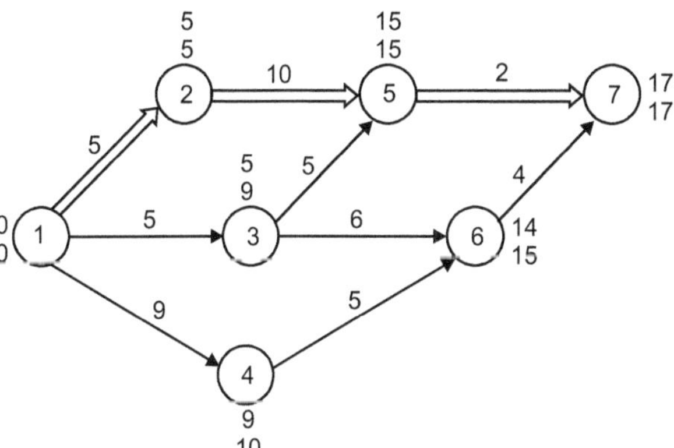

Fig. 8.79

5. As activity 5 – 7 has minimum slope crash it by 1 day.

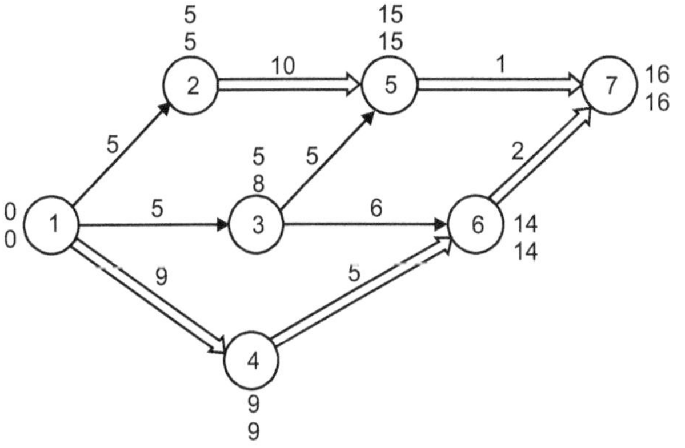

Fig. 8.80

Critical path:

(1) $1 - 2 - 5 - 7$

(2) $1 - 4 - 6 - 7$

   $$T_C = 6900 + 16 \times 300 + 3 \times 50 + 2 \times 50 + 2$$
   $$\times 200 + 1 \times 100$$
   $$= 12450$$

6. Crash activity 2 – 5 which has minimum slope crash activity by 2 days.

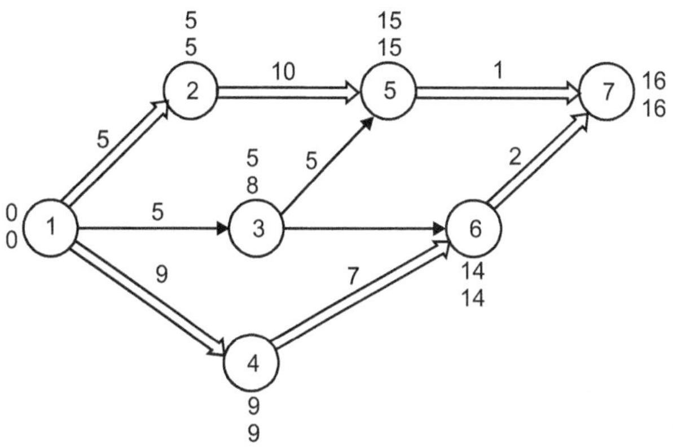

Fig. 8.81

Critical path: $1 - 4 - 6 - 7$

   $$T_C = 6900 + 16 \times 300 + 3 \times 50 + 2 \times 50 + 2$$
   $$\times 200 + 100 + 2 \times 200$$
   $$= 12850$$

7. Now crash activity 1 – 4 by 3 days.

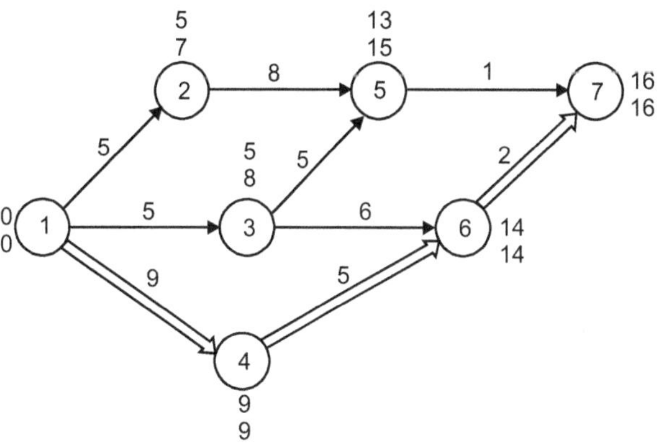

Fig. 8.82

Critical path: $1 - 2 - 5 - 7$

All the activities are crashed.

   $$T_C = 6900 + 14 \times 300 + 3 \times 50 + 2 \times 50$$
   $$+ 2 \times 200 + 100 + 2 \times 200 + 3$$
   $$\times 216.67 = 12900$$

Example 8.29: *The interact costs are given below:*

Table 8.63

Days	15	14	13	12	11	10	9	8
Indirect cost	600	100	400	250	175	100	75	50

Crash the activity to minimum total cost.

Solution:

Table 8.64

Activity	Predecessor	N_T	C_T	N_C	C_C
A	–	4	3	60	90
B	–	6	4	150	250
C	–	2	1	38	60
D	A	5	3	150	250
E	C	2	2	100	100
F	A	7	5	115	175
G	D, B, E	4	2	100	240
				Σ 713	

Table 8.65

Activity	N_T	C_T	N_C	C_C	ΔT	ΔC	$\Delta C/\Delta T$
A (1 – 2)	4	3	60	90	1	30	30
B (1 – 3)	6	4	150	250	2	100	50
C (1 – 4)	2	1	38	60	1	22	22
D (2 – 3)	5	3	150	250	2	100	100
E (4 – 3)	2	2	100	100	0	0	0
F (2 – 5)	7	5	115	175	2	60	30
G (3 – 5)	4	2	100	240	2	140	70

Table 8.66

Network No.	Critical path	Activity crashed	Crash days	Slope	T_C (₹)
1.	1 – 2 – 3 – 5 1 – 2 – 5 1 – 3 – 5	–	–	–	1113
2.	1 – 2 – 3 – 5 1 – 2 – 5 1 – 3 – 5	1 – 2	1	30	993
3.	1 – 2 – 3 – 5 1 – 2 – 5 1 – 3 – 5	2 – 5	2	30	1053
4.	1 – 2 – 3 – 5 1 – 3 – 5	1 – 3	2	50	1153
5.	1 – 2 – 3 – 5 1 – 3 – 5	3 – 5	2	70	1143
6.	1 – 2 – 3 – 5	2 – 3	2	100	1293

Crashing the activity upto 12 weeks the optimum cost is ₹993, then the cost increases.

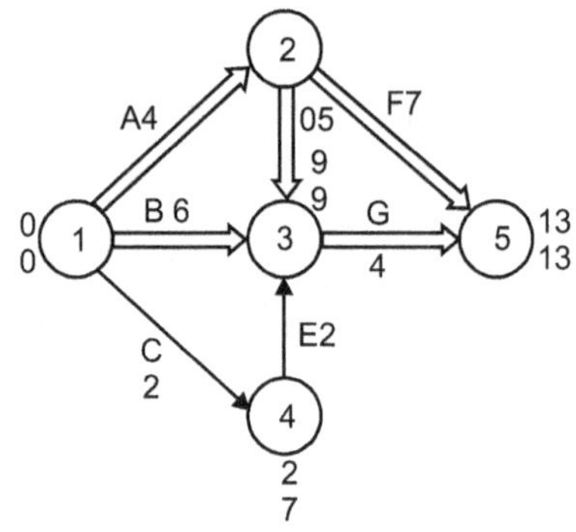

Fig. 8.83

Critical path:

1.   1 – 2 – 3 – 5
2.   1 – 2 – 5
3.   1 – 3 – 5

$$T_C = 713 + 400 = 1113$$

2.   Select activity 1 – 2 which should be crashed as it has minimum slope 30 by 1 day.

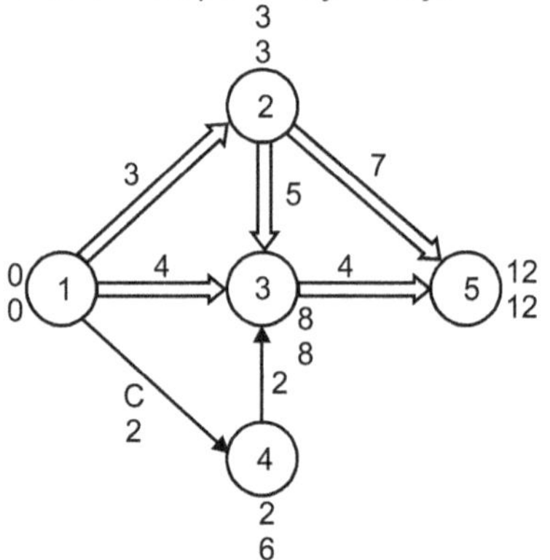

Fig. 8.84

3.   Crash activity 2 – 5 which has minimum slope = 30 by 2 days.

Critical path:

(i)   1 – 2 – 3 – 5
(ii) 1 – 2 – 5 – all crashed
(iii) 1 – 3 – 5

$$T_C = 317 + 250 + 30 + 60 = 1053$$

Fig. 8.85

4. Crash activity 1 – 3 having slope 50 and crash by 2 days

Critical path:

(i) $1 - 2 - 3 - 5$

(ii) $1 - 3 - 5$

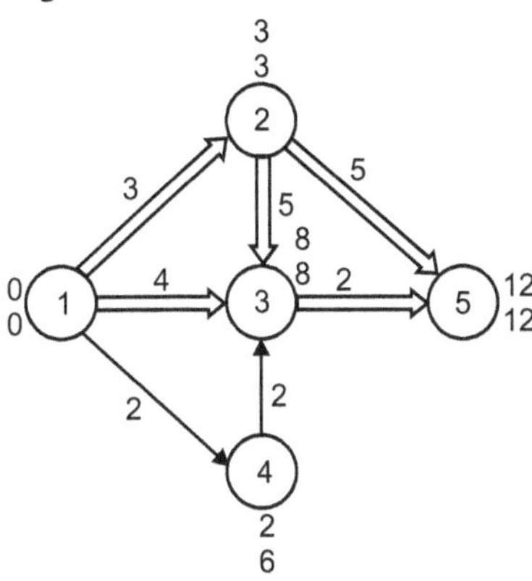

Fig. 8.86

$T_C = 713 + 250 + 30 + 60 + 100 = 1153$

5. Crash activity 3 – 5 having slope 70 by 2 days.

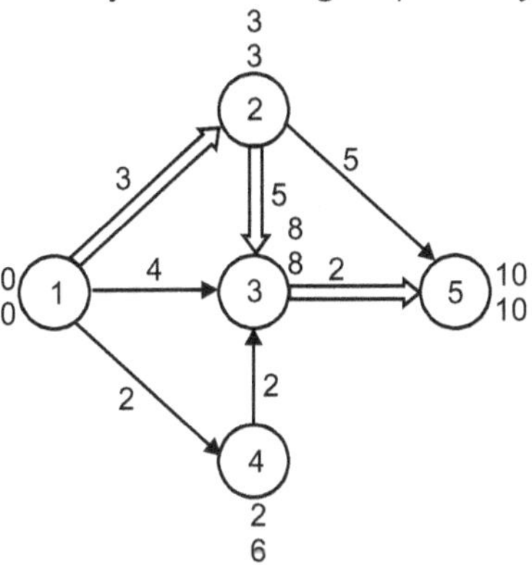

Fig. 8.87

Critical activity: $1 - 2 - 3 - 5$

$$T_C = 713 + 100 + 30 + 60 + 100 + 140$$
$$= 1143$$

6. Crash activity 2 – 3 having slope 100 by 2 days. All activities are previously crashed.

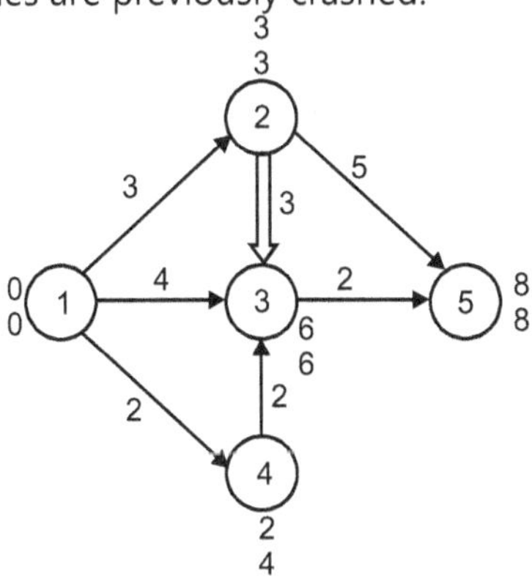

Fig. 8.88

$$T_C = 713 + 50 + 30 + 60 + 100 + 140 + 200$$
$$= 1293$$

Example 8.30: *The following Table 8.67 shows activity of project along N_T, C_T, N_C, C_C, indirect cost ₹60.*

Table 8.67

Activity	N_T	C_T	C_C	N_C
1 – 2	20	17	720	600
1 – 3	25	25	300	300
2 – 3	10	8	440	300
2 – 4	12	6	700	100
3 – 4	5	2	350	200
4 – 5	10	5	650	350

Draw network and identify C.P. Calculate N_T and corresponding cost. Crash the activities and determine optimal duration of project and corresponding cost.

Solution:

Critical path $= 1 - 2 - 3 - 4 - 5$

$$T_C = 1850 + 45 \times 60$$
$$= ₹ 4550$$

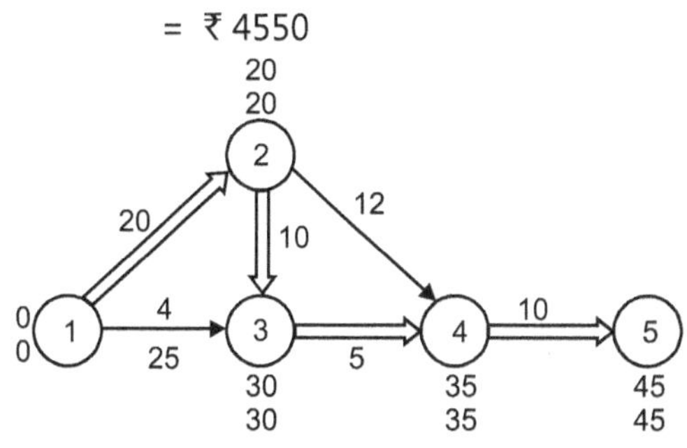

Fig. 8.89

Table 8.68

Activity	N_T	N_C	C_T	C_C	ΔC $(C_C - N_C)$	ΔT $(N_T - C_T)$	$\dfrac{\Delta C}{\Delta T}$
1 – 2	20	600	17	720	120	3	40
1 – 3	25	300	25	300	0	0	0
2 – 3	10	300	8	440	140	2	70
2 – 4	12	100	6	700	600	6	100
3 – 4	15	200	2	350	150	3	50
4 – 5	10	350	5	650	300	5	60
			$\Sigma = 1850$				

Activity 1 – 2 has less cost slope. Hence, crash activity 1 – 2 by 3 days.

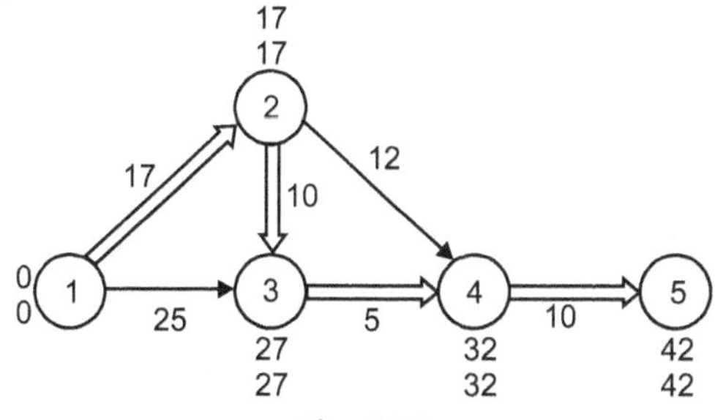

Fig. 8.90

Critical path = $1 - 2 - 3 - 4 - 5$

Direct cost = $1850 + 3(40) = ₹ 1970$

Indirect cost = $60 (42) = ₹ 2520$

∴ Total cost = $1970 + 2520 = ₹ 4490$

Now activity 3 – 4 has less cost slope.

Hence, crash 3 – 4 by 3 days.

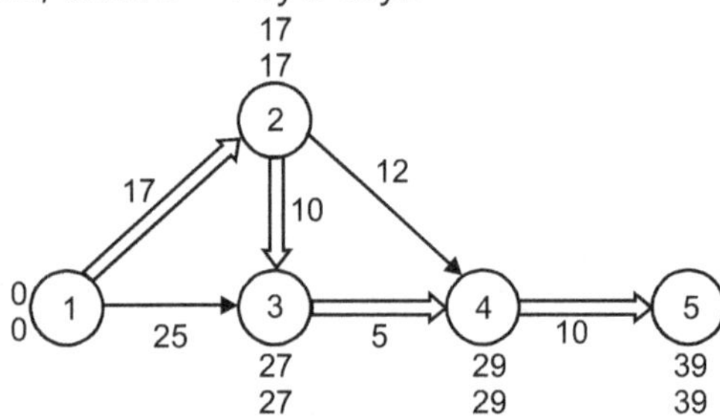

Fig. 8.91

Critical path = $1 - 2 - 3 - 4 - 5$

Direct cost = $1970 + 3(50)$

= ₹ 2120

Indirect cost = $39 (60) = ₹ 2340$

Total cost = ₹ 4460

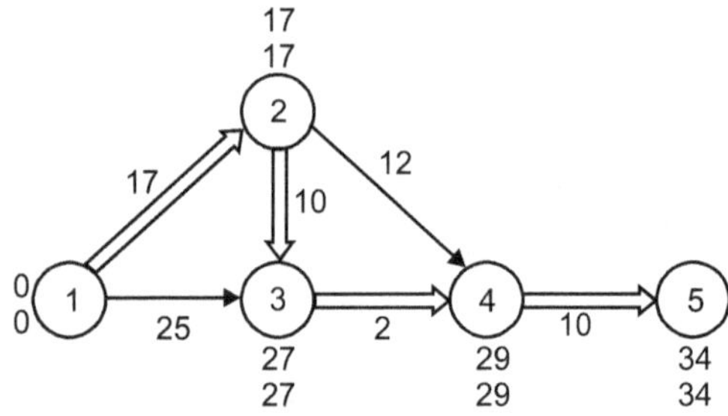

Fig. 8.92

Now, crash activity 4 – 5 by 5 days.

$C_p = 1 - 2 - 3 - 4 - 5$

Direct cost = $2120 + 5 (60)$

Indirect cost = $60 (34) = ₹ 2040$

Total cost = ₹ 4460

Activity 2 – 3 has left crash 2 – 3 by 2 days.

Critical path = $1 - 2 - 4 - 5$

Direct cost = $1970 + 2 (70) = ₹ 2110$

Indirect cost = $60 (34) = ₹ 2040$

∴ Total cost = ₹ 4150

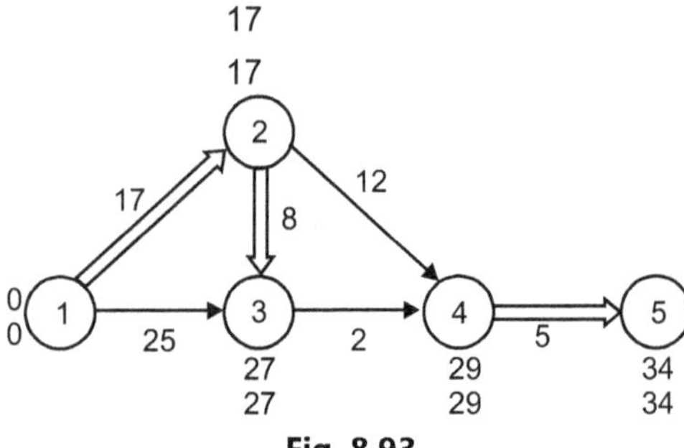

Fig. 8.93

Crash activity 2 – 4 by 6 days.

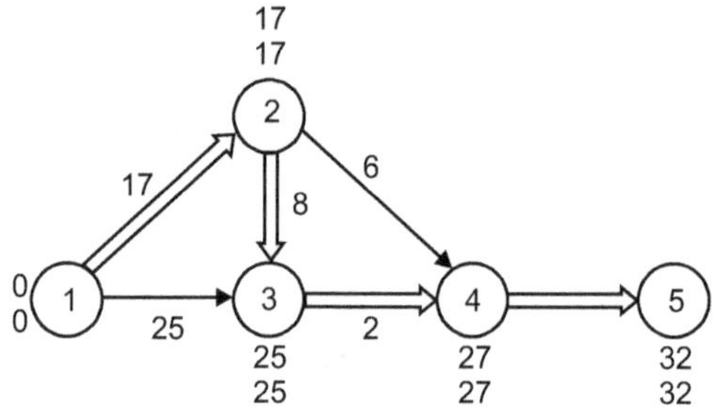

Fig. 8.94

$C_p = 1 - 2 - 3 - 4 - 5$

Direct cost = $2110 + 6 (100) = ₹ 2710$

Indirect cost = $60 (32) = ₹ 1920$

∴ Total cost = ₹ 4630

Example 8.31: *The project of laying of optical fibre cable in a particular town has the following activities. Draw the network and find critical path. Estimate total project cost if total direct cost of project is ₹ 90,000 and indirect cost of ₹ 8000/week. The deviation of activity 1 – 2, 1 – 4, 2 – 3, 2 – 4, 3 – 4, 4 – 5 can be reduced by 2, 1, 3, 8, 2 weeks respectively at an direct additional cost of ₹ 11000, 7000, 12300, 8900 respectively. Crash network step by step to find optimal solution. What will be minimum project cost at optimum project duration?*

Normal deviation : 4, 11, 3, 9, 5, 5.

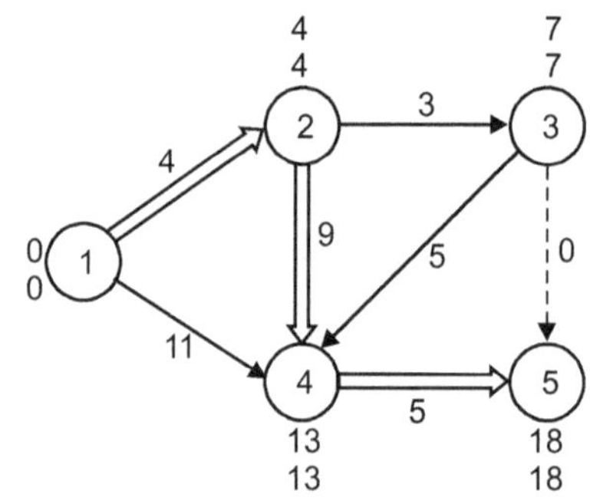

Fig. 8.95

Table 8.69

Activity	Duration	Crashing Weeks	Cost
1 – 2	4	2	11000
1 – 4	11	–	–
2 – 3	3	1	7000
2 – 4	9	3	12300
3 – 4	5	–	–
4 – 5	5	2	8400

Solution:

1. Crashing activity 1 – 2 by (4 – 2) = 2 weeks.

    Total cost = 90000 + 8000 × 18 = 234000

    Hence, crashing activity 1 – 2 and redrawing network diagram.

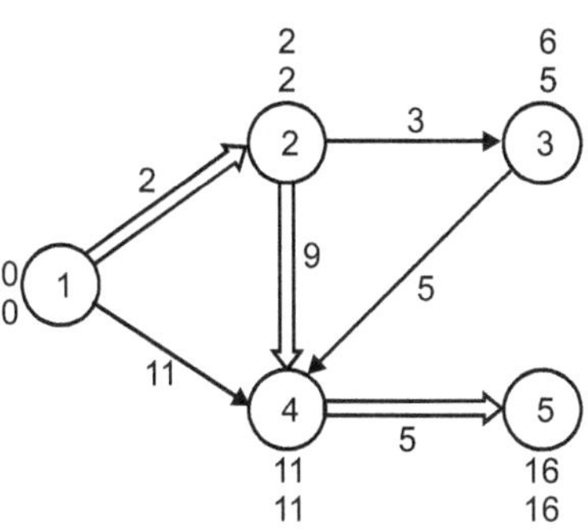

Fig. 8.96

2. Critical path: 1 – 2 – 4 – 5

    Crashing activity 2 – 4 by (9 – 3) = 6 weeks

                        Total cost = 9000 + 3000 × 16

                                    = 229000

3. Now crashing 2 – 4 by 6 weeks and redrawing network.

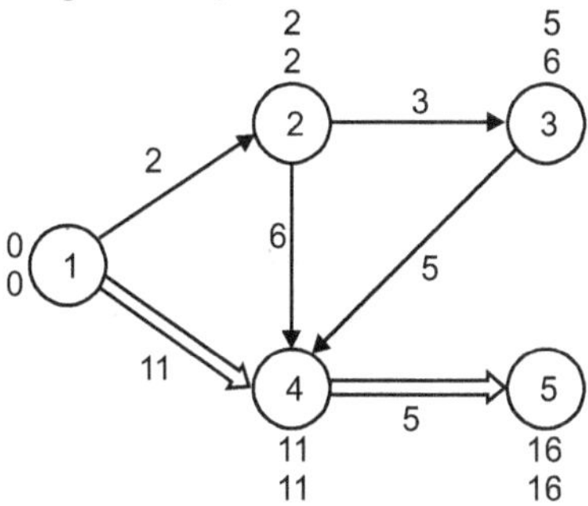

Fig. 8.97

Critical path: 1 – 4 – 5

Crashing activity 4 – 5 by 3 weeks

    Total cost = 90000 + 8000 × 16 + 123000

                = 2303000

4. No crashing 4 – 5 by 3 weeks and redrawing network

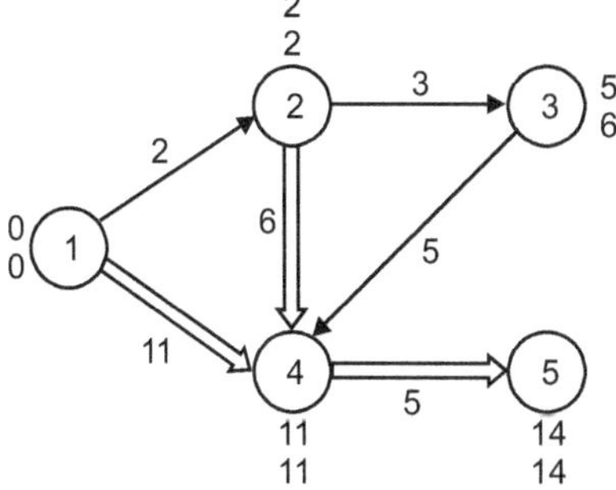

Fig. 8.98

Critical path: 1 – 4 – 5

As 2 – 3 does not lie on critical path then crash by days (3 – 3) = 0 weeks

Crashing activity 2 – 3 by (3 – 3) = 0 week

    Total cost = 90000 + 8000 × 14 + 8400

                = 2,10,400

5. Now crashing 2 – 3 by 0 week and redrawing network diagram.

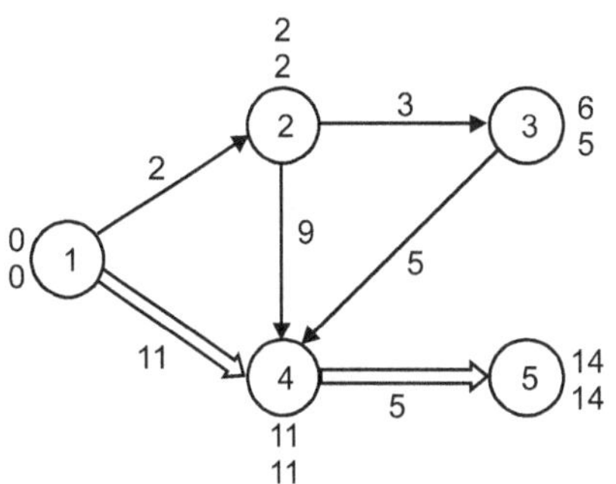

Fig. 8.99

Critical path: 1 – 4 – 5

    Total cost = 90000 + 80000 × 14 + 700 = 209000

Table 8.70

Sr. No.	Critical Path	Crash Activity	Duration	Total
1.	1 – 2 – 4 – 5	1 – 2	2	2,34,000
2.	1 – 2 – 4 – 5	1 – 2	6	2,29,000
3.	1 – 4 – 5	2 – 4	3	2,30,300
4.	1 – 4 – 5	4 – 5	3	2,10,400
5.	1 – 4 – 5	2 – 3	0	2,09,000

Example 8.32: *Project of laying the optical fibre cables in a particular town has the following details.*

Table 8.71

Activity	Duration in week
1 – 2	4
1 – 4	11
2 – 3	3
2 – 4	9
3 – 4	5
4 – 5	5

Solution: Draw network diagram, find critical path, estimate the total project cost if total direct cost of project is ₹ 80000 and indirect cost is ₹ 5000/ week. Duration of activities 1-2, 2-3, 2-4 and 4-5 can be reduced by 2 weeks, 1 week, 3 weeks and 2 weeks respectively at an additional direct cost of ₹ 10000, ₹ 6000, ₹ 11000 and ₹ 8000 respectively.

Table 8.72

Network No.	Critical Path	Activity Weeks	Crash Weeks	Additional Direct Cost	T_C
1.	1 – 2 – 4 – 5	–	–	–	1,70,000
2.	1 – 2 – 4 – 5 1 – 4 – 5	1 – 2	2	10000	1,70,000
3.	1 – 4 – 5	2 – 4	3	11000	1,71,000
4.	1 – 4 – 5	4 – 5	2	8000	1,58,000

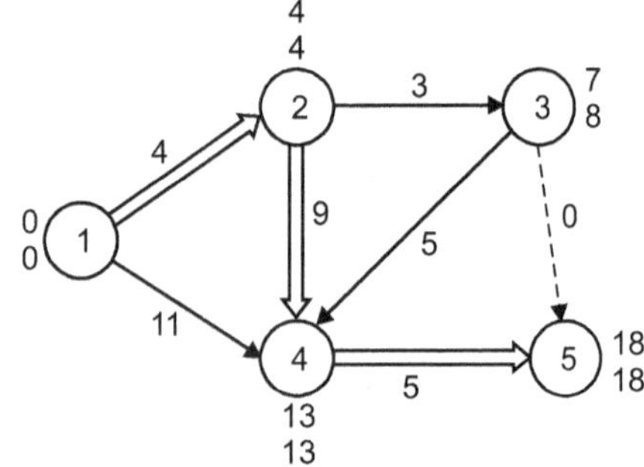

Fig. 8.100

Critical path: 1 – 2 – 4 – 5

$$T_C = 80000 + 5000 \times 18 = 1700000$$

2. Crash the activity 1 – 2 by 2 weeks:

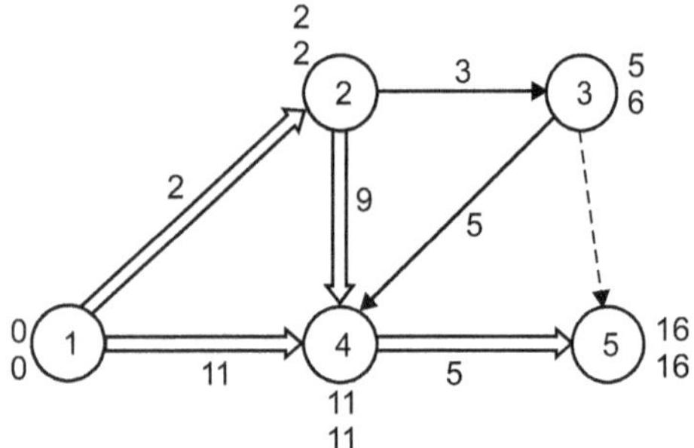

Fig. 8.101

Critical path: 1 – 2 – 4 – 5

$$T_C = 80000 + 5000 \times 16 + 10000 = 170000$$

3. Crash the activity 2 – 4 by 3 weeks:

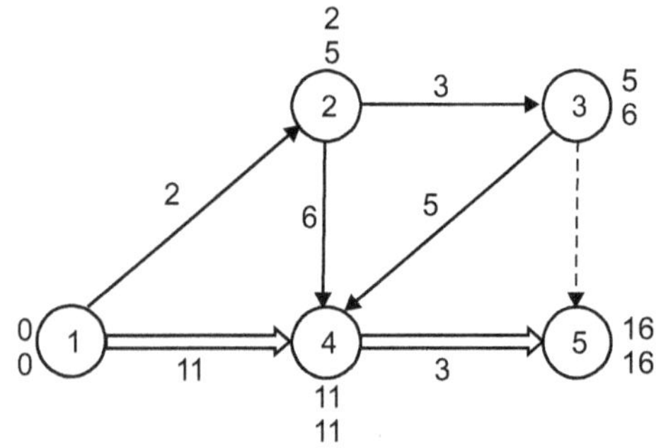

Fig. 8.102

Critical path: 1 – 4 – 5

$$T_C = 80000 + 5000 \times 16 + 110000$$
$$= 171000$$

4. Crash the activity 4 – 5 by 2 weeks:

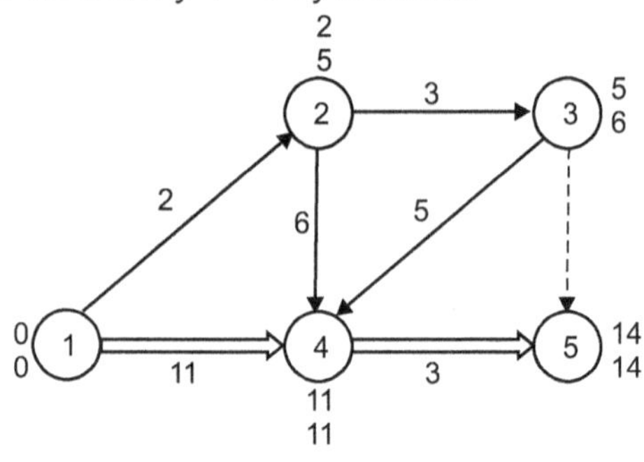

Fig. 8.103

$$T_C = 80000 + 5000 \times 14 + 8000 = 158000$$

Crash the activity upto 14 weeks the optimum cost is ₹ 158000.

Example 8.33: *A small project consists of 13 activities. Their precedence relationships and duration in days is given in table 8.73.*

Table 8.73

Activity	Predecessor	Duration (Days)
A	–	6
B	A	4
C	B	7
D	A	2
E	D	4
F	E	10
G	–	2
H	G	10
I	J, H	6
J	–	13
K	A	9
L	C, K	3
M	I, L	5

(i)   Construct the project network.

(ii) Find the critical path.

(iii) Find total completion time of the project.

[Dec. 2010]

Solution:

Network diagram for the given project is

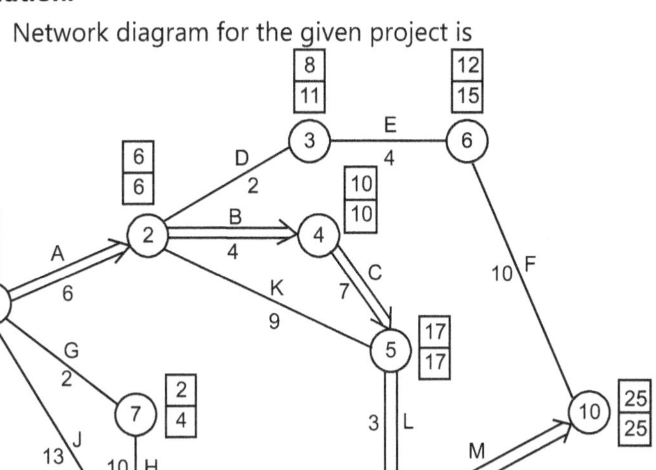

Fig. 8.104

Critical path: A – B – C – L – m

Total completion time of project = 25.

QUESTIONS FOR PRACTICE

Q.1. A list of activities, precedence relations and activity completion times is given in the following table.

Activity	Predecessor Activities	Time (Days)
A	–	5
B	a	4
C	b	2
D	a, c	6
E	d	8
F	e	5
G	c	4
H	d, e, g, i	13
I	C	2
J	g, h	1
K	f, h, j	6

(i)   Draw an activity on node network diagram.

(ii) What is the critical path and expected completion time for the project ?

Q.2. A project schedule has the following characteristics :

Table 8.74

Activity	Time (Weeks)	Activity	Time (Weeks)
1 – 2	4	5 – 6	4
1 – 3	1	5 – 7	8
2 – 4	1	6 – 8	1
3 – 4	1	7 – 8	2
3 – 5	6	8 – 10	5
4 – 9	5	9 – 10	7

(i)   Construct the network.

(ii) Find the critical path.

Q.3. The utility data for a project are given. Draw a network and find the critical path.

Table 8.75

Activity	0-1	1-2	1-3	2-4	2-5	3-4	3-6	4-7	5-7	6-7
Duration	2	8	10	6	3	3	7	5	2	8

Q.4. A project has the following data of offer.

Table 8.76

Head Activity	Tail Activity	Estimated Duration (Months)		
		Smaller	Middle	Highest
A	–	1	1	7
B	–	1	4	7
C	–	2	2	8
D	A	1	1	1
E	B	2	5	14
F	C	2	5	8
G	D, E	3	6	15
H	F, G	1	2	3

(i)   *Draw network and determine project duration.*

(ii) *Calculate floats .*

(iii) *What duration of project will have 95% confidence on project completion?*

Q.5. A project has the following activities, precedence relationships, and activity durations:

Table 8.77

Activity	Immediate Predecessors	Activity Duration (Weeks)
A	-	3
B	-	4
C	-	3
D	C	12
E	B	5
F	A	7
G	E, F	3

(a)   Draw a Gantt chart for the project.

(b)   Construct a CPM network for the project.

(c)   Identify those activities comprising the critical path.

(d)   What is the project's estimated duration?

(e)   Construct a table showing for each activity, its activity duration, earliest start time, latest start time, earliest finish time, latest finish time, and the activity slack.

Answers:

(c)   C, D , (d) 15 weeks

Q.6. A project designed to refurbish a hospital operating theatre consists of the following activities, with estimated times and precedence relationships shown. Using this information draw a network diagram, determine the expected time and variance for each activity, and estimate the probability of completing the project within sixty days.

Table 8.78

Activity	Immediate Predecessors	Optimistic Time	Most Likely Time	Optimistic Time
A	-	5	6	7
B	-	10	13	28
C	A	1	2	15
D	B	8	9	16
E	B, C	25	36	41
F	D	6	9	18

Q.7. An activity has these time estimates: optimistic time o = 15 weeks, most likely time m = 20 weeks, and pessimistic time p = 22 weeks.

(a) Calculate the activity's expected time or duration t.

(b) Calculate the activity's variance v.

(c) Calculate the activity's standard deviation.

Q.8. A project has the following activities, precedence relationships, and time estimates in weeks:

Table 8.79

Activity	Immediate Predecessors	Optimistic Time	Most Likely Time	Optimistic Time
A	-	15	20	25
B	-	8	10	12
C	A	25	30	40
D	B	15	15	15
E	B	22	25	27
F	E	15	20	22
G	D	20	20	22

(a) Calculate the expected time or duration and the variance for each activity.

(b) Construct the network diagram.

(c) Tabulate the values of ES,EF,LS,LF and slack for each activity.

(d) Identify the critical path, and the project duration.

(e) What is the probability that the project will take longer than 57 weeks to complete?

Q.9. The project detailed below has the both normal costs and "crash" costs shown. The crash time is the shortest possible activity time given that extra resources are allocated to that activity.

Table 8.80

Activity	Immediate Predecessors	Normal Time	Normal Time Cost (£)	Crash Time	Crash Time Cost (£)
A	-	5	2 000	4	6 000
B	A	8	3 000	6	6 000
C	B	2	1 000	2	1 000
D	B	3	4 000	2	6 000
E	C	9	5 000	6	8 000
F	C, D	7	4 500	5	6 000
G	E, F	4	2 000	2	5 000

Assuming that the cost per day for shortening each activity is the difference between crash costs and normal costs, divided by the time saved, determine by how much each activity should be shortened so as to complete the project within twenty-six days and at the minimum extra cost.

Table A$_1$

Proportion of total area under the normal curve from $-\infty$ to z.

Where z = normal variate

Z	p(z)	Z	p(z)	Z	p(z)	Z	p(z)
0.00	0.5000	0.65	0.7422	1.30	0.9032	1.95	0.9744
0.01	0.5040	0.66	0.7454	1.31	0.9049	1.96	0.9750
0.02	0.5080	0.67	0.7486	1.32	0.9066	1.97	0.9756
0.03	0.5120	0.68	0.7517	1.33	0.9082	1.98	0.9761
0.04	0.5160	0.69	0.7549	1.34	0.9099	1.99	0.9767
0.05	0.5199	0.70	0.7580	1.35	0.9115	2.00	0.9772
0.06	0.5239	0.71	0.7611	1.36	0.9131	2.02	0.9783
0.07	0.5279	0.72	0.7642	1.37	0.9147	2.04	0.9793
0.08	0.5319	0.73	0.7673	1.38	0.9162	2.06	0.9803
0.09	0.5359	0.74	0.7703	1.39	0.9177	2.08	0.9812
0.10	0.5398	0.75	0.7734	1.40	0.9192	2.10	0.9821
0.11	0.5438	0.76	0.7764	1.41	0.9207	2.12	0.9830
0.12	0.5478	0.77	0.7794	1.42	0.9222	2.14	0.9838
0.13	0.5517	0.78	0.7823	1.43	0.9236	2.16	0.9846
0.14	0.5557	0.79	0.7852	1.44	0.9251	2.18	0.9854
0.15	0.5596	0.80	0.7881	1.45	0.9265	2.20	0.9861
0.16	0.5636	0.81	0.7910	1.46	0.9279	2.22	0.9868
0.17	0.5675	0.82	0.7939	1.47	0.9292	2.24	0.9875
0.18	0.5714	0.83	0.7967	1.48	0.9306	2.26	0.9881
0.19	0.5753	0.84	0.7995	1.49	0.9319	2.28	0.9887
0.20	0.5793	0.85	0.8023	1.50	0.9332	2.30	0.9893
0.21	0.5832	0.86	0.8051	1.51	0.9345	2.32	0.9898
0.22	0.5871	0.87	0.8078	1.52	0.9357	2.34	0.9904
0.23	0.5910	0.88	0.8106	1.53	0.9370	2.36	0.9909
0.24	0.5948	0.89	0.8133	1.54	0.9382	2.38	0.9913
0.25	0.5987	0.90	0.8159	1.55	0.9394	2.40	0.9918
0.26	0.6026	0.91	0.8186	1.56	0.9406	2.42	0.9922
0.27	0.6064	0.92	0.8212	1.57	0.9418	2.44	0.9927
0.28	0.6103	0.93	0.8238	1.58	0.9429	2.46	0.9931

0.29	0.6141	0.94	0.8264	1.59	0.9441	2.48	0.9934
0.30	0.6179	0.95	0.8289	1.60	0.9452	2.50	0.9938
0.31	0.6217	0.96	0.8315	1.61	0.9463	2.52	0.9941
0.32	0.6255	0.97	0.8340	1.62	0.9474	2.54	0.9945
0.33	0.6293	0.98	0.8365	1.63	0.9484	2.56	0.9948
0.34	0.6331	0.99	0.8389	1.64	0.9495	2.58	0.9951
0.35	0.6368	1.00	0.8413	1.65	0.9505	2.60	0.9953
0.36	0.6406	1.01	0.8438	1.66	0.9515	2.62	0.9956
0.37	0.6443	1.02	0.8461	1.67	0.9525	2.64	0.9959
0.38	0.6480	1.03	0.8485	1.68	0.9535	2.66	0.9961
0.39	0.6517	1.04	0.8508	1.69	0.9545	2.68	0.9963
0.40	0.6554	1.05	0.8531	1.70	0.9554	2.70	0.9965
0.41	0.6591	1.06	0.8554	1.71	0.9564	2.72	0.9967
0.42	0.6628	1.07	0.8577	1.72	0.9573	2.74	0.9969
0.43	0.6664	1.08	0.8599	1.73	0.9582	2.76	0.9971
0.44	0.6700	1.09	0.8621	1.74	0.9591	2.78	0.9973
0.45	0.6736	1.10	0.8643	1.75	0.9599	2.80	0.9974
0.46	0.6772	1.11	0.8665	1.76	0.9608	2.82	0.9976
0.47	0.6808	1.12	0.8686	1.77	0.9616	2.84	0.9977
0.48	0.6844	1.13	0.8708	1.78	0.9625	2.86	0.9979
0.49	0.6879	1.14	0.8729	1.79	0.9633	2.88	0.9980
0.50	0.6915	1.15	0.8749	1.80	0.9641	2.90	0.9981
0.51	0.6950	1.16	0.8770	1.81	0.9649	2.92	0.9982
0.52	0.6985	1.17	0.8790	1.82	0.9656	2.94	0.9984
0.53	0.7019	1.18	0.8810	1.83	0.9664	2.96	0.9985
0.54	0.7054	1.19	0.8830	1.84	0.9671	2.98	0.9986
0.55	0.7088	1.20	0.8849	1.85	0.9678	3.00	0.99865
0.56	0.7123	1.21	0.8869	1.86	0.9686	3.20	0.99931
0.57	0.7157	1.22	0.8888	1.87	0.9693	3.40	0.99966
0.58	0.7190	1.23	0.8907	1.88	0.9699	3.60	0.999841
0.59	0.7224	1.24	0.8924	1.89	0.9706	3.80	0.999928
0.60	0.7257	1.25	0.8944	1.90	0.9713	4.00	0.999968
0.61	0.7291	1.26	0.8962	1.91	0.9719	4.50	0.999997
0.62	0.7324	1.27	0.8980	1.92	0.9726	5.00	0.999997
0.63	0.7357	1.28	0.8997	1.93	0.9732		
0.64	0.7389	1.29	0.9015	1.94	0.9738		

Model Question Paper for
End - Semester Examination (60 Marks)

Marks : 60　　　　　　　　　　　　　　　　　　　　　　　　　　　　　**Time : 3 Hour**

Instruction to the candidate:

(i)　Each questions carries 12 marks.

(ii)　Attempt any five questions to the following.

(iii)　Illustrate your answers with neat sketches, diagram etc., wherever necessary.

(iv)　If some part or parameter is noticed to be missing, you may appropriately assume it and should mention it clearly.

1. **(a)** Explain in brief how operation research has been evolved. **[06]**

　　(b) Solved L.P.P. by suitable method. **[06]**

$$\text{Maximize:} \qquad Z = -x_1 + 2x_2$$
$$\text{Subject to:} \quad -x_1 + 3x_2 \leq 10$$
$$x_1 + x_2 \leq 6$$
$$x_1 - x_2 \leq 2$$
$$\text{where,} \qquad x_1, x_2 \geq 0$$

2. **(a)** Differentiate between Transportation Problem and Assignment. **[06]**

　　(b) Solve the following Transportation Problem involving three sources and four destinations. The cell entries represent the cost of transportation per unit. Obtain solution by VAM method. **[06]**

		Destination 1	2	3	4	Supply ↓
	1	3	1	7	4	300
Source	2	2	6	5	9	400
	3	8	3	3	2	500
Demand →		250	350	400	200	

3. **(a)** In large maintenance department, fitters draw parts from the parts stores, which is at present staffed by one storekeeper. The maintenance foreman is concerned about the time spent by fitters in getting parts and wants to know if the employment of a stores helper would be worthwhile. On investigation it is found that: **[06]**

　　(i) A simple queue situation exists,

　　(ii) Fitters cost ₹ 2.50 per hour,

　　(iii) The storekeeper costs ₹ 2/- per hour and can deal on an average with 10 fitters per hour.

　　(iv) A labour can be employed at ₹ 1.75 per hour and would increase the capacity of the stores to 12 per hour.

　　(v) On an average 8 fitters visit the stores each hour.

　　(b) A super market has two girls ringing up sales at the counters. If the service time from each customer is exponential with a mean of 4 minutes, and if people arrive in a Poisson fashion at the rate of 10 an hour, then find: **[06]**

　　(i) what is the probability of having an arrival has to wait for service?

　　(ii) what is the expected percentage of idle time for each girl?

4. **(a)** Define types of cost in inventory model . **[06]**

　　(b) The production department for a company requires 3600 kg of raw material for manufacturing a particular item/year. It has been estimated that cost of placing an order is ₹ 36 and cost of carrying an inventory is 25% of investment in inventories. The price is ₹ 10/kg. The purchase manager wishes to determine an ordering policy for raw material. **[06]**

(P.1)

5. (a) Differentiate between CPM and PERT.   **[06]**

(b) A small maintenance project consists of following 12 jobs. Draw the network of the project. Summarize CPM calculations in tabular form. Calculating the three types of floats for jobs and hence determine critical path.   **[06]**

Job	1-2	2-3	2-4	3-4	3-5	4-6	5-8	6-7	6-10	7-9	8-10	9-10
Duration	2	7	3	3	5	3	5	8	4	4	1	7

6. (a) The project of laying of optical fibre cable in a particular town has the following activities. Draw the network and find critical path. Estimate total project cost if total direct cost of project is ₹ 90,000 and indirect cost of ₹ 8000/week. The deviation of activity $1-2, 1-4, 2-3, 2-4, 3-4, 4-5$ can be reduced by 2, 1, 3, 8, 2 weeks respectively at an direct additional cost of ₹ 11000, 7000, 12300, 8900 respectively. Crash network step by step to find optimal solution. What will be minimum project cost at optimum project duration?   **[06]**

Normal deviation : 4, 11, 3, 9, 5, 5.

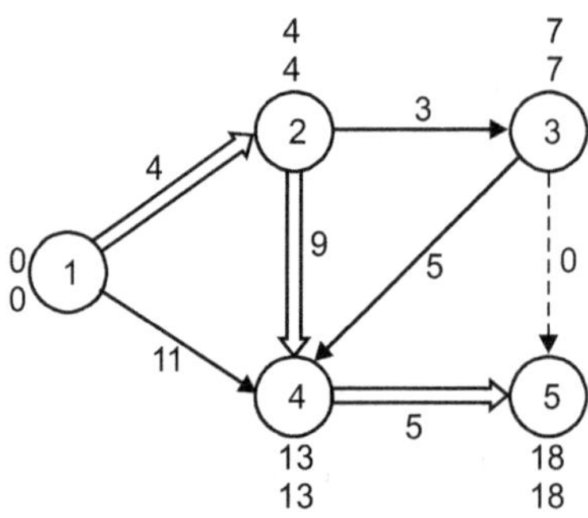

Fig. 1

Activity	Duration	Crashing Weeks	Cost
1 – 2	4	2	11000
1 – 4	11	–	–
2 – 3	3	1	7000
2 – 4	9	3	12300
3 – 4	5	–	–
4 – 5	5	2	8400

(b) A small maintenance project consist of following 12 jobs. Draw the network of the project. Summarize CPM calculations in tabular form. Calculating the floats for jobs and hence determine critical path.   **[06]**

Job	1-2	2-3	2-4	3-4	3-5	4-6	3-8	6-7	7-9	6-10	8-9	9-10
Duration	2	7	3	3	5	3	5	8	4	4	1	7

◈ ◈ ◈

9 789389 686845